Agricultural Meteorology and Climatology: Principles and Applications

Agricultural Meteorology and Climatology: Principles and Applications

Levis Green

STATES
ACADEMIC PRESS
www.statesacademicpress.com

States Academic Press,
109 South 5th Street,
Brooklyn, NY 11249, USA

Visit us on the World Wide Web at:
www.statesacademicpress.com

ISBN: 978-1-63989-032-3 (Hardback)

Cataloging-in-Publication Data

Agricultural meteorology and climatology : principles and applications / Levis Green.
p. cm.
Includes bibliographical references and index.
ISBN 978-1-63989-032-3
1. Meteorology, Agricultural. 2. Climatology. 3. Meteorology.
4. Atmospheric science. I. Green, Levis.

S600.5 .A37 2022
630.251 5--dc23

Table of Contents

Preface

Agricultural meteorology or agrometeorology is a sub-discipline of applied meteorology which is concerned with the study of the relationship between weather and agricultural production. It is a biophysical science which focuses on making use of weather information to enhance the quantity and quality of crop production. Various other disciplines contribute to this subject such as soil physics, chemistry, hydrology, phenology and agronomy. Agricultural climatology or agroclimatology is a branch of science which deals with the study of the influence of climate on conditions of agricultural plants and animal husbandry. This book attempts to understand the multiple branches that fall under the discipline of agricultural meteorology and climatology and how such concepts have practical applications. The various studies that are constantly contributing towards advancing technologies and evolution of this field are examined in detail. As this field is emerging at a rapid pace, the contents of this book will help the readers understand the modern concepts and applications of the subject.

The researches compiled throughout the book are authentic and of high quality, combining several disciplines and from very diverse regions from around the world. Drawing on the contributions of many researchers from diverse countries, the book's objective is to provide the readers with the latest achievements in the area of research. This book will surely be a source of knowledge to all interested and researching the field.

In the end, I would like to express my deep sense of gratitude to all the authors for meeting the set deadlines in completing and submitting their research chapters. I would also like to thank the publisher for the support offered to us throughout the course of the book. Finally, I extend my sincere thanks to my family for being a constant source of inspiration and encouragement.

Levis Green

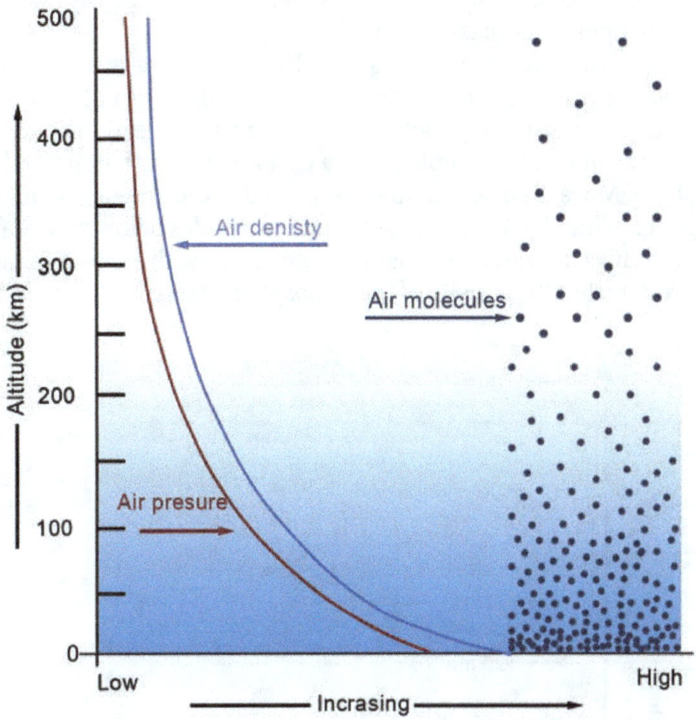

Figure 1.1: The atmospheric density and pressure distribution.

This is the dual nature of O_3, usually described as "good" and "bad" O_3, with the latter represented by tropospheric O_3. Near the surface, O_3 is created in complex photochemical reactions involving NO_2 and hydrocarbons, most of which are volatile organic compounds (VOCs). Tropospheric O_3 has the strongest impact on plants. In highly polluted air, O_3 will cause more damage to plant tissue than all other pollutants together. Together with other gases, ozone enters leaves through open stomata, where it attacks cells as a strong oxidant. Exposure symptoms include visible lesions on the leaf surface, reduced growth and a hypothesized reduction in the allocation of carbohydrates to roots. These symptoms can significantly impact plant development and crop yield. The part of the atmosphere in which concentration of "good" ozone is reduced – i.e., when the ozone layer becomes thinner with respect to its normal value – is called an ozone hole.

Finally, aerosols are important constituents of the atmosphere that appear in the form of small particles or drops. Even if their concentration is very small (0.01 ppm), aerosols significantly affect physical and chemical processes of the atmosphere. When they increase in size and/or concentration, aerosols can affect atmospheric transparency by producing the effect of light cloudiness or fog.

Atmospheric density and pressure. Due to the presence of sources and sinks, the concentration of trace gases changes with altitude. Photochemical processes at high altitudes (above 10 km) lead to concentrations of ozone and monoatomic oxygen that are higher than in the lower atmosphere. More than 50% of mass is located in the lowest 5.5 km of the atmosphere, while 99% of it can be found in altitudes below 30 km above the ground (Figs. 1.1 and 1.2). This is in accordance with experimental evidence proving that the density of the atmosphere decreases with altitude.

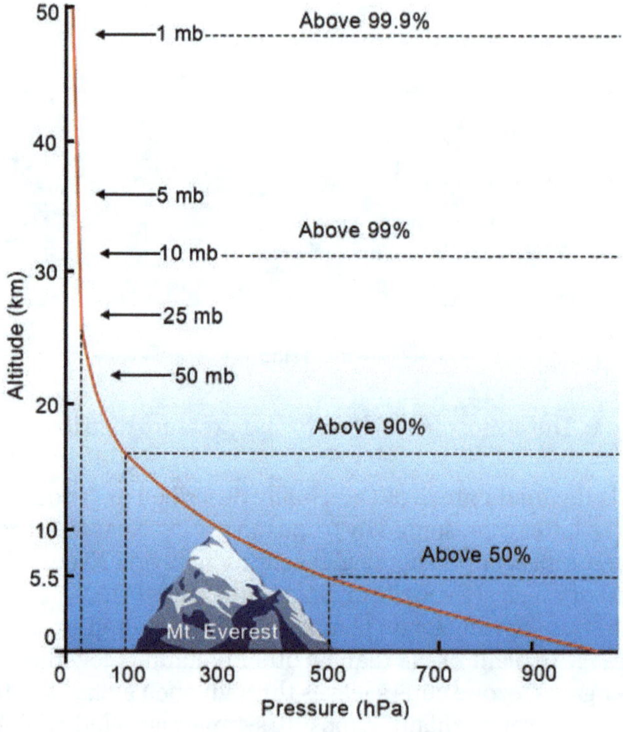

Figure 1.2: Distribution of atmospheric pressure and mass (in percent) by altitude.

Atmospheric stratification. Atmospheric pressure and mass distribution decrease with altitude quite rapidly in the lower atmosphere, while above 20 km the decrease is very slow. Air temperature as a very prominent atmospheric characteristic does not vary uniformly with altitude (Fig. 1.3). Measurements have shown that starting from the Earth's surface air temperature decreases with altitude at an average lapse rate of 6.5 °C/1000 m up to 11 km. This part of the atmosphere is called the *troposphere*. Of course, the depth of this layer of decreasing temperature is not constant over the globe and throughout the year. It varies from 7 km above the poles during winter to 20 km above the Equator. Since a major source of energy for the troposphere is heat from the Earth's surface, moving away from it brings a temperature decrease. The troposphere contains 75–80% of the mass of the atmosphere responsible for almost all processes which form and affect weather. At the upper limit of the troposphere, there is a thin layer of air called the tropopause. Here the temperature lapse rate starts changing from 6.5 °C/1000 m to 2 °C/1000 m or even less for the next few kilometres (from 2 to 4 km depending on latitude) (WMO, 1992), forming a layer of almost constant temperature. For thermodynamic reasons, the constant temperature of the layer dampens all vertical motions through the layer. In effect, the tropopause tends to function as an imperfect lid separating the troposphere from the higher reaches of the atmosphere. In the first layer above the tropopause, temperature increases with altitude. The *stratosphere*, as it is called, owes its temperature profile to its gas content and distance from the ground. Specifically, at heights above 20 km, the concentration of atmospheric ozone is highest. This gas intensively absorbs solar radiation in the UV part of the spectra and increases the temperature of this part of the atmosphere. Because of the stable stratification of stratosphere, vertical mixing and convection are almost negligible and particles that get into the stratosphere (in volcano eruptions or nuclear accidents, for example) can stay there for decades. In addition, the stratosphere is very dry. Most water vapour stays in the troposphere, while 20 % of the atmosphere's mass is distributed over a layer measuring more than 30 km. At an altitude of about 50 – 55 km we find the top of stratospheric inversion layer, called the stratopause. On average, the temperature at this altitude is –15 °C, while atmospheric pressure of 100 Pa (1 mb) indicates that 99.9% of the atmosphere is located below this level. At an altitude of approximately 50 km, air temperature again starts to decrease, reaching at 85 km the absolute atmospheric minimum temperature of –90 °C. This atmospheric layer, which has a very low density of gas molecules, rich in UV radiation and almost no oxygen, is called the *mesosphere*. Due to its low density, absorption of solar radiation is negligible, while heating from the stratosphere is reduced with increasing altitude. Since the position of the mesosphere is too high for research aircraft and balloons and too low for satellites, it remains an almost "mysterious" part of the atmosphere. The mesopause divides the mesosphere from the highest atmospheric layer – the *thermosphere* – which extends up to 100 km. Air in this layer is so thin that even atomic and molecular collusions are very rare. In the upper part, the main components are atomic nitrogen, oxygen and helium. Gas content and temperature are strongly affected by the Sun's activities. Even if temperature is believed to increase with altitude here, this cannot

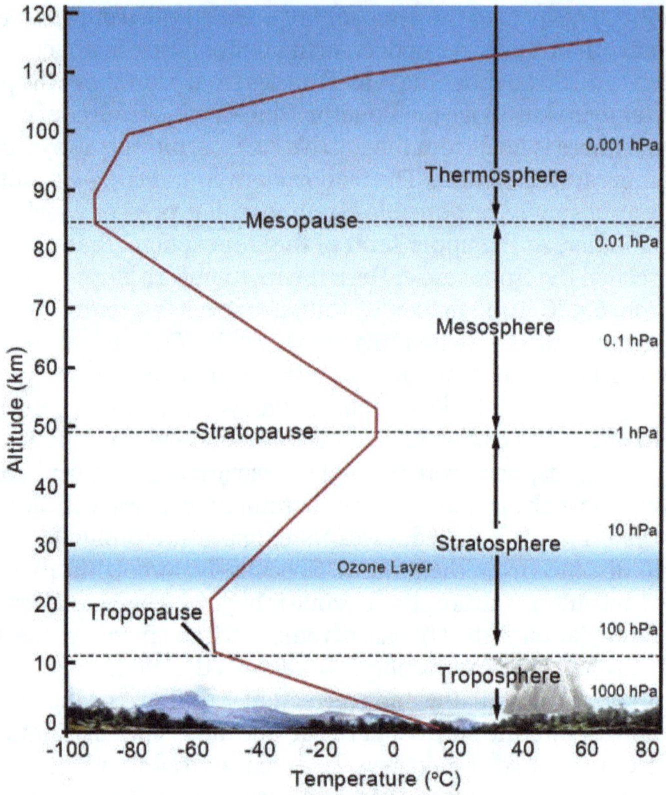

Figure 1.3: Atmospheric layers with respect to vertical variation of air temperature.

be taken as fact, as in this part of the atmosphere temperature is, literally, the average kinetic energy of a limited number of molecules.

1.2 Impact of atmospheric composition on plants

Atmosphere and soil comprise the entire plant environment. Atmospheric content and conditions crucially affect plant development and growing conditions. The balance of atmospheric gases drives plant physiology, chemistry and complete growth.

The impact of CO_2 increase on plants has been and continues to be a subject of great concern since the last decades of the 20th century. Its impact mostly concerns the intensity of photosynthesis, as under current atmospheric concentrations CO_2 is a limiting factor of photosynthesis (Fig. 1.4) and transpiration for many plants. Under well–watered condi-

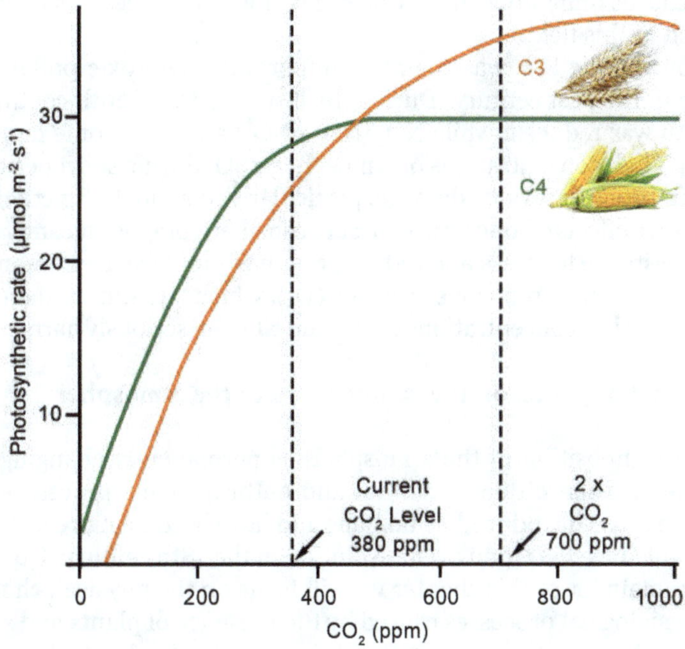

Figure 1.4: Short-term response of photosynthetic rate on changes of CO_2 concentration for C_3 and C_4 plants.

tions, the respective responses of C_3 and C_4 plant[1] photosynthesis on CO_2 elevation differ significantly. For gas concentrations varying from 180 to 700 ppm, rates of photosynthesis were changed by 30% and 142% for C_4 and C_3 plants, respectively (Ward *et al.*, 1999). Net photosynthesis of wheat is about 15.9 mol m^{-2} h^{-1} ($4.4167 \cdot 10^{-3}$ mol m^{-2} s^{-1}), while in the case of maize it is about 12.5 mol m^{-2} h^{-1} ($3.47 \cdot 10^{-3}$ mol m^{-2} s^{-1}), for photosynthetic active radiation (PAR) of 279.12 W m^{-2}. Furthermore, for maize, a C_4 crop, the photosynthetic rate becomes constant at lower CO_2 concentrations (450 ppm) than, for example, wheat, a C_3 crop (850 ppm). Transpiration response at elevated CO_2 takes place at the stomata level, resulting in increased stomatal resistance and, therefore, reduced intensity of the transpiration process. It should be noted that plants of the same type can react very differently to changing CO_2 levels, depending on other

[1] C_3 plants are temperate weather crops like wheat, barley, rice, soybean, sunflower and all trees, while C_4 are plants from arid climate zones with hotter or tropical weather such as corn, sugarcane, millet and sorghum.

environment conditions such as stress, nutrition and genetic constitution (i.e. cultivar effects). Moreover, long–term adaptation is uncertain, which causes difficulties in making predictions and necessitates further research in this field.

Atmospheric SO_2 was the most important phytotoxic pollutant in Europe in the past century. During the last decades of 20th century, its emission was reduced, while NO_x ($NO + NO_2$) and O_3 become more important. High concentrations of atmospheric SO_2 during short periods of time can produce serious damage, particularly in stomata functionality. Cumulative effects of long term exposure can, therefore, significantly affect plant resistance to stress and reduce growth without clear symptoms. In combination with tropospheric ozone (O_3), which has similar damaging effects, even low concentrations of SO_2 and NO_x can seriously harm plants.

1.3 Impact of plants on the composition of the atmosphere

The composition of the atmosphere is permanently changing over the time and space due to natural and anthropogenic processes. The latter are those introduced by humans and have become more and more important since first industrial revolution in the 18th century. However, in the remainder of this chapter we will focus on the physical, chemical and physiological processes caused by the presence of plants and soil.

Let us begin with one of the most important, namely, the balance in the production and destruction of oxygen and nitrogen in the atmosphere. Nitrogen is produced and released from soils to the atmosphere through the processes of nitrification (biological oxidation of ammonia to nitrites) and denitrification (the microbial process of nitrate reduction), which release N_2, N_2O, NO and NH_3. On the contrary, biological activity of the soil (especially N fixation) is mainly responsible for nitrogen removal from the atmosphere. The most important source of atmospheric oxygen is plant photosynthesis, while the oxidation process of organic (vegetation, animal and bacterial respiration) and inorganic matter as well as the aerobic decay of organic matter represent sinks of oxygen.

Plants affect the composition of the atmosphere on both mechanical and physiological levels. Plant canopy, particularly forest, increases surface roughness, changes air flow above the canopy and influences gas exchange from the atmosphere to the canopy air space. Atmospheric gases such as O_2, CO_2 and H_2O are permanent components of important physical and physiological processes related to plant growth and canopy environmental interactions. Some other gases can be assimilated by plants (O_3) or deposited on the surface of leaves (NO_X, SO_X,...). The presence of plant canopy affects carbon balance by means of biomass distribution, respiration, assimilation of CO_2, decomposition of organic matter and nutrient mineralization. In the case of H_2O, vegetation always increases atmospheric humidity by transpiration and reduction of the transfer of water vapour from the ground surface. This is why plant canopy is considered globally as an important source and sink of most atmospheric constituents.

2

Atmospheric Energy Balance

Three major mechanisms of energy transfer in nature are conduction, convection and radiation. Energy is transferred by *conduction* either through solid materials or between them. It occurs when atoms or molecules interact with each other, transferring part of the kinetic or vibrational energy from particles having higher energy (the warmer part) to those with lower energy (the cooler part). *Convection* is typical for energy transfer within fluids or through the interface between a solid surface and a streaming fluid. *Electromagnetic* radiation is the mechanism of energy transfer by electromagnetic waves which can propagate through free space or a material medium at the speed of light.

2.1 Shortwave and longwave radiation in the atmosphere

Wave length, l and frequency, n are characteristics commonly used to describe electromagnetic radiation. Since electromagnetic waves are just a special case of energy transfer by wave motion, the relation between wave length and frequency – as for all waves – can be written in the form

$$c = \lambda \cdot \nu \tag{2.1}$$

where c is the speed of wave propagation. In the case of electromagnetic waves in the air or a vacuum, this speed is commonly known as the speed of light, with a value of $3 \cdot 10^8$ m s^{-1}. According to Planck's quantum theory, energy is not emitted or absorbed continuously but in the form of quanta. In the case of electromagnetic radiation, this package of energy is called a photon, and its energy, E is proportional to frequency, ν

$$E = h \cdot \nu \tag{2.2}$$

where h is Planck's constant ($6.62 \cdot 10^{-34}$ J s).

Every body with a temperature above 0 K emits radiation. The energy of electromagnetic radiation emitted from each square meter of body surface in each second, B is proportional to the fourth power of its absolute temperature, T according to Stefan–Boltzman's law

$$B = \varepsilon \cdot \sigma \cdot T^4 \tag{2.3}$$

where σ is Stefan–Boltzman's constant ($5.67 \cdot 10^{-8}$ W m^{-2} K^{-4}) and ε is the surface emissivity. At the same temperature, different bodies radiate differently. Therefore, body energy emission is not just a matter of temperature but also of their material characteristics. Surface emissivity varies from 0 for a so–called "absolute white body" to 1 for a so–called "absolute black body". Of course in nature there is no such thing as an absolute white or black body, but some natural surfaces behave very similarly in their emissivity (Tab. 2.1).

Land cover	Albedo	Emissivity
Tropical forest	0.13	0.99
Woodland	0.14	0.98
Farmland/natural grassland	0.20	0.95
Semi–desert/stony desert	0.24	0.92
Dry sandy desert/salt pans	0.37	0.89
Water	<0.08	0.96
Sea ice	0.25–0.60	0.90
Snow–covered vegetation	0.20–0.80	0.88
Snow–covered ice	0.80	0.92

Table 2.1: Mean annual albedo (see Section 2.2) and emissivity of different land covers (Thompson, 1998).

Wave length (or frequency) and electromagnetic spectra are the most important characteristics of electromagnetic radiation. The energy of the latter is inverse to wave length, with the term "shortwave" used for high energy and "longwave" for low energy radiation. The electromagnetic spectra of an emitting body represents the distribution of the monochromatic irradiance/intensity of radiation over a range of wavelengths (Fig. 2.1). According to wavelength referring to a maximum in electromagnetic spectra, radiation of body in case is considered as shortwave or longwave. The most important source of shortwave radiation for the Earth is the Sun (maximum 500 nm), while all forms of electromagnetic radiation originating from the much cooler surfaces of the Earth belong to the longwave part of the spectra.

The Sun is a star, a ball of hot gases and plasma whose surface radiates like a black body at a temperature of approximately 6000 °C. Under clear

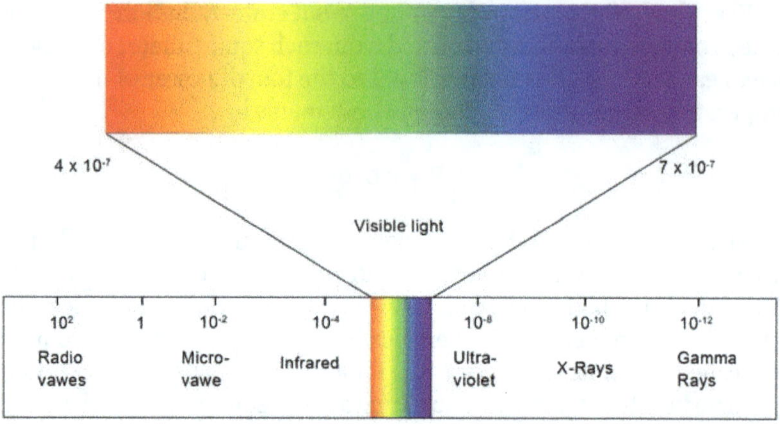

Figure 2.1: Electromagnetic radiation spectra.

sky conditions, atmospheric components perform only minor absorption of solar radiation in the visible part of the spectra (380 – 760 nm), allowing high energy radiation to reach the Earth's surface and to play the role of the most important driving force in the whole atmosphere. However, absorption is not the only mechanism of solar radiation attenuation in the atmosphere. *Diffuse reflection* and *selective absorption* are processes which can significantly change the quality and quantity of solar radiation reaching the surface (expressed as changing colours of the sky, for example). Diffuse reflection occurs when radiation reaches a surface (drops, clouds, etc.) which is not flat (Fig. 2.2). In the atmosphere this is common situation. An incident ray reflects at many angles other than at the incident one, thus changing its original spectral characteristics. Selective absorption refers to different intensities of absorption in different parts of the spectra. The absorption spectra of atmospheric components are responsible for selective absorption of solar radiation in the atmosphere (Fig. 2.3). The presence and concentration of particular atmospheric components determine the intensity of absorption at certain wavelengths.

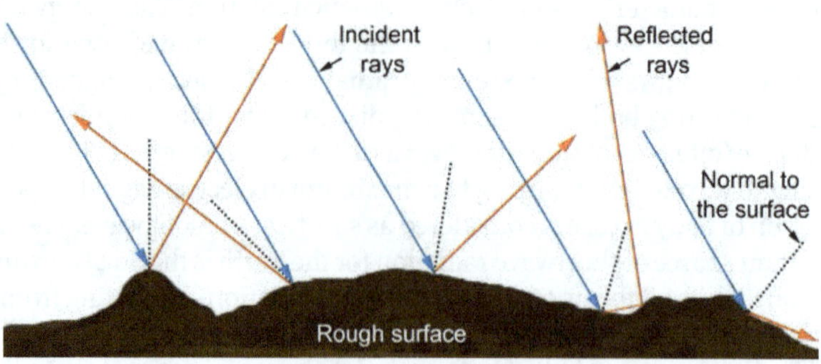

Figure 2.2: Diffuse reflection of radiation from rough surface.

The Earth is a massive body with changing temperature and emissivity over its surface. The intensity of the Earth's radiation can be calculated using Eq. (2.3), taking into account ground surface temperature and emissivity. In the case of the atmosphere, radiation emission is much more complex. Specifically, every gas component, droplet or cloud part in the atmosphere is a small body which emits energy according to Stefan–Boltzman's law. Since for each of these the surrounding air represents an infinite heat reservoir, it can be assumed that the temperatures of all of these small bodies are equal to that of the surrounding air. In that case, radiation emitted by the atmosphere is proportional to the air temperature and the number of small sources of radiation in the atmosphere. Obviously, the exact relation for calculating atmospheric radiation is almost impossible to formulate. But, there are many empirical relations taking into account impact of clouds on global radiation or correlation between sunshine duration and global radiation (see Chapter 14).

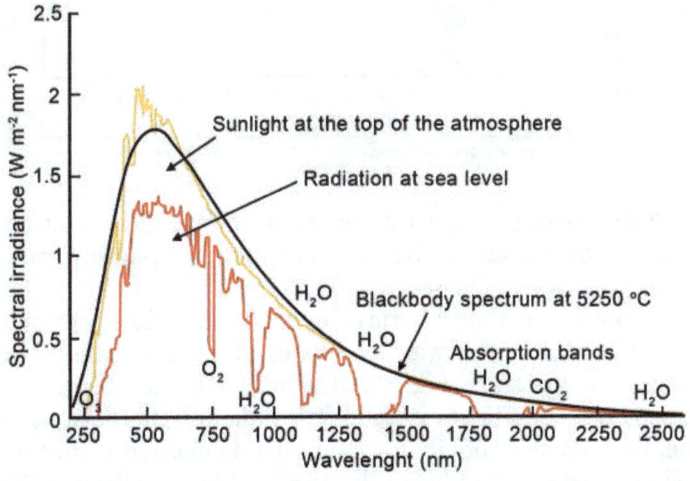

Figure 2.3: Solar radiation spectra at the top and the bottom of the atmosphere.

Since the Earth and the atmosphere have temperatures in the range of 200–300 K, emitted radiation belongs to the thermal infrared part of electromagnetic spectra (4–100 μm) with a maximum intensity of 10 μm (Fig. 2.4). In the subject literature, the commonly used term for both of these radiation fluxes is *terrestrial radiation*. This radiation is the most powerful source of longwave radiation in the Earth's climate system and is responsible for surface energy balance at night. In the case of high humidity and/or cloudiness, atmospheric radiation is higher.

The presence of clouds additionally rises surface warming by increased reflection of radiation coming from the ground, therefore making the surface energy balance positive even at night, in the absence of short wave radiation.

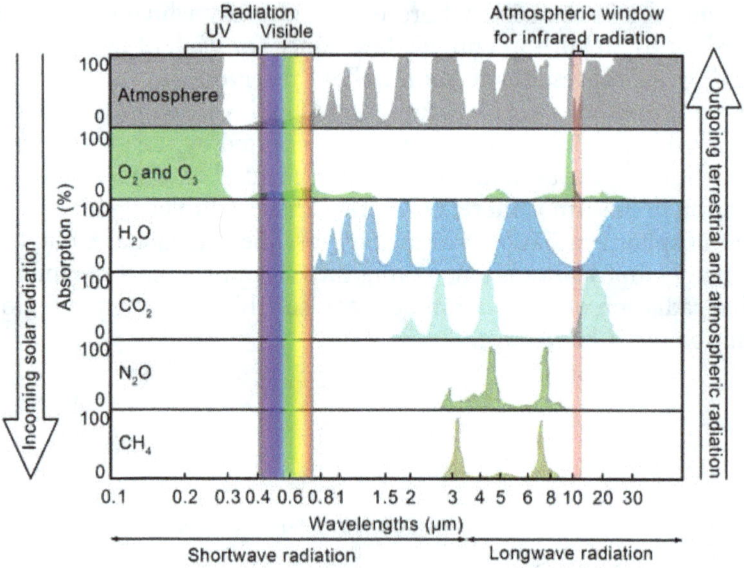

Figure 2.4: Absorption spectra for major natural "greenhouse gases" (GHGs) in the Earth's atmosphere of shortwave and longwave radiation.

A large portion of atmospheric components absorbs electromagnetic radiation, but at different wavelengths or absorption windows of the electromagnetic spectrum (Fig. 2.4). The most powerful absorber of electromagnetic radiation below 0.4 µm (UV radiation) in the stratosphere is ozone. With an absorptivity for UV radiation close to 1, stratospheric ozone is the "guardian" of life on Earth. In addition, nitrogen (N_2) and oxygen (O_2) also absorb electromagnetic radiation, but at wavelengths below 0.1 µm and at 0.245 µm, respectively. At wavelengths between 0.39 µm and 0.70 µm, the absorption spectrum of the atmosphere has a gap, with absorptivity close to 0, called the "atmospheric window" (Fig. 2.4). It allows solar radiation, in the so–called visible part of the spectra, to reach the Earth's surface. However, with a certain reduction of intensity, the Earth's surface receives solar radiation of a broader range of wavelengths (0.3–3 µm), which include the near–ultraviolet (NUV), visible and near–infrared (NIR) parts of the spectra–in other words, shortwave radiation. Absorption of this radiation is mostly affected by water vapour and carbon dioxide (Fig. 2.4).

Outgoing terrestrial radiation belongs to longwave radiation spectra with wavelengths above 4 μm. In atmospheric absorption of longwave radiation, the most important role is played by a group of gases which were once called "greenhouse gases" (GHGs)[2]: water vapour, carbon dioxide, ozone, nitrous oxide and methane. GHGs in the atmosphere are responsible for the absorption of radiation mostly in the infrared "window" between 8 and 14 μm. Only CO_2 has an excessive absorption line beyond 13 μm.

2.2 Energy balance

Solar radiation reaches the Earth's surface in the form of *direct* and *diffuse* radiation. Direct radiation is a part of solar radiation which comes to a horizontal surface along with the Sun's ray at one narrow spatial angle (up to 5). After diffuse reflection and scattering in the atmosphere, solar radiation falls to the surface in the form of diffuse radiation. The measure of total solar radiation coming to the Earth's surface is so–called *global* radiation, defined as the sum of both direct and diffuse radiation in the range of 0.3–3 μm wavelengths. Partitioning global radiation between direct and diffuse parts is mostly effected by the presence of clouds and the solar angle. In the case of clear sky, both direct and diffuse radiation are present. However, when thick clouds cover the sky and the Sun is completely obscured, global radiation is equal to the diffuse measure. The same situation occurs at sunset, when the Sun is below the horizon and only the reflection of solar radiation from the sky to the ground produces diffuse radiation on the Earth's surface. For the region of Central Europe, typical values of clear–sky daily global radiation vary from about 5 to 30 MJ m^{-2}, depending on the season and the latitude.

At the surface, global radiation can be separated into two basic types of radiation: *absorbed* (energy absorbed by the surface and transferred deeper into the soil) and *reflected* (energy reflected from the surface back to the atmosphere). Shortwave absorbed radiation varies with latitude and season (the dominating effect of the sunbeam's angle) and the type of surface. A portion of reflected radiation is determined by the surface characteristic *albedo*. The magnitude of the albedo of natural surfaces varies from 0.13 for tropical forest to 0.8 for snow–covered ice (Tab. 2.1) and can be calculated as the ratio of reflected (R) and global (G) radiation intensities. Albedo changes with colour, wetness, structure and surface cover. Dark and moist ground after deep ploughing has a low albedo

[2] The name "greenhouse gases" is kept for historical reasons even if this term is scientifically incorrect. Specifically, the main reason for temperature increase in a greenhouse is the prevention of heat conduction due to the presence of glass walls and a ceiling. In the case of warming related to GHGs, the cause of the rise in temperature is the increased absorption of longwave radiation by GHGs.

because of an increase in absorbed solar radiation. In other words, dark bodies absorb better than white ones, while the presence of water in the soil increases its heat capacity because air (with a lower heat capacity) is replaced by water (with a higher one). Deep ploughing increases multiple reflection of solar radiation on the ground surface before its final reflection towards the atmosphere. An important lesson to be learned in this regard is that a component of the energy budget of the Earth's surface–reflected radiation–is affected by surface characteristics, which are changing dynamically. More about the impact of vegetation on albedo can be found in Section 3 of this Chapter.

Before going into a detailed analysis of energy balance, some important sources and sinks of energy should be mentioned, namely, *sensible* and *latent* heat, which are transferred by convection and turbulence. Sensible heat is energy transferred from warmer to cooler places in the presence of a gas (air) or fluid, thereby producing a temperature change of all bodies that come into contact with it (sensation). In 1884 Joule described it as energy indicated by a thermometer. Latent (hidden) heat, on the other hand, is energy involved in phase changes at a constant temperature. At the Earth's surface and in the atmosphere, this energy is associated with changes in the state of water from water vapour to liquid water and ice and *vice versa*. Sensible heat flux is positive if it is directed from the surface to the atmosphere, while latent heat flux is positive if, for example, a surface evaporates. That convention is reasonable, since in both cases the atmosphere becomes warmer.

At this point, let us shed more light on two aspects of the analysis of incoming solar radiation which sometimes produce a great deal of confusion. One is related to the amount of solar radiation that passes through a plane normal to the direction of the Sun. The greatest amount of solar energy that can reach 1 m^2 of surface above the Earth's atmosphere in 1 s is called the solar constant; on average its value is 1367 W m^{-2}. A second analysis relates to the average amount of radiation that the Earth (at the top of the atmosphere) receives per unit area. It differs from the previous definition because there is always a "light" and "dark" part of the Earth. The average amount of received radiation refers to the energy which each part of the atmosphere receives if the incoming energy is uniformly distributed all over the planet at a given moment. A quick calculation proves that this equals one quarter of the solar constant, or 342 W m^{-2}.

The shortwave energy balance at the Earth's surface and the atmosphere as well as the partition of radiation fluxes, is commonly made with respect to incoming solar radiation (whose total is set to 100 units). We will express the amount of each different radiation flux as a portion of this total radiation. On its path through the atmosphere, then, solar radiation goes through the processes of: a) *absorption* in the stratosphere (2) and the troposphere (17); b) *reflection* from clouds (20) and the ground surface

(4), including scattering in the atmosphere (6), and c) *transmission* in the form of direct (28) and diffuse (23) radiation (Fig. 2.5). It produces an overall effect of 30 units of reflected radiation, which corresponds to a planetary albedo of 0.3. Another calculation brings us to the conclusion that the atmosphere absorbs only 19% of incident solar radiation, while about 51% reaches the Earth's surface.

Absorption of solar radiation by the atmosphere and the surface below produces their heating. Let us suppose that the Earth's surface at 18.55 °C (291.7 K) and with an emissivity of 0.95 is in thermodynamic equilibrium with the atmosphere. According to Stefan–Boltzman's law, this surface emits 390 W m^{-2}, which is approximately 114% of the incoming solar radiation (Liou, 2002).

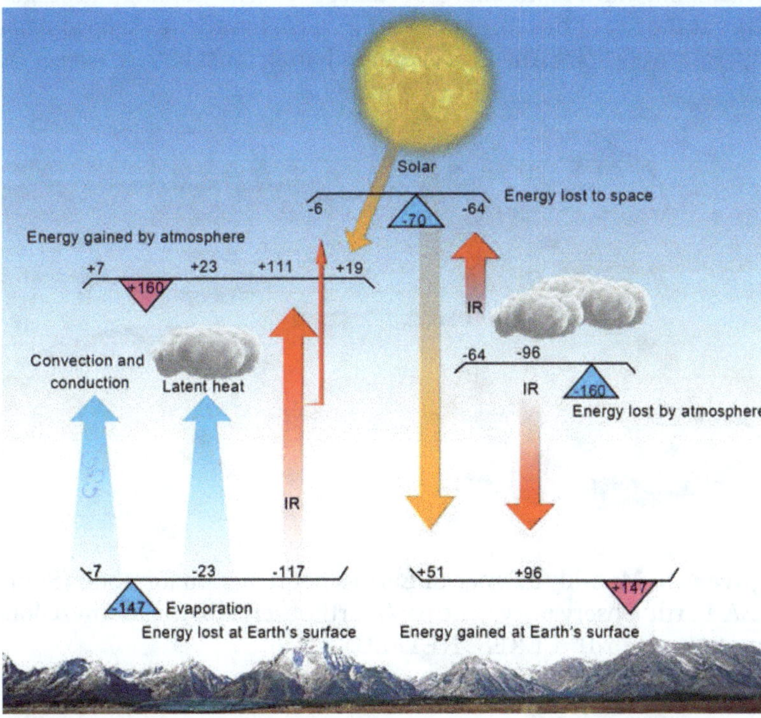

Figure 2.5: Shortwave and longwave energy balance of the Earth's surface and the atmosphere.

If we partition the flux as was done for shortwave radiation, longwave radiation fluxes at the Earth's surface and the atmosphere are as follows (Fig. 2.5): from 117 units of terrestrial longwave radiation coming from surface, 111 will be absorbed by atmospheric gasses, while 6 units will be lost to outer space. Additionally, 30 units on average come from the surface to the atmosphere in the form of latent (23) and sensible (7) heat. The atmosphere therefore gains 19 units from solar radiation, 111 from terrestrial and 30 from sensible and latent heat fluxes, giving us a total of 160 units. But the atmosphere emits as well. According to its average temperature and current content, atmospheric downward emission is 96 units, while 64 units are lost to space.

These 96 units of reemitted energy (i.e., 111 units of absorbed terrestrial radiation) produce the so–called "greenhouse gas (GHG) effect". Of course in the case of a higher concentration of gases absorbing longwave radiation, absorption of surface emission will be above 111, which will increase the equilibrium temperature and consequently the amount of energy reemitted to the ground.

To better understand this effect, we should consider the global impact of all these processes. One of the best indicators in this regard is the Earth's net radiation, which represents the balance between incoming and outgoing energy at the top of the atmosphere (Fig. 2.6). In order to keep global air temperature constant, the Earth's net radiation must be zero over a year.

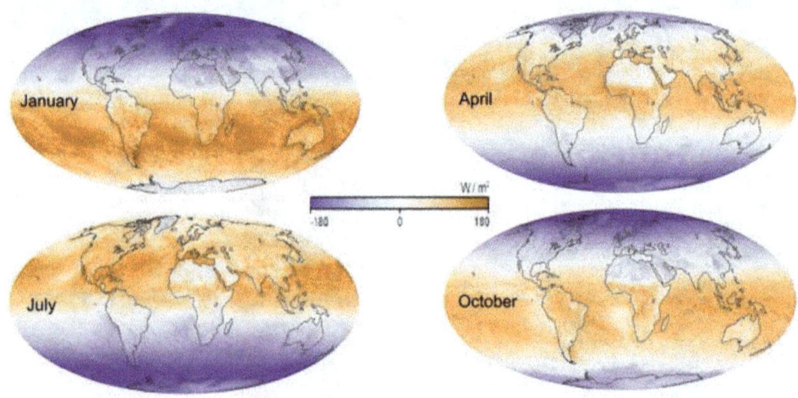

Figure 2.6: Monthly average of Earth's net radiation for 2006 (Source: NASA Earth Observatory, <https://earthobservatory.nasa.gov/GlobalMaps/view.php?d1=CERES_NETFLUX_M>).

2.3 Impact of radiation on plants

Radiation impacts plants through wavelength, energy and duration. From the point of view of plant physiology, the most important sections of the electromagnetic spectra are the visible ones, also known as the light and UV–B ranges. Light is responsible for chlorophyll formation, the functioning of stomata, photosynthesis, growth and phenology, and in the case of mixed vegetation, such as forest and grassland, for species stratification and the size and structure of leaves. For photosynthesis, the range of 400–700 nm (photosynthetic active radiation – PAR) is crucial. Daily variation of PAR intensity and the partition between direct and diffuse components is significant (Fig. 2.7). After 14 h, direct PAR intensity is negligible, and photosynthesis completely relies on diffuse radiation.

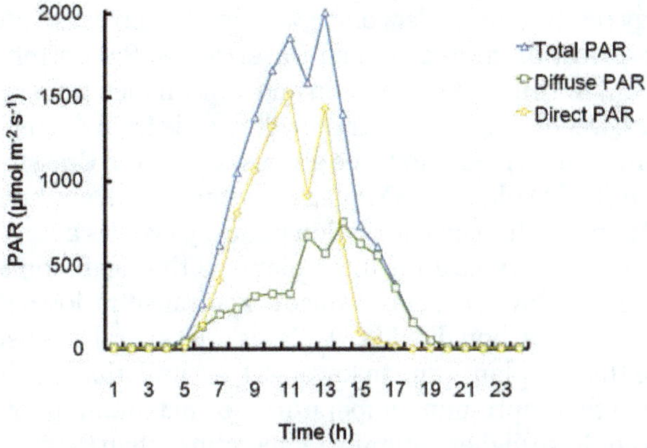

Figure 2.7: PAR partitioning above forest on 1st June 2012 at Prospect Hill Tract, Harvard Forest, Petersham (MA, USA)[3] (Munger and Wofsy, 2017).

UV radiation, particularly UV–B radiation (280–315 nm), has negative effects on photosynthesis, reducing plant size and destroying enzymes and deoxyribonucleic acid (DNA), which leads to plant tissue damage and increased sensitivity to disease. Additionally, UV–B radiation affects the productivity of phytoplankton, which are responsible for carbon storage in the ocean.

Infrared radiation (IR) is electromagnetic radiation with a wavelength between 700 nm and 1 mm. Infrared radiation is emitted by every "hot" body; once absorbed by a body, it gives rise to the phenomenon of heat. These thermal properties are common to all electromagnetic radiation, but they occur more particularly with infrared radiation, which is therefore also called thermal radiation. Infrared emission is the result of transitions between vibrational energy levels of molecules. It opens a window of opportunity for investigating molecular structures and reticular bonds of plants through analysis of their IR emission and absorption spectra.

All physiological and biochemical processes important for the development of plants are triggered and/or limited by temperature (the lower and upper thresholds, respectively). The rate of process is often proportional to effectively invested energy, i.e., the energy transferred between an organism and its environment at temperatures between certain

[3] The tower at Prospect Hill Tract, Harvard Forest, Petersham (MA, USA) was installed in 1989. The eddy–flux measurements taken there constitute the longest running record of net–ecosystem carbon exchange in a North American forest.

thresholds. Although photosynthesis is the most important physiological process involving radiation, only 50% of the incident portion is employed by the plant to perform photosynthesis. It is also worth remembering that from energy received by the leaf only 5% is finally converted into biomass, while 60% is lost on non–absorbed wavelengths, 8% on reflection and transmission, 8% on heat dissipation and 19% on metabolism (Campillo *et al.*, 2012). Depending on how carbon dioxide is fixed during photosynthesis, plants are divided into three groups: C_3, C_4 and Crassulacean acid metabolism (CAM) plants. CAM plants are those with crassulacean acid metabolism in which a carbon fixation pathway evolves as an adaptation to arid conditions. Typical for these plants is that leaf stomata remain closed during the day in order to reduce evapotranspiration but stay open at night to collect carbon dioxide. C_3 plants are generally considered less productive than C_4 plants due to lower CO_2 assimilation efficiency. They also differ in their optimum temperatures for maximum photosynthesis rates: C_4 crops have higher optimum temperatures than C_3 crops. Legumes need additional metabolism for N–assimilation, which accounts for their lower radiation use efficiency, except under conditions where nitrogen availability is a limiting growth factor.

Plants as physical systems are always in a state of energy balance with their environment. Even if some organisms can autonomously change temperatures, in the long term their temperatures are stable, the result of energy exchange with the environment. A common assumption is that the net radiation, R_{net} received by a plant is balanced by the turbulent fluxes of latent, LE and sensible heat, H together with storage in the ground, Q_G and in the biomass, Q_S:

$$R_{net} = LE + H + Q_G + Q_S \tag{2.5}$$

Plant energy storage which is used for some small, autonomous temperature regulations and internal processes does not exceed 20-30 W m^{-2}. Therefore, it is thought that the first three terms on the right–hand side of the balanced equation account for more than 95% of net radiation.

Latent heat flux from the leaf surface, LE_l, is proportional to the intensity of evapotranspiration, E_l. The coefficient of proportionality is the latent heat of evaporation, L which depends on the temperature of the evaporating liquid. In the case of leaves, the temperature of the water inside leaves and on the leaf surface is equal to its temperature, and L can be considered as a function of leaf temperature, T_l. It is useful to keep in mind that for water, $L = 2.43 \cdot 10^6$ J kg^{-1} at 30 °C and $L = 2.501 \cdot 10^6$ J kg^{-1} at 0 °C. Evapotranspiration is a complex process which includes the physical process of evaporation from the soil and leaf surfaces and also the physiological process of transpiration through the plant's stomata (Fig. 2.8). Since water potentially evaporates from the leaf surface only during short periods of time after precipitation or dew formation, considerations of leaf energy balance include only transpiration.

Commonly used parameterisation of transpiration is based on Fick's law of diffusion and resistance representation of vegetation, implying analogy with Ohm's law. According to Fick's law, water vapour flux between the leaf interior and the air is directly proportional to the difference in its concentration (commonly expressed through the partial pressure of water vapour in the air, e) and inversely proportional to transfer resistance. A similar concept can be applied to CO_2 diffusion between the plant and the surrounding air using the CO_2 concentration difference between the inside and outside of the leaf stomata and its transfer conductivity. This is why CO_2 uptake and water vapour release from the plant surface are very closely correlated. This is usually expressed using the so-called transpiration coefficient, which represents a very close relation of transpiration to the photosynthesis rate.

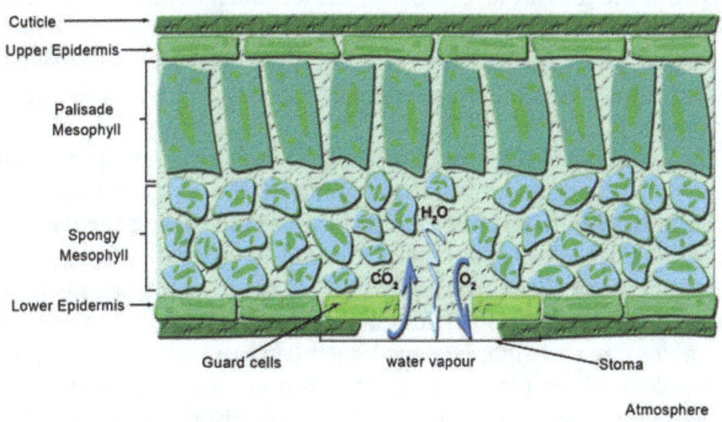

Figure 2.8: Stomata scheme.

It is a reasonable assumption that water vapour within a leaf is in saturation at the leaf temperature and therefore that water vapour concentration can be expressed using the saturation vapour pressure at the leaf temperature, $e_s(T_l)$. If the relative humidity of the air, ρ is known, then the vapour pressure of the surrounding air can be expressed as $r \cdot e_s(T_a)$, where T_a is the air temperature. Since water vapour is transferred through plant stomata and air, two resistances–stomatal, r_s and aerodynamical, r_a – should be taken into account. Water vapour flux can then be expressed in the form of latent heat

$$L(T_l)E_l = \frac{\rho c_p}{\gamma} \frac{e_s(T_l)\text{-}re_s(T_a)}{r_s + r_a} \tag{2.6}$$

where ρ is the air density, c_p is the specific heat at constant pressure, ρ is the relative humidity, e_s is the saturated water vapour pressure and g is the psychrometric constant. Since the difference in concentrations at the end of the stomata and between the stomata and the air is not equal, then according to Ohm's law the stomatal, r_s and the aerodynamical, r_a resistances should be treated as a series of connections with equivalent resistance, equal to the sum of the resistances ($r_s + r_a$).

Sensible heat flux, H from the leaf surface is a consequence of the temperature difference between the leaf and its environment. Applying Fick's law and the resistance representation on sensible heat transfer, this flux can be written in the form

$$H = \rho C_p \frac{T_l - T_a}{r_a}. \quad (2.7)$$

Within the thin air layer close to the leaf surface and on the interface between the plant and the ground, molecular conduction is an important mechanism of energy transfer. This energy, however, is very difficult to calculate, and it is common practice to express it as a percentage of net radiation R_{net}.

The duration of radiation is commonly related to the period during the day when solar radiation can be measured, i.e., the period between sunrise and sunset. The photoperiod is a period during the day when a plant or animal is exposed to light. The day length will differ at different latitudes because of the Earth's axis tilt. At the Equator, the photoperiod is constant (day length is 12 h and night length is 12 h), but moving closer to the poles day and night become longer (up to a 24 h day or a 24 h night at the poles). Many plant species are photoperiodic, which means that they will flower only under certain light conditions. According to their photoperiod sensitivity, plants are classified into three groups: short–day plants (SDPs), long–day plants (LDPs) and day–neutral plants (DNPs). SDPs generate response when the day is shorter than the night, LDPs when the day is longer than the night, while DNPs do not respond to day/night length changes during the year. When plants react to the balance between day length and night length, they are programming their development stages to respond to certain environmental conditions likely to develop in that period of the year. Plants have chosen this phenomenon over the annual temperature cycle since it is predictable and consistent from year to year. Plants detect the light signal on their leaves (absorption of the photon by a chromoprotein phytochrome) and produce mobile signals which trigger certain processes leading to the next stage of development. The most important developing stages triggered by the photoperiod in light–responsive plants are flowering, tuberization and bud set.

2.4 Impact of plants on radiation

The presence of vegetation can make a significant difference in the radiation spectra and the energy balance of the Earth's surface.

The absorption, reflection and transmission spectra of vegetation are affected by a series of factors covering its morphological and physiological characteristics that change–sometimes more, sometimes less–over the year. Absorption of radiation in the photosynthetically active part of the spectra (PAR) (0.4 – 0.7 μm) is about 90%, and 15–20% in the nearby infrared (IR) spectra (0.7 – 2.5 μm). Reflectance of radiation by plant leaves is mostly affected by the presence of atmospheric water (0.7 – 1.9 μm) and by water present in the leaf and canopy structure (0.7 – 1.5 μm). Changes in the absorption and reflection spectra of plants are good indicators of water content and different degrees of stress and plant health, which find practical use in the field of remote sensing, for example.

Optical properties of green leaves compiled for PAR, near IR and solar shortwave radiation are shown in Table 2.2.

	PAR	NIR	Solar shortwave
Reflectance	0.09	0.51	0.30
Transmittance	0.06	0.34	0.20
Absorbance	0.85	0.15	0.50

Table 2.2: Typical optical properties for green (living) leaves.

Moving from the leaf to the canopy level opens up the possibility of analyzing the impact of plant canopies or vegetation on the energy balance of the Earth's surface. For example, under the same atmospheric conditions, the emissivity and temperature of bare soil differ considerably from those of the vegetation surface, affecting terrestrial longwave radiation.

The intensity of radiation reflected from the surface changes in the presence of vegetation due to changed surface albedo. In contrast to many other natural surfaces, vegetation changes its albedo autonomously during the seasons, producing differences in reflected PAR and global radiation and, consequently, in surface albedo (Fig. 2.9). The most important canopy characteristic commonly used to describe this feature is leaf area index (LAI), which is defined as the surface of all leaves per unit area of ground surface covered by plants. Monitoring changes in LAI allows tracking of vegetation and specifically of changes in surface albedo over the year, and vice versa.

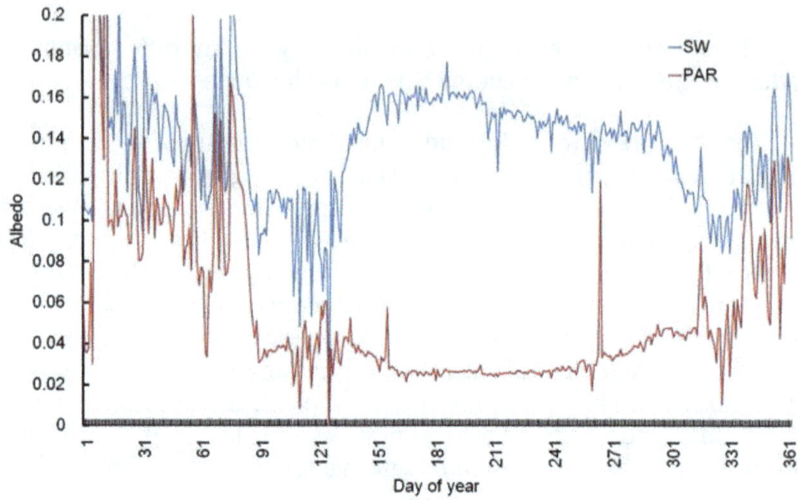

Figure 2.9: Mean midday (10 to 14h) values of global (SW) and PAR albedo measured during 2004 above forest at Prospect Hill Tract, Harvard Forest, Petersham (MA, USA) (Munger and Wofsy, 2017).

Similar values of solar and PAR albedo are typical for the leafless period when a strong increase in albedo is most likely caused by the presence of snow. After DOY 120, when leaves appear, PAR albedo decreases because absorption of PAR increases. At the same time, solar radiation albedo increases because of the extension of the leaf surface which reflects radiation. This situation continues until DOY 270, when leaves start to fall and both albedo values converge toward each other. Low albedo values in November and December are the result of radiation reflection from steam and, probably, from wet ground which is not too high.

Plant canopy absorbs radiation very efficiently. From radiation which enters the closed canopy, after a part of incident radiation is reflected, less than 10% reaches the ground. This portion, of course, strongly depends on canopy type, density and structure but also on leaf surface and orientation. Intensity of absorbed PAR greatly changes over the growing season. When the canopy is fully leafed (maximum LAI), absorption of PAR by the tree crown can reach 80% (Fig. 2.10).

In the Harvard forest, leaf emergence starts typically in April. Extension of the plant surface increases absorption of PAR. This is why PAR intensity within the for-est reaches its maximum level in April and then decreases until October when all the leaves fall.

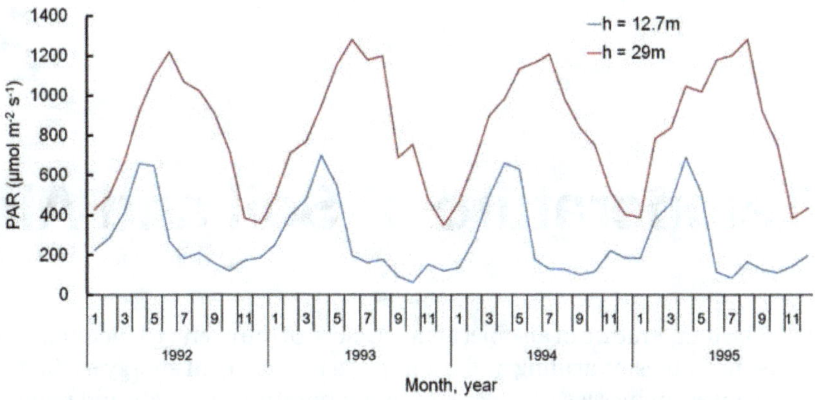

Figure 2.10: PAR monthly mean of midday values measured above and inside Prospect Hill Tract, Harvard Forest, Petersham (MA, USA) (Munger and Wofsy, 2017).

Under the same incoming radiation, plant type and canopy structure can significantly affect radiation balance at the surface (Fig. 2.11). Even during winter when plants are in hibernation, they affect energy balance components by their presence and variety specifics. For example, in the case of miscanthus plants, there is less negative balance because of the lower surface temperatures of the miscanthus, while in a barley field there is significant soil heat flux to the surface, producing a more negative balance. Additionally, the ground surface of a miscanthus canopy is covered with miscanthus leaves, which creates an insulation layer and reduces thermal emission from the soil surface, producing lower daily variations of net radiation (Eitzinger and Koessler, 2002).

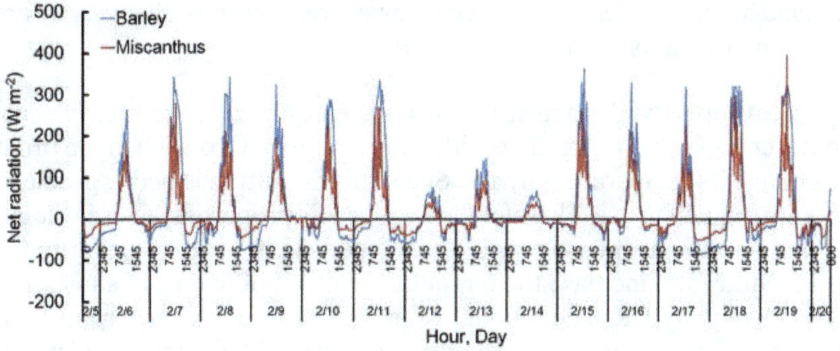

Figure 2.11: Daily variation of net radiation balance of miscanthus and winter barley during 5th to 20th February 1998 in Marchfeld, Austria (Source: BOKU-Met Weinviertel, Austria).

3

Temperature of Soil and Air

The temperature of an object is a measure of the average kinetic energy of its molecules. Warming and cooling are results of an energy exchange that causes an increase or decrease in temperature. The rate and magnitude of temperature change depend not only on the absorbed or emitted amount of energy but also on the thermal properties of the object.

Soil and atmosphere make up the plant growing environment, in which the plant life cycle occurs. Plant temperature is the result of the radiation balance at the plant surface and the energy exchange with the surrounding air – mostly by convection and with the soil – by conduction.

3.1 Heat, thermal properties and temperature

Heat is a form of energy. It can be absorbed or emitted by a body, resulting in a change of temperature.

Thermal properties are related mostly to physical quantities describing the ability of a body to treat heat: when a body stores heat, we speak of its *heat capacity*; when it transfers heat, we are dealing with its *thermal conductivity*; and when it transfers temperature changes, we are concerned with its *thermal diffusivity*.

Heat capacity, C is a measurable physical quantity numerically equal to the energy which should be added to or removed from a body in order to change its temperature by 1 K. Since this definition is body specific, it is quite difficult to use. Therefore, more specific properties, such as molar, C_n and specific heat capacity, C_h are introduced in order to quantify the heat required to increase the temperature of 1 mol or 1 kg of substance, respectively, by 1 K. The amount of energy required for a certain temperature change depends on whether energy is added or removed under constant pressure or constant volume. In the case of an ideal gas, all energy added under constant volume is transformed into the kinetic energy of the gas molecules, leading to a temperature increase. On the other hand, if energy is added under constant pressure, it will be partitioned into work for volume increase (in order to keep pressure constant) and the kinetic energy of gas molecules, i.e. a temperature increase. Hence, to achieve the same temperature increase, under constant pressure more energy is required than under constant volume; in other words, heat capacity at constant pressure, C_p (J kg^{-1} K^{-1}), is larger than heat capacity at constant volume, C_v (J kg^{-1} K^{-1}). The ratio between these two heat capacities is usually denoted by g, which for an ideal gas is equal 5/3. In the case of real gases, g changes slightly with temperature.

	ρ (10^3 kg m^{-3})	C_h (10^3 J kg^{-1} K^{-1})	K_b (J m^{-1} s^{-1} K^{-1})	D_h (10^{-6} m^2 s^{-1})
Air	0.00116	1.007	0.025	21.4
Quartz	2.65	0.84	8.8	3.95
Humus	1.4	1.9	0.25	0.094
Water	1.00	4.22	0.57	0.14
Ice	0.91	2.11	2.2	1.15

Table 3.1: Typical values for density, r, specific heat capacity, C_h, thermal conductivity K_h and thermal diffusivity, D_h, of some constituents of soils and geologic formations.

Thermal conductivity, K_h is a property of a substance describing its ability to conduct heat. It is numerically equal to the energy transferred in 1 s through the volume of unit bases (1 m^2) and height (1 m) when the temperature difference between opposite sites is 1 K.

Thermal diffusivity, D_h is the measure of thermal inertia. It is a material–specific heat property which measures its ability to conduct thermal energy relative to its ability to store energy. It can be calculated as the ratio of thermal conductivity, K_h and volumetric heat capacity, VHC where the latter is actually the product of density, ρ and specific heat capacity, C_h.

Soil is a composite environment, consisting of solids, air and moisture. In dry state, sandy soil has the highest thermal conductivity. However, water has much higher heat capacity than soil solids or air (Tab. 3.1). Because of this, overall thermal properties of soil vary with its moisture content. Thermal conductivity should always be reported together with moisture content.

3.2 Heating and cooling of the soil

Heating and cooling of the soil is the result of the energy balance of Earth's surface (see Chapter 2.2). Since electromagnetic radiation cannot penetrate through soil, a portion of energy left at the surface after reflection becomes transformed into heat. Intensive heating of the surface soil layer and energy conduction govern the soil temperature profile. Absorption and conduction of energy is greatly affected by surface and soil characteristics such as *colour and roughness*, *specific heat capacity* and *thermal conductivity*.

- *Colour and roughness*: A dark surface has high absorptivity, which reduces its albedo. A rough surface has a lower albedo than a smooth surface of the same soil due to intensive absorption of radiation caused by multiple reflections and to absorption of radiation on elements of roughness (see, Chapter 2).
- *Specific heat capacity*: In comparison to water, soil has a lower heat capacity, resulting in faster heating and cooling of the soil surface. However, moistening of the soil increases heat capacity.
- *Thermal conductivity*: Soil wetness affects thermal conductivity. Energy transfer from the soil surface to deep soil layers, and vice versa, is more efficient in wet than in dry soils.

Surface temperature, or so–called "skin" temperature, is a key factor affecting the temperature of the thin overlying part of the atmosphere, which is responsible for dew and frost formation. Subsurface soil temperature depends on heating of the surface layer and the soil characteristics affecting energy transfer. Heating and soil characteristics both change during the day and year and produce the diurnal and annual cycles of the soil temperature profile.

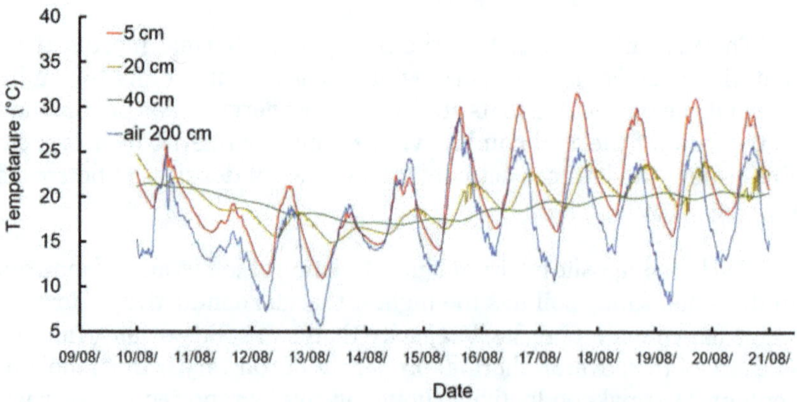

Figure 3.1: Daily variation in soil (grass cover) and air temperature during the summer 2016 in Goggendorf (Austria) (Source: BOKU–Met).

During the day, the surface soil layer attains its maximum temperature approximately one hour after maximum solar radiation, while its minimum is reached just before sunrise. However, a time lag in the occurrence of maximum and minimum soil temperatures increases with depth. That time is required for surface layer heating and energy transfer through the soil column, which greatly depends on the heat capacity of the soil. Daily variation in soil temperature decreases with depth until it reaches a constant temperature level (Fig. 3.1). The depth of this level depends on soil type and wetness, season and latitude. The daily course of soil temperature has a seasonal pattern, which is particularly evident at the surface layer (Fig. 3.2).

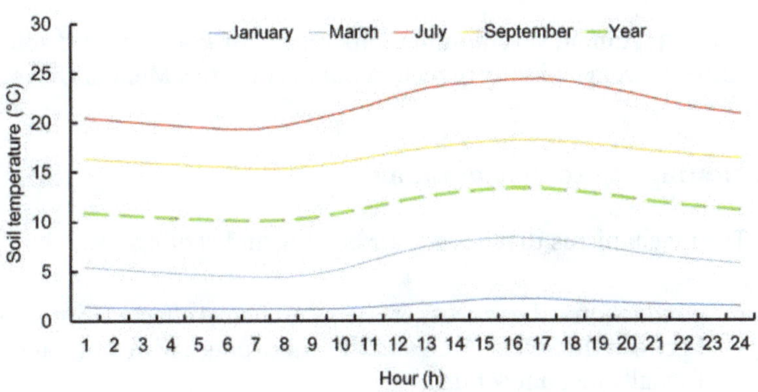

Figure 3.2: Seasonal variation in the daily course of soil temperature at 10 cm depth in Ridjica (Serbia) (2013–2017) (Source: PIS Vojvodina, Serbia).

In winter, surface heating during the day is lower than in summer, due to less daylight and the lower intensity of solar radiation. It produces a smaller daily variation in soil temperature during the winter and a greater variation during the summer. Additionally, daily variation in temperature increases at lower latitudes, due to the intensive diurnal heating and nocturnal cooling that take place as one nears the Equator. Consequently, at low latitudes the depth at which constant daytime soil temperature is achieved is greater during the summer.

Soil temperature (Fig. 3.3) in the Northern Hemisphere reaches its annual minimum in January and its maximum in July. Over the course of a year, variation in soil temperature decreases with depth until the level of constant temperature is reached. Annual changes are the cumulative effect of daily changes and because of that penetrate much deeper into the soil than daily variations.

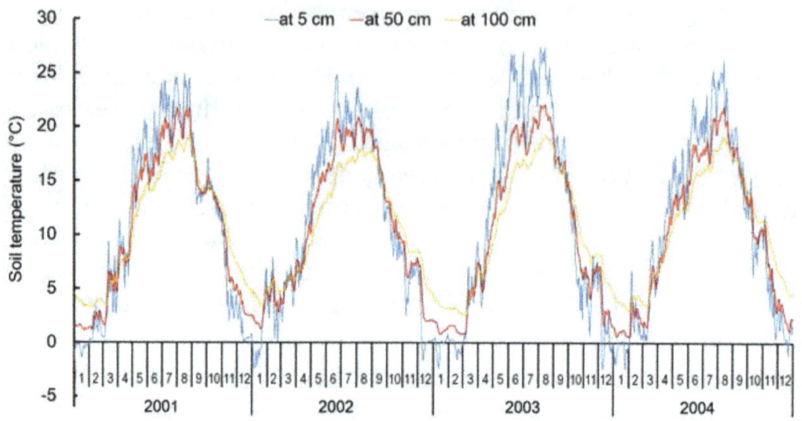

Figure 3.3: Annual variation in temperature for grass covered soil during 2001–2004 in Doksany (Czech Republic) (Source: Mendel University, Brno).

3.3 Heating and cooling of the air

The mechanisms that govern air heating and cooling are:

- *conduction* – the process of energy exchange effected by molecules between the soil surface and the atmosphere with a thin air layer (usually just a few mm);
- *turbulent mixing* – the exchange of energy (and substance) between the soil surface and the atmosphere and between different atmospheric layers by turbulent eddies within an atmospheric boundary layer. The thickness of this layer varies during the day, with a typical value of 1 km;
- *convection* – the rising of warmer and the descending of colder air, governed by buoyancy force;
- *radiation* – energy transfer by electromagnetic waves, which in the case of the atmosphere is dominated by the absorption of solar and terrestrial radiation (see Chapter 2.2). The content of the atmosphere can significantly affect atmospheric warming by radiation;
- *advection* – in the atmosphere, this is energy transfer governed by wind blowing from one region to another and bringing air of a different temperature;
- *evaporation and condensation* – phase changes of water in the atmosphere followed by the release or engagement of energy in the form of latent heat flux. This heat transfer is an important source and sink of energy for the atmosphere (23 % of the solar constant: see Chapter 2.2).

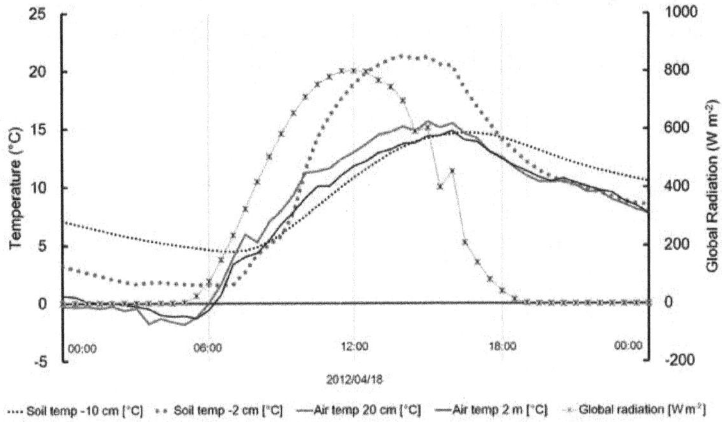

Figure 3.4: Daily variation of solar radiation, soil and air temperature in Groß–Enzersdorf (Austria) (Source: BOKU–Met).

During the day, at the height of a few meters above the ground, air temperature reaches its maximum 2–3 hours after the time of maximum solar radiation and 1–2 hours after that of maximum soil surface temperature (Fig. 3.4). This lag in time is required to produce soil heating by solar radiation and air heating by soil (and solar) radiation. The measurements of solar radiation, soil and air temperature presented in Fig. 3.4 are perfect examples of this phenomenon. The high intensity of solar radiation (800 W m^{-2} at noon) produces intensive soil heating at depths of 2 cm and 10 cm with a lag in time (the time needed for energy transfer) and a decrease in amplitude (the amount of energy left in the layer above 10 cm). Air temperature at 20 cm is higher than at 2 m during the day, since the 20 cm level is closer to the ground surface. At night, however, this difference disappears and in the case of intensive cooling after sunset can even be the opposite. We should also note that energy transfer is faster in air than in the soil: the time lag is therefore shorter and the temperature difference lower between 20 cm and 2 m when compared to the variance between these two measurements at depths of 2 cm and 10 cm in the soil.

Daily variation in air temperature depends on: latitude (higher if closer to the Equator), season (highest in summer), land use (highest in the case of sandy bare soil or rock), landscape (higher in lowlands), altitude (higher at lower elevations), cloudiness (higher under a clear sky) and vegetation cover (higher in the case of bare soil). However, daily variation in air temperature is lower above a water surface (never exceeding 1.7 °C).

The annual cycle of air temperature follows annual cycle of the underlying surface temperature (Fig. 3.5). The highest annual variation in maritime air is up to 20 °C, while for places deeper inland it reaches 60 °C. Annual variation in air temperature depends on: latitude (lower if closer to the Equator), land use (highest in the case of sandy bare soil or rock), altitude (higher at lower elevations), cloudiness (higher under a clear sky) and vegetation cover (higher in the case of bare soil).

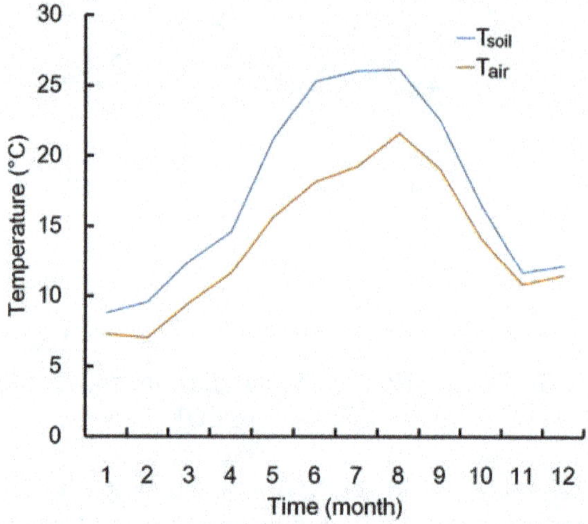

Figure 3.5: Average annual variation in air and soil temperatures for 2005–2011 in Elba (Italy) (Source: Regional Hydrologic Service of Tuscany).

Marine environments can significantly affect daily and annual variation in air temperature. Due to the higher heat capacity of water, oceans accumulate 16 times more energy than landmasses over the year. Therefore, bodies of water warm up more slowly during spring and summer and cool down more slowly in autumn and winter. During the winter, air temperature above water and above land regions affected by bodies of water is always warmer than above areas farther inland; on the other hand, this temperature is cooler in summer, with much smaller daily and annual variations (Fig. 3.6).

3.4 Impact of soil and air temperature on plants

Soil temperature and growth. From sowing till harvesting, heat and/or temperature are catalysts of many physiological processes governing plant development. Soil temperature affects plant growth directly and indirectly. In the case of crops, direct impact occurs through the effects of environmental temperature on germination and emergence of crops. For every species there is an optimal soil temperature for seed germination. At that temperature, the maximum number of seeds will germinate in the shortest time. The minimum temperature for germination of a cereal crop is 4 °C and the preferred temperature is 20 °C. Indirect effect is the result

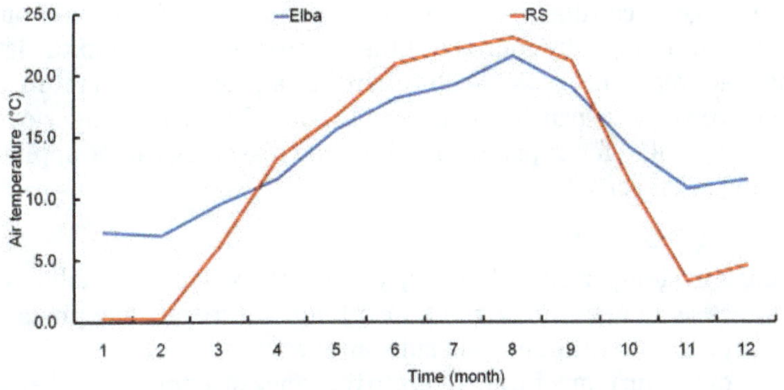

Figure 3.6: Annual variation in air temperature during 2011 in Elba (Italy) and Rimski Sancevi (RS) (Serbia) (Source: Regional Hydrologic Service of Tuscany and Republic Hydrometeorological Service of Serbia).

of soil temperature impact on root growth and water and nutrient uptake from the soil. Soil cooling decreases the rate of physiological processes while increasing the risk that the future plant will lack proper vigor. Acting together, soil temperature and humidity initiate and determine the germination of crops, the budding of trees and the emergence of leaves in a forest. The lower soil temperature thresholds for the germination of some crops are shown in Table 3.2.

Type	Minimum (°C)	Optimum (°C)	Maximum (°C)
Barley	3–4	20	28–30
Rye	1–2	25	30
Wheat	3–4	25	30–32
Sunflower	3	28	35
Corn	8–10	32–35	40–44

Table 3.2: Germination threshold soil temperatures of some major crops.

Air temperature and growth. In accordance with the typical processes of their development, the life cycle of living organisms is divided into growth stages. In general, the start of these phases is triggered and their duration limited by temperature. Typical temperatures, which are usually associated with plant production but completely applicable to all living organisms, are cardinal (base, optimum and ceiling point) and lethal (minimum and maximum) temperatures. The difference between these two ranges regards a plant's potential for recovery. With regard to the range of cardinal temperatures, a plant can revive even if out of the optimum range. By contrast, if the temperature is beyond the lethal limits, recovery is not possible. Both cardinal and lethal temperatures are specific to species, growing stage and locality. In the remainder of this chapter we will refer to plants, but the same reasoning can be applied to all living organisms.

Cardinal temperatures (Tabs. 3.3 and 3.4) are defined as follows:

a. base temperature is the lowest temperature at which metabolic processes result in a net substance gain;
b. optimum temperature refers to the range of temperatures between the top (upper threshold) and the bottom (lower threshold). The optimum temperature provides favorable conditions for plant development. The lower threshold temperature is the minimum temperature required for growth stage initiation. At temperatures below the lower threshold, plant growth is interrupted. After the temperature rises above this minimum, a plant continues to grow without damage or yield losses;
c. ceiling point temperature is the so–called failure temperature, at which grain yield falls to zero. This temperature is above the optimum but below the maximum lethal temperature.

		Components			Phenophases				
Temperature	Lethal limits	Leaf initiation	Shoot growth	Root growth	Sowing to emergence	Vernalisation	Terminal spikelet	Anthesis	Grain filling
Lethal minimum	-17.2			-20					
Lethal maximum	47.5								
Cardinal		-1.0	3.0	2.0^b	3.5	-1.3	1.5	9.5	9.2
Optimum		22.0	20.3	$<16.3^b$	22.0	4.9	10.6	21.0	20.7
Failure point T_{max}		24.0	>20.9	$>25.0^b$	32.7	15.7	>20.0	31.0	35.4

Table 3.3: Air temperature thresholds (°C) for wheat. With superscript "b" are denoted soil temperature values (Porter and Gawith, 1999).

Lethal maximum and minimum temperatures (Tabs 3.3 and 3.4) refer to critical temperatures beyond which a particular plant cannot recover. If air temperature–i.e., the plant temperature–crosses either lethal threshold, the plant will wither away or experience significant damage. Cold (or low–temperature) stress, either in the form of chill (0–15 °C) or frost (below 0 °C), is one of the major environmental factors affecting crop growth and productivity. For all plants, temperatures below 0 °C are critical because of water freezing in the plant tissue.

Crops	Lethal minimum	Cardinal	Optimum	Failure point T_{max}
Barley	−17.3 ~ −12.9			
Maize		33 ~ 38 (leaf photosynthesis rate)	35 (TLmean, pollen viability)	
Tomato		7 (leaf appearance rate, rate of progress to anthesis) 6–8 (leaf photosynthesis) 5.7 (fruit development and maturation)	22 (leaf appearance rate, rate of progress to anthesis) 30 (leaf photosynthesis) 26 (fruit development and maturation) 22 ~ 25 (individual fruit growth rate) 17 ~ 18 (fruit size) < = 26 (fruit set rate)	35[a] (vegetative growth)
Soybean		6 (time to anthesis) 13.2 (pollen germination) 12.1 (pollen tube growth)	26 (Time to anthesis) 23 (post–anthesis, SSGR, SS, yield) 30.2 (pollen germination) 36.1 (pollen tube growth)	39 (lethal, single seed growth rate, seed size, Seed harvest index) 47.2 (pollen germination) 47 (pollen tube growth)

Table 3.4: Air temperature thresholds (°C) for different crops. Superscript "a" denotes cardinal temperatures which can be applied to other phenophases (Luo, 2011).

Since only water increases its volume as its temperature decreases, ice has a larger volume than water. As it forms, ice damages plants from the inside. But for some plants, even temperatures far above 0 °C can be critical and cause chill injury. More about cold impact on plants can be found in Chapter 9.3, which discusses frost and its effects.

At temperatures above the upper threshold, plant development is interrupted and in some cases injury can be expected. Heat stress is defined as the rise in temperature beyond a critical threshold for a period of time sufficient to cause irreversible damage to plant growth and development. The scale of damage depends greatly on the plant growth stage, the duration of high temperatures, the variation in daily temperature and the rate of temperature change. A number of experimental studies have found that the effects of high temperatures are greater if the plant is exposed to high daily temperature variation and with sudden rather than gradual temperature increase (6 °C h^{-1}, for example).

The effect of high temperatures with respect to the growing stage is plant specific. In the case of wheat, for example, the effect of exposure at pre–anthesis on grain yield is greater than at post–anthesis. In extreme cases, up to 97% of variation in yield can be attributed to differences in grain number per square meter, which is closely related to a maximum temperature during the 4–day period in which 50% anthesis is reached. Air tempera-tures above 30 °C are particularly dangerous for maize 8–10 days after anthesis, reducing grain sink strength and yield. Soybean growth under high temperatures (30–38 ° C) in comparison to a lower temperature regime (22–30 °C) can lead to a 34% reduction in pollen production, a 56% reduction in pollen germination and pollen tube elongation of 33% (Salem *et al.*, 2007). Soybean seed yield at harvest maturity appears to be particularly sensitive to high temperatures at the late flowering/early pod filling stage. Researchers have recorded a yield reduction of 29% for a 10 °C increase over an eight–day span during this period (Ferris *et al.*, 1998). If tomatoes are subjected to high temperatures (26–32 °C) during their growing season, the number of fruits per plant can be reduced by 90% compared to growth at moderate temperatures (22–28 °C).

Crop (variety, location)	Base temperature	Growing degree days to maturity
Beans (Snap, South Carolina)	10.0	650–705
Corn (Sweet, Indiana)	10.0	1200–1500
Cotton (Delta Smooth leaf, Arkansas)	15.5	1040–1380
Peas (Early, Indiana)	4.4	600–650
Rice (Vegold, Arkansas)	15.5	930–1150
Wheat (Indiana)	4.4	1150–1320

Table 3.5: Accumulated degree days during growing period of different plants (Ahrens, 2010).

Accumulated degree days (DD) or growing degree days (GDD) are commonly used to describe the requirements of a living organism for heat but also to assess the rate of organism development under certain thermal conditions. For a given plant and growing stage, DDs are equivalent to the energy required for completing the "full package" of processes typical for a specific stage. This measure represents the sum of temperature differences between the average daily temperature and the base temperature (T_b) (Tab. 3.5). In plant production and protection, DDs are a common tool for assessing the developmental dynamics of plants and harmful organisms, for scheduling management activities such as irrigation, fertilization and sampling for pests and diseases, and for taking protective actions. The "Numerical examples" explains more about DD calculation.

Plant temperature and growth. The energy balance of the canopy and the resulting plant temperature are affected by plant type and growth stage as well as by the morphological and aerodynamic characteristics of the canopy (LAI, vegetation cover, roughness length, displacement height).

Plant type and growth stage affect plant temperature mostly through sensible and latent heat fluxes (the evapotranspiration rate). Adapting to its environment, a plant develops the size and angle of its leaves (which affect incoming solar radiation and outgoing sensible heat flux) as well as its stomata resistance (affecting transpiration). These characteristics all significantly affect canopy energy balance. Additionally, during the phase of intensive growth and biomass production, a plant tends to increase its rate of all exchange processes, especially transpiration.

The morphological and aerodynamic characteristics of canopy, such as LAI and vegetation cover (the portion of land covered by plants), affect energy partitioning by increasing bare soil fluxes for small LAI and vegetation cover. On the other hand, large LAI values increase latent heat flux in the energy balance equation. Roughness length and displacement height are aerodynamic characteristics of the canopy; they are described in more detail in Chapter 6.

For different plant canopies and plant structures, different temperatures can be expected under the same incoming radiation. More importantly, plant temperature can significantly differ from air temperature, even more so as the latter increases. In the course of the day, differences in air temperature within and above the canopy can also be significant. Leaf area development over the seasons plays an important role in causing this difference (Fig. 3.7). During winter, when trees are leafless, temperature difference is a result of the common variation of the vertical air temperature profile. In July, when a forest is fully leafed and the crown behaves as a barrier between the free atmosphere and the forest canopy air space, daily courses of air temperature are almost parallel, indicating negligible mixing between these two environments.

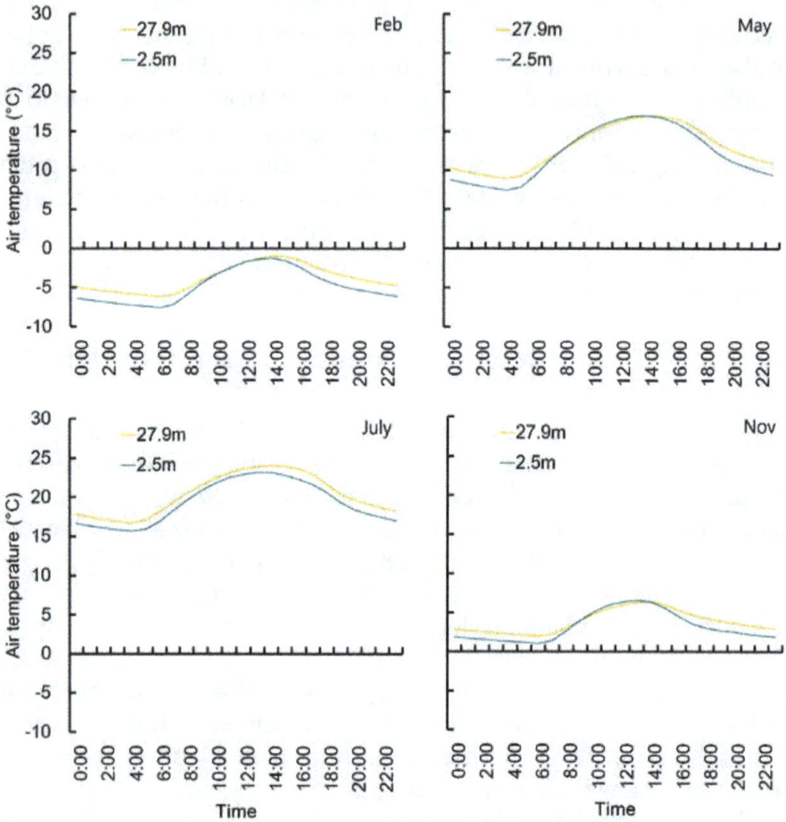

Figure 3.7: Daily variation of average air temperature (1991–2015) within and above canopy at Prospect Hill Tract, Harvard Forest, Petersham (MA, USA) (Munger and Wofsy, 2017).

Plant physiological processes and temperature. According to van Hoff's rule, the rate of chemical processes doubles with each 10 °C temperature increase. In the case of plants, this rule holds until the upper threshold temperature is reached. Beyond this temperature, some processes are diminished or interrupted. The most important plant physiological processes governed by temperature are photosynthesis, dissimilation, transpiration and the assimilation of nutrients.

Photosynthesis. Occurs in chloroplast where a plant manufactures its food. Using water and nutrients from the soil and CO_2 from the air, in the presence of chlorophyll and solar radiation (PAR) a plant produces sugars, starches, carbohydrates and proteins, and releases O_2. The intensity of photosynthesis is greatly affected by temperature and PAR intensity. An important feature of this process is saturation. At a certain PAR intensity, a threshold temperature is reached corresponding to a break in the rate increase. An increase in air temperature beyond the threshold causes a decrease in the photosynthetic rate.

Respiration. Using O_2 from the air, a plant converts sugars into energy, releasing CO_2 and water. Respiration occurs in the dark as well as in the light, and its rate increases with temperature. The consumption of the photosynthesis products during respiration reduces plant growth potential. During the day, photosynthesis overcomes respiration if the temperature is below critical (Fig. 3.8). Above this temperature, losses of photosynthesis exceed production, interrupting growth and reducing plant biomass.

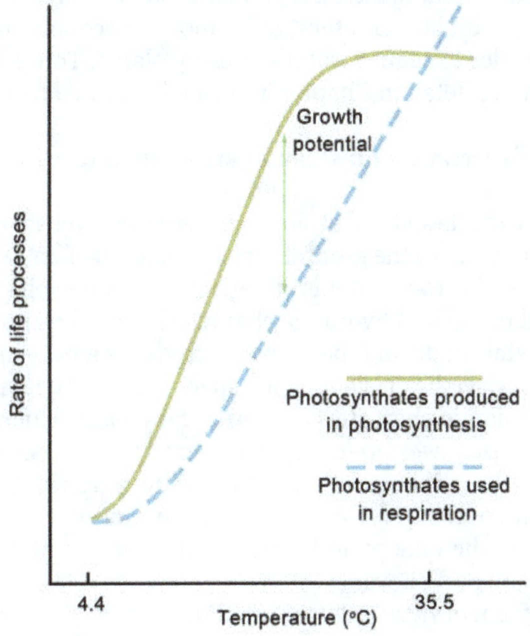

Figure 3.8: Example of temperature impact on the rate of photosynthesis and respiration.

Assimilation of nutrients. Soil temperature can significantly influence the movement of water and nutrients from the soil through the root system by affecting the hydraulic conductance of the soil and root permeability. The most important root functions, water and nutrient absorption, are temperature mediated. Temperature affects both passive and active water absorption by increasing permeability of the cell membrane as well as metabolic activity. A temperature increase from 13 °C to 28 °C causes water absorption by the root to double. A temperature increase from 5 °C to 22 °C is accompanied by a sevenfold increase in water uptake by the pea. An increase in crop root decreases hydraulic resistance; it also increases water uptake and exudation and the extrusion of osmotically active substances. Studies of the influence of temperature on nitrogen uptake suggest that ammonium is absorbed more readily than at low temperatures. An increase in K content is found with the increase in temperature. The same holds true for phosphorus and other nutrients.

Transpiration. It governs the process of water exchange between the plant interior and the atmosphere through the stomata, which control mineral uptake and transport, turgor pressure and plant cooling. Each step in this process is greatly affected by soil or air temperature. However, the most profound impact is related to stomata resistance and water deficiency of the surrounding air. At high plant temperatures and under dry air conditions, plant stomata tend to be more closed and increase their resistance in order to reduce water losses by plants. This subject will be discussed in more detail in Chapter 4, which is devoted to air humidity.

3.5 Impact of vegetation on soil and air temperature

Vegetation affects soil and atmospheric surface layer structure as well as the energy balance of the ground surface; hence it affects both soil and air temperature. The root systems of vegetation change soil structure and therefore its thermal and hydraulic characteristics. The appearance and growth of vegetation during the season introduces a new entity into the atmospheric surface layer–canopy air space. Changing its morphological and aerodynamic characteristics on a daily basis, canopy air space is the atmospheric lower boundary, with energy and turbulent transfer characteristics that differ greatly from the canopy atmosphere above. It is a significant source and sink of heat, water and momentum, whose presence affects the energy and water balance of the atmosphere and increases friction and buoyancy.

From the point of view of soil and air temperature, the most important impact of canopy air space is that on energy balance; this occurs through:

- changing the partitioning of energy components by reducing soil heat flux and reflected radiation (the albedo depends on the vegetation season), and
- introducing the short– and long–wave radiation which is reflected and absorbed by plants, the sensible and latent heat flux exchange between plant surface and the surrounding air, and the emission of long–wave radiation by plants.

Soil, plant and within–canopy air temperatures are the result of energy balance at the canopy top, the canopy air space and the ground surface. In

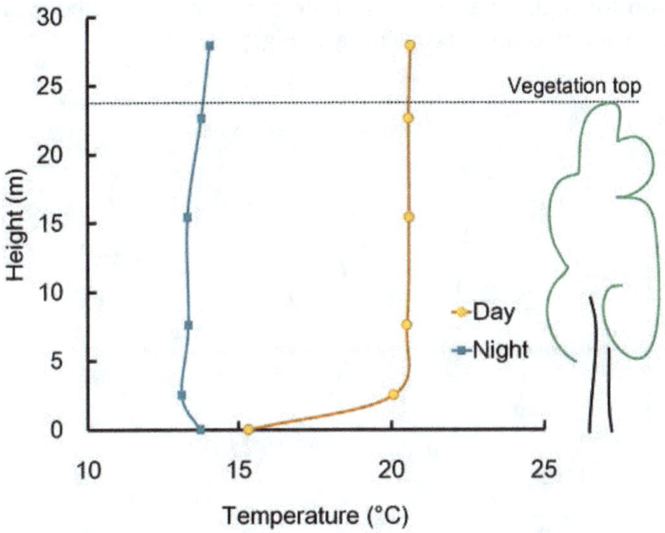

Figure 3.9: Vertical profile of average canopy air temperature during the day (06:00-18:00) and night (21:00-04:00) in June (1991-2015) at Prospect Hill Tract, Harvard Forest, Petersham (MA, USA) (Munger and Wofsy, 2017).

contrast to the soil, electromagnetic radiation can penetrate through the canopy top. Intensively attenuated over canopy height, soil surface reaches only 2–10 % of incoming solar radiation and heats it up. Significantly, reduced soil heat flux (in comparison to bare soil), combined with permanently changing soil structure and vegetation–related conditions on the ground surface, decreases soil temperature and its daily and annual variation.

Within the canopy, most solar radiation is absorbed by vegetation elements and used for plant warming and photosynthetic processes. This absorption produces a decreasing vertical air temperature profile for the canopy from top to bottom (Fig. 3.9). At night, when only long–wave radiation contributes to the energy balance, the Earth's long–wave radiation commonly overcomes atmospheric radiation, producing a temperature profile opposite to its daily course. Higher temperatures can be found closer to the ground, while at the canopy top air temperature reaches its minimum, because of intensive radiation.

The presence of vegetation reduces daily and annual variations in air temperature. A reduction in vegetation density and height increases annual variation in soil temperature, while producing a much lesser impact on air temperature amplitude. However, slight differences in extreme air temperatures may still be noticed. For example, over the years it is quite common for higher variations in daily temperatures to be measured in orchard canopy than in crop canopy (Fig. 3.10).

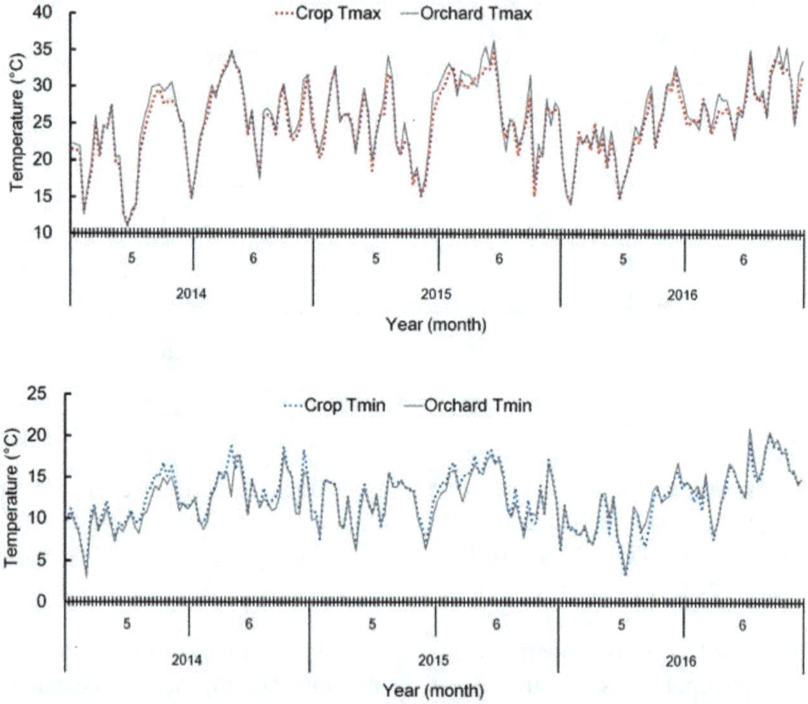

Figure 3.10: Variation of daily maximum and minimum air temperatures in apple orchard (Cenej, Serbia) and crop canopy (Kac, Serbia). (Source: PIS Vojvodina).

4

A Study of Air Humidity

Air humidity is a result of the atmospheric water cycle. As a consequence of numerous phase–changing processes, water transforms from ice to liquid, from liquid to water vapour and vice versa, providing sources and sinks of water in the atmosphere. The role of plants in the atmospheric water cycle is as important as the role of water in the plant life cycle.

4.1 Quantifying air humidity

Variables commonly used to describe air humidity are: (i) water vapour pressure, (ii) saturation water vapour pressure, (iii) vapour pressure deficit (VPD), (iv) dew point temperature, (v) relative humidity, (vi) specific humidity and (vii) absolute humidity.

Water vapour pressure (e). In a gas mixture, as in the atmosphere, water vapour pressure is the partial pressure of water vapour when equilibrium with ice or liquid water is established. This depends on the amount of water vapour in the atmosphere, temperature and atmospheric pressure.

Saturation water vapour pressure (e_s). If the amount of water vapour in the air is enough to initiate condensation at a given temperature, then the air is saturated and the water vapour pressure in the air reaches its saturation value. This depends only on temperature.

Vapour pressure deficit (VPD). The difference between saturated and unsaturated water vapour pressure at a given temperature is commonly used to describe the dryness of air, i.e. how much additional water vapour (expressed in pressure units) moist air can hold before becoming saturated.

Dew point temperature (T_d). If the air is saturated, its temperature is called the dew point temperature because any further temperature drop causes condensation and dew formation. This temperature varies according to atmospheric pressure and humidity.

Relative humidity (r). Commonly used to describe air humidity with respect to saturation, it is defined as the ratio of water vapour pressure to its saturated value at a given temperature.

Specific humidity (q) *and absolute humidity* (a). The mass of water vapour in a unit mass or unit volume of humid air is used to define specific humidity (g kg^{-1}) or absolute humidity (kg m^{-3}), respectively.

Annual and diurnal variations in energy balance as well as air and soil temperature affect the water cycle and therefore air humidity variations. Typically, water vapour pressure and relative humidity variations are analysed on different time and spatial scales.

Location–maritime or continental–determines vapour pressure diurnal variation. Above an open water surface, which is an unlimited source of water vapour, it is in accordance with the variation in air temperature which governs the water vapour supply through evaporation. Consequently, in maritime conditions (and in continental conditions during winter) the water vapour pressure maximum is at 2 pm and the minimum at sunrise. During the summer, far inland at mid–latitudes, diurnal variation of water vapour pressure has two minimums (4 am and 3 pm) and two maximums (9 am and 9 pm). What causes this phenomenon? In the morning, just before sunrise (4 am), the energy balance of the Earth's surface and the air temperature are at their daily minimums, leading to a minimum of evaporation and a minimum of water vapour supply to the atmosphere. As incoming energy increases, the intensity of evaporation and the amount of water vapour in the air above the evaporating surface also increase. During the first few hours after sunrise, when vertical transfer is still quite weak, accumulation of water vapour takes place, increasing its pressure to the maximum value, usually at 9 am. At midday, the surface and air warm up significantly, thereby enhancing the vertical (turbulent) transfer of water vapour and reducing it to the minimum level, which coincides with the maximum of atmospheric turbulent transfer in the early afternoon (3 pm). As vertical transfer decreases towards sunset, accumulation of water vapour takes place again, producing the second maximum of water vapour pressure (9 pm), even if the intensity of evaporation drops due to a decrease in incoming radiation and in surface and air temperatures. On an annual scale, water vapour pressure is in accordance with air temperature variation.

Diurnal variation of relative humidity is a direct consequence of the definition of relative humidity and the fact that saturation water vapour pressure depends only on air temperature. For this reason, diurnal (Fig. 4.1) and annual (Fig. 4.2) variations in relative humidity are inversely proportional to temperature variation. Figure 4.1 illustrates some of the points discussed in the previous paragraph regarding daily changes in air humidity. At night, differences between air humidity within and above the canopy are not significant. After sunrise, the relative air humidity at a height of 2 m begins to decrease due to the increase in air temperature; by contrast, at 20 cm humidity continues to increase even when the air temperature at this level begins to rise. This is the result of intensive evaporation of the ground (and potentially formation of dew), which can greatly increase air humidity in the plant canopy in the morning, thereby overcoming the effect of temperature on relative humidity.

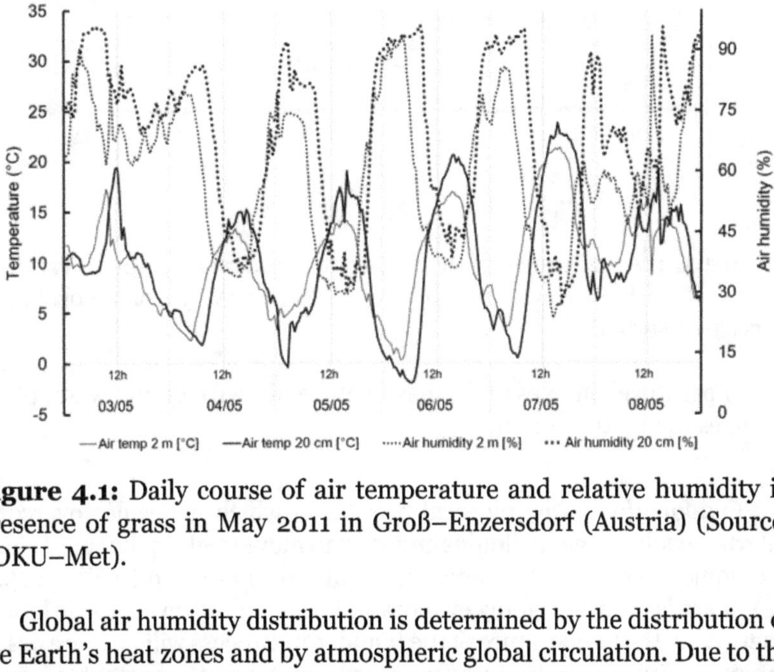

Figure 4.1: Daily course of air temperature and relative humidity in presence of grass in May 2011 in Groß–Enzersdorf (Austria) (Source: BOKU–Met).

Global air humidity distribution is determined by the distribution of the Earth's heat zones and by atmospheric global circulation. Due to the high intensity of evaporation at low latitudes close to Equator, relative humidity is high over the year, decreasing toward the north and south up to 30° latitude. At the poles, because of low temperatures, the maximum water vapour pressure is very low and therefore relative humidity is high.

4.2 Processes following the water cycle

Water is a triatomic polar molecule. Whether it is found as ice, liquid water or vapour is determined by the balance of attraction between the molecules and their kinetic energy. The attraction of intermolecular forces acts to keep molecules close to each other, as occurs in the water ice phase. However, an increase in kinetic energy (directly proportional

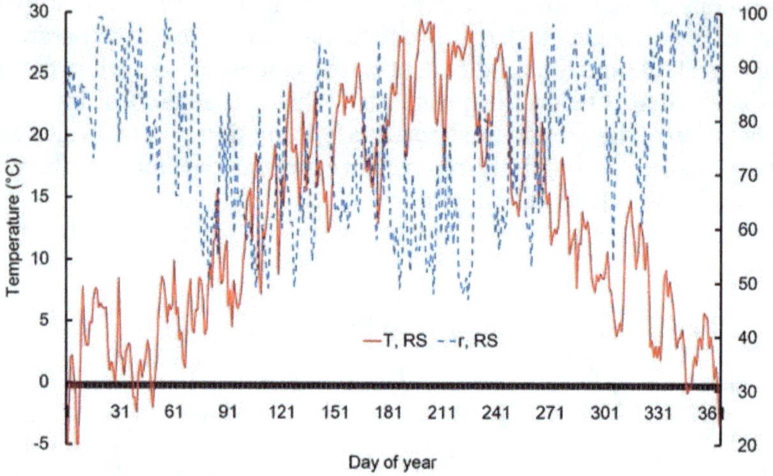

Figure 4.2: Anual course of air temperature and relative humidity during 2015 on Rimski Sancevi (Serbia) (Source: Republic Hydrometeorological Service of Serbia).

to temperature) enables molecules to move more freely and water phase changes to liquid or vapor.

Evaporation is the physical process in which molecules overcome intermolecular forces in liquids and start to move freely in the space above the liquid surface. At temperatures above absolute zero (0 K = -273.15 °C), there is always ongoing evaporation. Depending on available kinetic energy, i.e. the temperature of the liquid, this process will be more or less intensive. Energy needed to evaporate 1 kg of water at a constant temperature is called the latent heat of vaporization; it varies from 2264.76 kJ kg^{-1} (at 100 °C) to 2500 kJ kg^{-1} (at 0 °C). The intensity of water evaporation is often related to the vapour pressure deficit (VPD), i.e. the ability of surrounding air to "accept" additional water molecules. The common conclusion is that when air is saturated and VPD is zero evaporation cannot take place–but this is not quite correct. Specifically, when the air above a liquid surface is saturated, molecules can leave this surface and evaporate, but some other molecule, with lower energy and close to the liquid surface, will be captured by surface molecules and transformed to liquid, giving rise to *condensation*. This means that evaporation and condensation are continuous processes: the difference between the number of molecules leaving and those returning to the liquid determines whether we are dealing with an evaporation or condensation process.

Under constant pressure and in a specific volume, the probability of aggregation with other gas molecules increases as average kinetic energy (temperature) decreases. This explains why maximum vapour pressure, i.e. the number of molecules required for condensation at a certain temperature, decreases with temperature and depends on temperature only. By contrast, under constant temperature an increase in atmospheric pressure reduces the mean free path of molecules and increases the probability of collisions, i.e. the intensity of condensation. Of course, the same effect occurs if, under constant air temperature and atmospheric pressure, water vapour is added.

Depending on the mechanism of the water–to–liquid transformation involved, condensation can occur as homogenous or heterogeneous. Homogenous condensation takes place in pure, supersaturated air without solid boundaries. Heterogeneous condensation, commonly found in the atmosphere, starts on small particles (typically 0.2 mm in diameter) called *condensation nuclei*. Condensation nuclei, such as dust, black carbon, sea salt, volcanic products and phytoplankton, have a different composition but are always hygroscopic, meaning that they can take up water at humidity levels well below saturation. Droplets formed on condensation nuclei play a major role in the formation of haze at a relative humidity as low as 70%.

Melting. The solid state of matter is characterized by the presence of intermolecular forces much stronger than the kinetic energy available for work against these forces. Solids retain fixed shape and volume. Melting is a physical process which occurs when the energy added to a solid state system is enough to overcome intermolecular forces and form a liquid state. The inverse process is *freezing*: as a result of removing energy from liquid, the kinetic energy of molecules is reduced and attractive intermolecular forces become dominant, forming a solid state. However, when a solid state turns directly into gas, this physical process is called sublimation. In the literature the same term is used to describe the opposite process as well.

4.3 Atmospheric evaporation and condensation

The Earth's surface is a permanent source and sink of energy and humidity for the atmosphere.

Evaporation from an open water surface is caused by net radiation, air temperature, humidity. It is also influenced by the wind, which reduces air humidity and increases the removal of water molecules from the surface, i.e. increases the intensity of evaporation.

Evaporation from bare soil is affected by the same meteorological conditions as evaporation from an open water surface, but also by soil characteristics such as: soil moisture, type and structure, salinity and the presence and level of groundwater. If soil is saturated with water, the intensity of evaporation reaches its maximum under given meteorological conditions; this is called *potential evaporation*.

Vegetation gives off water into the atmosphere as a result of two processes: the physical process of evaporation from the vegetation surface and the physiological process of transpiration, which is actually the diffusion of water vapour from the plant interior to air space through the stomata. A common name for evaporation from a plant surface and transpiration from its interior is evapotranspiration; this phenomenon will be explained in more detail in the sections below. If the ground surface is fully covered by plants and if there is no water deficit in the soil, evapotranspiration reaches its maximum intensity with respect to actual meteorological conditions: this is called *potential evapotranspiration*.

Condensation, followed by the release of the latent heat of condensation, is a powerful source of energy for the atmosphere. Condensation of water in the air occurs when:

a. moist air flows over a colder surface, cooling significantly as it comes into contact with cold objects;
b. air temperature decreases due to intensive irradiation of an underlying surface;
c. the mixing of warm and cold air masses occurs;
d. air is forced upward, greatly reducing its temperature due to pressure decrease, while keeping the initial amount of unsaturated water vapour. At a certain point, air will cool enough to reach the dew point temperature for a given amount of water vapour.

Products of condensation – fog. Fog is a result of water vapour condensation in the near–surface air. When small water droplets or ice crystals are suspended in the air, visibility is reduced to 1000 m in the case of aviation fog, to 180 m in the case of thick fog and to 50 m in the case of dense fog. Fog is commonly called a (stratus) cloud on the Earth's surface, since the mechanism of its production is very close to that of cloud formation.

Depending on whether a process leads to condensation or near–condensation conditions (a difference between air and dew point temperatures of less than 2.5 °C), different types of fog are produced.

Advection fog is the result of the horizontal transfer of warm moist air over a much cooler land or water surface, which causes it to cool below its dew point temperature. This type of fog is often found above a cold water surface. In addition, when cold air moves over a warm and moist surface, steam fog can be formed. In some areas of the world, advection fog in particular is a significant source of water, which has given rise to a new technique called fog harvesting.

Evaporation fog appears when cool air lies above a warmer water surface. In this case, water evaporates into cold air and increases its water vapour content, resulting in more pronounced condensation conditions.

Radiation fog is the result of condensation which takes place in near–surface air due to the Earth's nocturnal cooling; it usually occurs under a clear sky and during calm winter nights. This fog dissipates after sunrise when the ground sufficiently warms, but the presence of inversion prolongs the duration of fog.

Frontal fog is related to changes in atmospheric condensation conditions due to the presence of frontal zones or a front passage. Fog appearing before a warm and after a cold front passage is called *precipitation* or rain fog. It is associated with light precipitation falling into cold stable air in a frontal zone, raising water vapour content and therefore the dew point of air. During a front passage, the mixing of warm and cold air masses leads to condensation conditions.

Upslope fog is formed when moist air is forced to rise by a steep hill or ridge. As a result, air cools to its dew point temperature, producing fog and clouds on the upwind side of the mountain range.

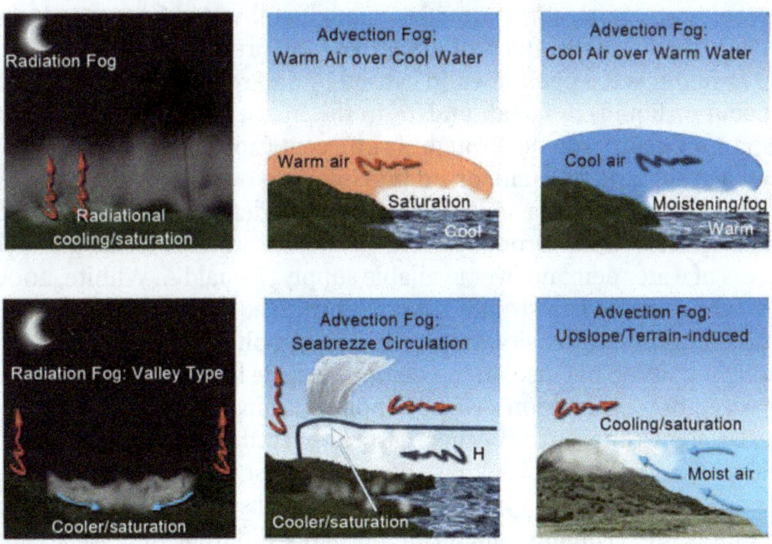

Figure 4.3: Types of radiation and advection fog.

4.4 Impact of air humidity on plants

Plants are very sensitive to air temperature and humidity variations. Both high and low air humidity affect plant development, pollination, disease resistance and final yield.

High relative humidity reduces the intensity of transpiration and can even stop it. Of course the slightest reduction of humidity produced by a light wind, for example, initiates the process again. Low transpiration brings higher turgor pressure in plant cells, which increases pressure on the elastic cell walls, causing cell expansion. This is one of the reasons for the intensive growth of tropical plant leaves. Pollination is highly sensitive to air humidity. High relative humidity can reduce pollen dispersal to distant locations. Warm and humid conditions are the ideal environment for the development of many different plant diseases as well as of some pests (thermophile insects) and weeds (crop–competing plants).

Low relative humidity increases stomata resistance and therefore decreases gas exchange between plants and their environment, decreasing the intensity of some of the most important physiological processes, such as photosynthesis and transpiration. In the case of low humidity, even irrigation effects are reduced. Specifically, under the same air temperature and photosynthetically active radiation (PAR) intensity, a lower yield is obtained if water for growth is provided by irrigation rather than precipitation. The reason is air humidity. The impact of irrigation on air humidity is much less pronounced than is the case with precipitation. Hence, even if irrigation provides enough soil water for normal plant growth, the negative effects of dry air are still present.

Drought is the phenomenon caused by and related to a lack of water (vapour or liquid) in the air and/or in the soil; it has the most profound impact on plant growth. Drought is an event (not only weather) following a period of inadequate or no precipitation over an extended period of time which creates a soil and air moisture deficit and a severe water balance disturbance. Drought is not the same as water scarcity, that is, an excess of water demand over available supply (Donald A. Wilhite, 2000). Drought severity is also difficult to determine, as it depends not only on the duration, intensity, timing and geographical extent of a specific drought episode but also on the demands made by human activities and by vegetation on a region's water supplies (Blain, 2012).

Drought differs from other natural hazards (e.g., floods, tropical cyclones, and earthquakes) in many ways. First of all, it is a "creeping" phenomenon (Gillete, 1950) that slowly sneaks up and impacts many sectors of the economy, operating on many different time scales. It is characterized in this way because its effects accumulate slowly over a long period of time and because it is difficult to determine the onset and the end of a drought episode (this is an objective of drought monitoring systems (DMS)).

Naturally, drought might have a specific definition in relation to a certain region and a certain type of human activity. In the first case, each region has specific climate characteristics, while in the second, agriculture, for example, could be the principal economic sector to be affected by drought, because soil moisture supplies are usually exhausted quickly. The impact of drought on crop yield may vary from crop to crop–from maize to wheat, for example–because of their different planting and growing times.

There are numerous classifications of drought based on: a) the system and source of available water; b) its time of appearance and duration; c) the medium of its cause.

According to the system in case and source of available water, drought can be meteorological, hydrological, agricultural or socioeconomical, following the most common classification of its types (Wilhite and Glantz,1987).

Meteorological drought is related to a deviation of air temperature and precipitation from normal values (30–year averages) during a specific period (month, season or year). However, the parameters selected and criteria chosen for the definition of meteorological drought depend on the climate characteristics of a particular region. If a climate is characterised by seasonal precipitation patterns, then deviations in the amount, intensity and distribution of precipitation are measures of precipitation deficit and drought. However, in regions regularly supplied with precipitation, like the west coast of Scotland, which has an average annual precipitation of 3000 mm during 250 rainy days, an increase in the number of consecutive days without precipitation is a measure of drought. With respect to severity, meteorological drought is classified according to the recorded precipitation during the period in question as follows: slight (11 – 25% less than the long–term average), moderate (25 – 50% less) and severe (more than 50% less).

Hydrological drought is determined by the propagation of meteorological drought through the terrestrial hydrological cycle: it is therefore influenced by the properties of the hydrological cycle (Van Lanen, 2006; Vidal *et al.*, 2010). This type of drought is expressed as the reduction in the amount of water in rivers, streams and lakes and also as ground water depletion. It appears with a certain time lag with respect to meteorological drought and affects both the power industry and agriculture. The impact of hydrological drought is not very easy to determine compared to that of agricultural drought: for example, a precipitation deficit is soon visible in a crop system, while the impact of this deficiency on groundwater level or reservoir levels may not affect hydroelectric power production or recreational uses for many months.

Agricultural drought occurs when air and soil moisture are inadequate to meet current plant water demand. Plant water demand depends on meteorological conditions, plant type, growing stage and soil characteristics (physical, chemical and microbiological). Agricultural drought should therefore be considered as a plant–specific event. More about agricultural drought and the different indices commonly used to describe the intensity of drought can be found in Chapter 9.

Socioeconomic drought is a result of meteorological, hydrological and agricultural drought (one of these or their combination) and appears when demand for an economic good exceeds supply as a result of a weather–related shortfall in water supply (Wilhite, 2005).

According to Thornthwaite, with respect to the time of its appearance and its duration, drought can be classified as:

- *permanent*–appears in regions with an arid climate where an inadequate amount of precipitation results in crop production becoming dependent on irrigation;
- *seasonal*– usually caused by large–scale seasonal circulation in regions with clearly defined wet and dry periods during the year (monsoon areas, for example);
- *contingent* – the result of high variability in the amount and intensity of precipitation, typical for regions of humid and sub–humid climates. If it coincides with the growing season of crops, it can produce significant yield reduction.

Drought can also be classified as atmospheric or soil drought. *Atmospheric drought* occurs when air humidity is very low. It is commonly the result of high temperatures and a reduced amount of precipitation, but not necessarily. When the quantity of precipitation is normal, this type of drought can be caused by strong warm winds, which increase the vapour pressure deficit in the air. A good example is Sukhovey wind, which typically blows in the Caspian region and in Kazakhstan. This is wind of moderate speed (5 – 20 m s^{-1}), accompanied by high temperatures (20 – 25 °C) and low relative humidity (less than 30 %). During the growing season it can trigger excessive evapotranspiration and significant water stress in plants. If atmospheric drought lasts long enough, it produces a significant reduction in soil moisture content, initially in the surface soil layer but afterwards spreading deeper into soil, affecting the root zone of crops–this is *soil drought*.

Leaf wetness refers to the presence of liquid water on the leaf surface. There are three sources of water for the surface of leaves: (i) precipitation (interception), (ii) overhead irrigation and (iii) dew. Three quantities commonly used to describe leaf wetness are the amount of water retained per unit leaf area, the portion of the leaf covered by liquid water and the duration of leaf wetness. These leaf wetness properties depend mostly on plant specific characteristics (leaf area and angle, surface wettability) and meteorological conditions, and can significantly affect the water balance of individual plants and canopy. The maximum amount of water which can be retained per unit leaf area before it starts to drip off varies among species, ranging from 0.1 to 500 ml m^{-2}. Plants can benefit from low as well as high water retention, i.e. leaf wetness. For example, low water retention improves water balance under dry conditions by allowing water to easily reach the soil before it evaporates from the plant surface. On the other hand, in the case of excessive rain, interception can delay precipitation inflow to the soil and provide optimal partitioning of water balance components. For many pathogens (apple scab, cedar apple rust, late blight potato, tomato early blight, etc.), the duration of leaf wetness plays a crucial role in the development of the pathogen infection rate.

Since water evaporates from the wet leaf surface as from any other open water (or water–covered) surface, air temperature and humidity as well as wind speed predominantly affect water removal from the leaf surface. It is important to note that a plant surface covered with water is not available for transpiration because retained water covers stomas and actually affects gas exchange between plants and their environment.

4.5 Impact of plants on air humidity

Plants can affect air humidity directly and indirectly. Directly, plants act as a source of water vapour coming from the plant interior (transpiration) or the plant surface (evaporation of water retained on the surface). Therefore is relative humidity measured in plant canopy commonly higher than above bare soil or meteorological shelter particularly during summer (Fig. 4.4).

Section 4.2 introduced the processes of transpiration and evapotranspiration. These processes are so important for plants as well as for the atmosphere that they deserve more attention. Since it is quite difficult to distinguish water vapour flux by transpiration from that by evaporation, in the remainder of the text only evapotranspiration (ET) will be considered.

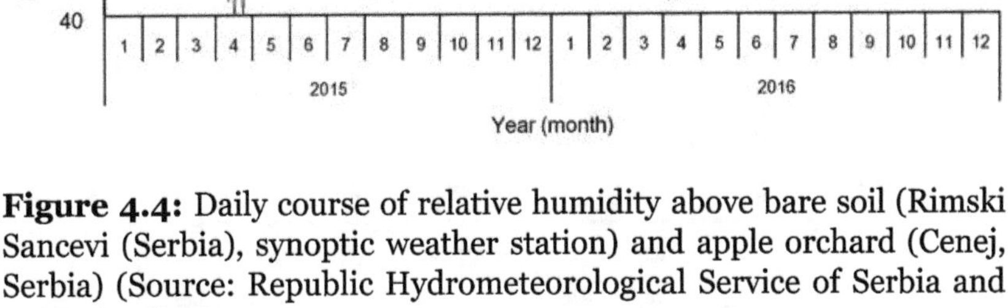

Figure 4.4: Daily course of relative humidity above bare soil (Rimski Sancevi (Serbia), synoptic weather station) and apple orchard (Cenej, Serbia) (Source: Republic Hydrometeorological Service of Serbia and PIS Vojvodina).

Meteorological conditions, crop and soil characteristics, environmental conditions and crop management affect the intensity of evapotranspiration. Depending on the level of complexity, i.e. the number of parameters involved, three types of evapotranspiration are commonly considered (Allen *et al.*, 1998):

a. reference ET,
b. ET under standard conditions and
c. adjacent ET corresponding to the actual ET from the vegetated surface.

Reference evapotranspiration, ET_0 is evapotranspiration from an optimally wetted soil surface fully covered by the reference vegetation, in most cases short grass. The intensity of reference ET depends only on meteorological conditions.

Evapotranspiration under standard conditions, ET_c often called *potential evapotranspiration*, is a process that occurs from an optimally wetted soil surface fully covered with the selected crop, which is disease–free and optimally fertilized (reference canopy conditions). In addition to meteorological conditions, the intensity of this type of evapotranspiration is strongly affected by characteristics of the soil (type, structure, salinity, fertility) and the crop (variety, growing stage, stomata number, distribution and openness). The calculation of ET_c is commonly made by the crop coefficient (K_c) approach, assuming that the potential evapotranspiration from any vegetated surface can be expressed with respect to the reference ET ($ET_c = K_c ET_0$). The crop coefficient (K_c) is a time dependent variable (Fig. 4.5), which introduces specific crop characteristics into the calculation of potential evapotranspiration and should be experimentally determined for at least three of the most important periods during the growing season: initial stage, mid–season and late season.

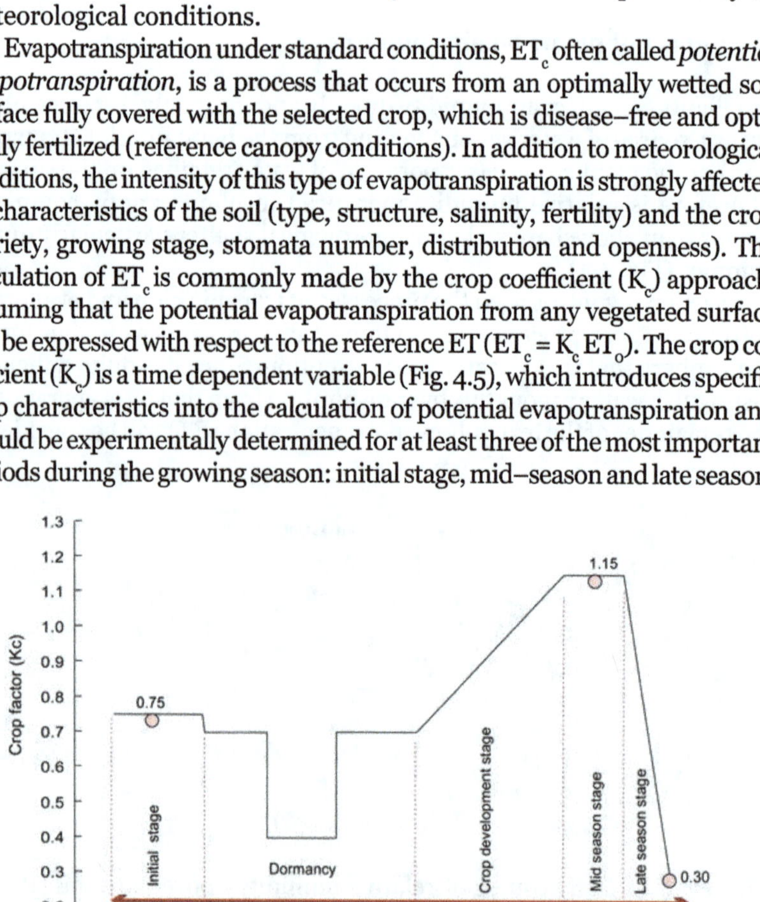

	Oct	Nov	Dec	Jan	Feb	Mar	Apr	May	Jun	Jul
Kc	0.75	0.75	0.70	0.40	0.70	0.78	0.97	1.12	0.92	0.30
Eto (mm day-1)	1.66	0.77	0.45	0.48	0.91	1.79	2.82	3.92	4.43	4.72
Etc (mm day-1)	1.25	0.58	0.32	0.19	0.64	1.40	2.74	4.39	4.08	1.42

Figure 4.5: Winter wheat crop coefficient K_c variation during growing season (Jacimovic, 2012).

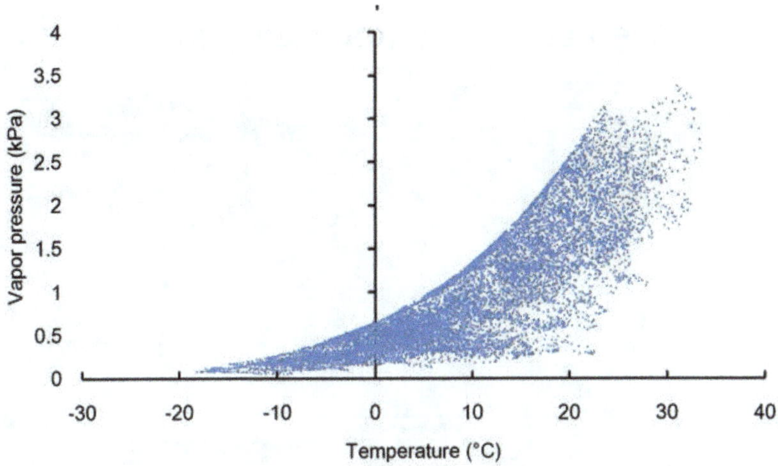

Figure 4.6: Measurements of water vapor pressure in the atmosphere during 2006 at Hemlock Tower, Prospect Hill Tract (Harvard Forest) (NY, USA) (Munger and Hadley, 2018.).

Actual ET is the real amount of water vapour obtained from a particular canopy under current meteorological, soil and plant conditions. The intensity of actual ET (ET_p) is determined by using the crop coefficient approach, whereby the effects of deviation from reference canopy conditions are taken into account using the adjustment factor (K_{adj}). In the case of a high water vapour deficit or a low quantity of available soil water content, plants can close their stomata (increase stomata resistance) and reduce ET_p. However, this is not an instantaneous process. If a high water vapour deficit appears suddenly, as in the case of Sukhovey wind, the stomata have no time to close and the plant loses water rapidly.

On all time and spatial scales, vegetation is an important source of water vapour for the atmosphere. It increases specific air humidity, but its impact on relative humidity also depends on air temperature and the capacity of the air to hold water. Measurements of water vapor pressure at the Hemlock Tower, Prospect Hill Tract (Harvard Forest) (NY, USA) (Fig. 4.6) depict two interesting facts related to air humidity and the impact of air temperature on it. First, at the same temperature, water vapor pressure is higher on the lowland than at the ridge top, which is, obviously, better ventilated. Secondly, there is a clear upper limit of water vapor pressure at a given temperature. This is, indeed, the definition of saturation water vapour pressure and argues in favour of the statement above (Chapter 4.1) that E depends only on air temperature. If tracing threshold values with respect to temperature produces an exponential curve, this describes the dependence of saturation water vapour pressure on temperature. Its mathematical formulation is well known the Clausius–Clapeyron relationship (see more in “Numerical examples”).

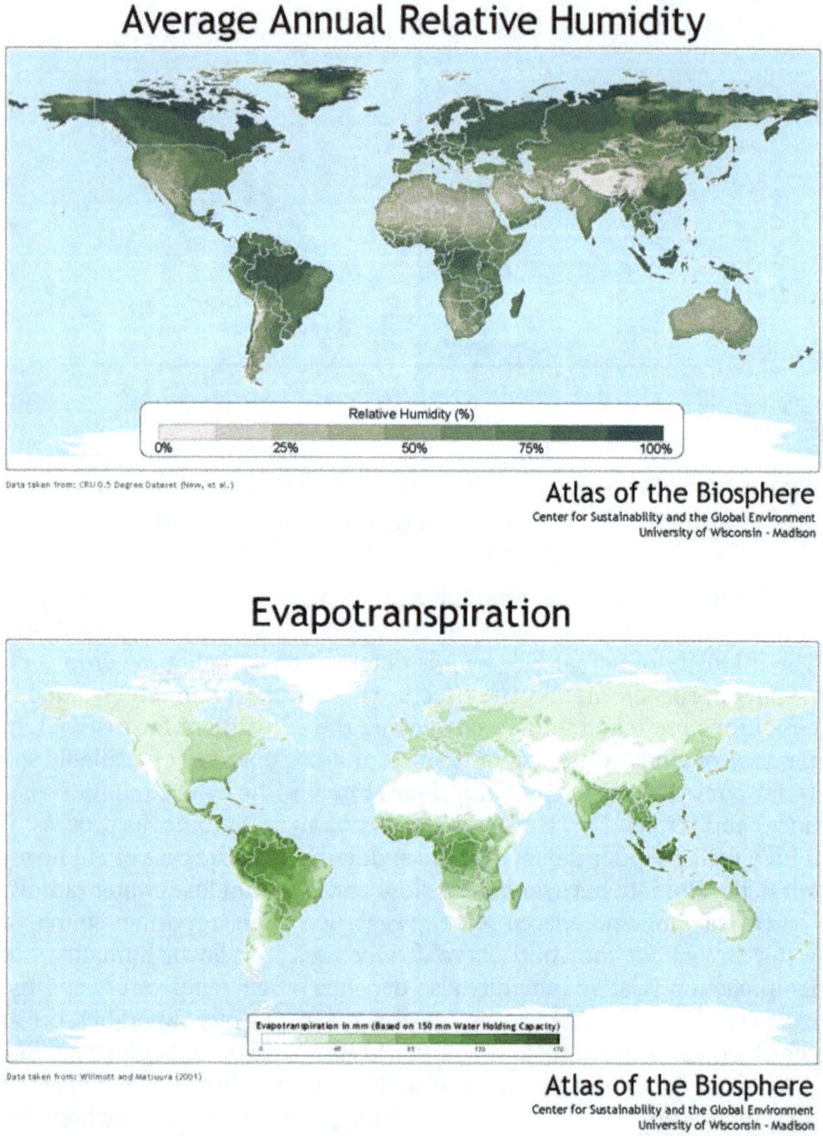

Figure 4.7: Global distributions of annual relative humidity and evapotranspiration (Source: Atlas of the Biosphere, https://nelson.wisc.edu/sage/data–and–models/atlas/maps.php?catnum=3&type=Ecosystems).

Vegetation over a large spatial extension, as in rain forests in Brazil, has a predominant impact (c. 50 – 75 %) on annual relative humidity and evapotranspiration (Fig. 4.7) and the formation of precipitation clouds (Fig. 4.8).

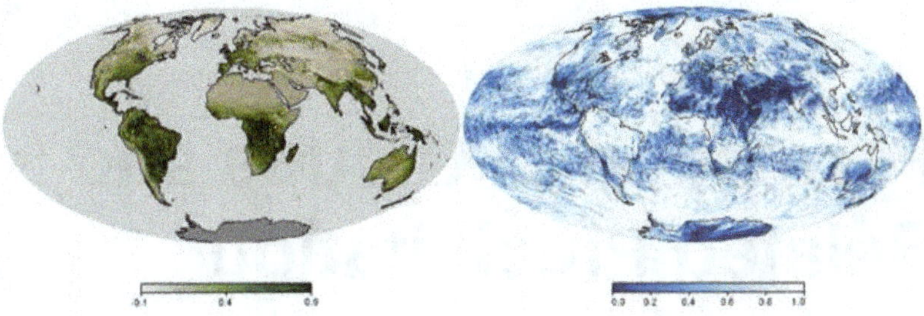

Figure 4.8: Global distributions of vegetation cover (left) and cloud fraction (right) for February 2000 (Source: NASA Earth Observatory, https://earthobservatory.nasa.gov/GlobalMaps).

Different plant types have significantly different impacts on air humidity. Under sufficient soil water conditions, the intensity of nocturnal transpiration affects air humidity and the dew point of canopy and the surrounding air space. Good examples are temperate C_3 plants like wheat and C_4 tropical plants like maize. In the case of wheat, it is typical for stomata to close at night to reduce transpiration. Without advection of humid air at night, the dew point will decrease by a few degrees °C. Dew can be formed in the wheat canopy only if air temperature drops to this temperature. By contrast, maize is a tropical plant which continues transpiration at night by increasing air humidity and leaving its dew point in the morning just 1–2 °C below its sunset value. This creates conditions for frequent due formation in the maize canopy.

5

Formation of Clouds and Related Precipitation

Whether thin as a spider's web and high in the sky, or dark as iron and almost at our "fingertips", clouds are the most impressive visible product of water vapour condensation as well as many other processes which define their type, content, height and scale. The presence of any cloud affects the energy balance of the atmosphere and the Earth's surface. Like snow, white cumulus clouds cause multiple reflections of solar radiation, producing much higher solar intensities in comparison to the top of the atmosphere. Precipitation is the fall of water, ice, or snow from the atmosphere to the surface of the Earth. But before considering how cloud formation and precipitation begin, some important points related to adiabatic raising and atmospheric stability have to be explained.

5.1 Adiabatic processes and atmospheric stability

In Chapter 4 we discussed how rising air affects the diurnal variation of water vapour pressure and upslope fog formation; we also saw how radiation fog can persist for several hours, keeping high humidity conditions near the surface. But it is not quite clear why this happens. In order to explain why air rises on some occasions and not on others, it is essential to introduce atmospheric adiabatic processes and explain the concepts of air parcel and atmospheric stability.

An *air parcel* is a volume element of air used to explain the behaviour of surrounding air. It is small enough to possess nearly uniform properties within its whole volume, maintaining all the basic thermodynamic and dynamic properties of the air which it represents.

An *adiabatic process* is one that occurs without the exchange of matter (energy and substance) between a thermodynamic system and its surroundings. According to the First Law of thermodynamics, energy brought to the thermodynamic system (Q) can be spent on changes in the internal energy of the system (ΔU), i.e. temperature change and work performed by the system (A), such as volume expansion. It can be expressed as

$$Q=\Delta U+A \tag{5.1}$$

In the case of adiabatic processes (Q = 0 J and A = -ΔU), there is no energy exchange with the surroundings. If any work is performed it is done at the expense of the internal energy of the system, expressed in the form of a temperature decrease.

In light of the definition of adiabatic processes, the question arises whether real world air parcel can be considered as an adiabatic system and its movement as an adiabatic process. Because predictions based on this model work, the answer is affirmative.

How does this manifest itself in the atmosphere? Let us suppose that due to the interaction of an air parcel with cooler and denser surrounding air, buoyancy force is exerted. Buoyancy will force the air parcel to rise as long as it is warmer and less dense in comparison to its surroundings. While rising, the air parcel will be exposed to reduced atmospheric pressure, which will lead to an increase in its volume and therefore a decrease in its temperature. At a lower temperature, the initial amount of water vapour will produce increased relative humidity and the air parcel will become more saturated as it ascends (see Chapter 4). As long as the relative humidity of the air parcel is less than 100%, its rise is called *dry adiabatic*, while the decrease in temperature with height is called the *dry adiabatic lapse rate* (Γ_s = 1 K/100 m). If the air parcel cools down to the dew point temperature, condensation occurs, and the latent heat of condensation warms up the air parcel, reducing its temperature decrease. This kind of rise is called *moist adiabatic*. Specifically, the *moist adiabatic lapse rate* (Γ_w), is always less than that of the dry adiabatic but closely depends on the air parcel's initial water vapour content. Commonly used values in examples and calculations are 0.4–0.8 K/100 m.

The *static stability* of any physical system, including an air parcel, is defined with respect to static equilibrium, the state in which forces are all balanced. Equilibrium can be stable, when a parcel tends to return to the previous state after being disturbed, or unstable, when a small disturbance produces significant changes and return to the initial state is no longer possible.

If the lapse rate of the surrounding air (environmental lapse rate) is less than the adiabatic lapse rate (Fig. 5.1b), then the air parcel that is forced to rise (by a steep hill, for example) will be permanently colder and denser than the surrounding air and will tend to return to the initial position. In that case the *atmosphere is absolutely stable*. A low environmental lapse rate occurs when the surface air is cooling or the air above is warming. Cooling of surface air is commonly due to cold air advection or intensive nocturnal cooling, which generally produces highly stable air just before sunrise when the temperature reaches its minimum. Since a stable atmosphere is strongly resistant to any vertical motion, it allows fog to stay close to the surface early in the morning (Fig. 5.2). Warming of the upper air is the result of warm advection or slow air sinking and consequent warming (a process opposite to cooling due to rising), which can produce subsidence inversion. In stable conditions, inversion behaves as an atmospheric lid, preventing vertical transfer and keeping all products of condensation, but also all pollutants, close to the ground.

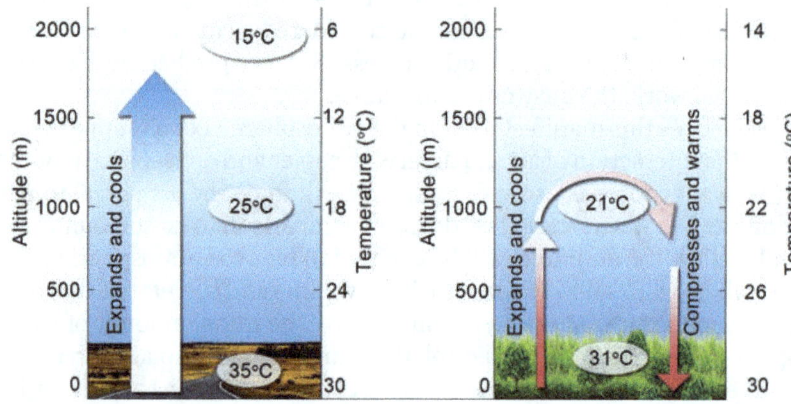

Figure 5.1: Air parcel in unstable (left-a) and stable (right-b) atmosphere.

Figure 5.2: Stable atmosphere in the morning keeps fog close to the ground.

If the environmental lapse rate is above the dry adiabatic lapse rate, a rising air parcel will cool off more slowly than its environment (Fig. 5.1a). Hence, once an air parcel starts to ascend, it will keep rising in this *absolutely unstable atmosphere*. Atmospheric instability increases when the environmental lapse rate increases, which is a common phenomenon when the surface air is warming or the air above is cooling. Surface air temperature increases as a result of warm advection or the daily heating of the Earth's surface by solar radiation, which produces intensive vertical mixing of the atmosphere. Accordingly, atmospheric stability can change from stable to unstable following temperature day cycle.

A stable atmosphere which holds water vapour, particularly in the morning before a dry day, is an important source of humidity for plants and animals. However, in the case of frost, stability associated with radiation inversion produces frost damage, whose severity depends on the intensity of the inversion and stability (which determine the duration of the frost) and on how far the temperature drops below 0 °C (which determines the intensity of the frost).

An unstable atmosphere encourages the vertical transfer of heat and water vapour from the ground and plant canopy to the atmosphere, greatly affecting the energy and water balance of the surface. It can boost the intensity of evapotranspiration and gas exchange between plants and the surface layer, which at the same time increases the intensity of all physiological processes involved.

5.2 Clouds and their formation

Clouds are visible products of atmospheric water vapour condensation and freezing. They can be formed close to the ground or at high altitudes; they can be thin with a large horizontal extension or thick up to boundary layer top but with "corn field size" horizontal scale.

Water vapour condensation (see Chapter 4) and cloud formation in the atmosphere generally occur when: a) air rises due to convection, b) air is forced to rise by mountains or weather fronts, and c) warm and cold air collide and mix in the atmosphere. The appearance of clouds and the type of precipitation that forms in them depends on whether clouds are formed by a slow rise in a stable atmosphere or fast convection in an unstable one.

Fully layered (stratus–like) clouds tend to form under conditions of large–scale subsidence and strong static stability. Convective clouds are usually produced by the atmospheric heating and convection of moist air from the ground to the top of the atmosphere or by fast moving cold fronts forcing upward motion. Mixed types are also possible with layer clouds having some convective elements if local instability occurs. In contrast to the great horizontal extension of layered clouds, convective clouds are characterised by their vertical scale. The faster the convection and the more water vapour introduced, the more pronounced will be the vertical thickness of the cloud.

As they are permanently changing, clouds appear in many forms, giving rise to the need for classification. The first cloud classification was made by amateur British pharmacist and meteorologist Luke Howard, who published his *Essay on the Modification of Clouds* in 1803. His classification is based on ten cloud classes; it was later extended and accepted by the World Meteorological Organisation (WMO) as the basis for the ten main groups, or "genera", of clouds: Cirrus (Ci), Cirrostratus (Cs), Cirrocumulus (Cc), Cumulonimbus (Cb), Altocumulus (Ac), Altostratus (As), Nimbostratus (Ns), Stratocumulus (Sc), Stratus (St) and Cumulus (Cu). Further classification is made according to the part of the atmosphere (troposphere) in which low, medium and high clouds commonly appear (Fig. 5.3).

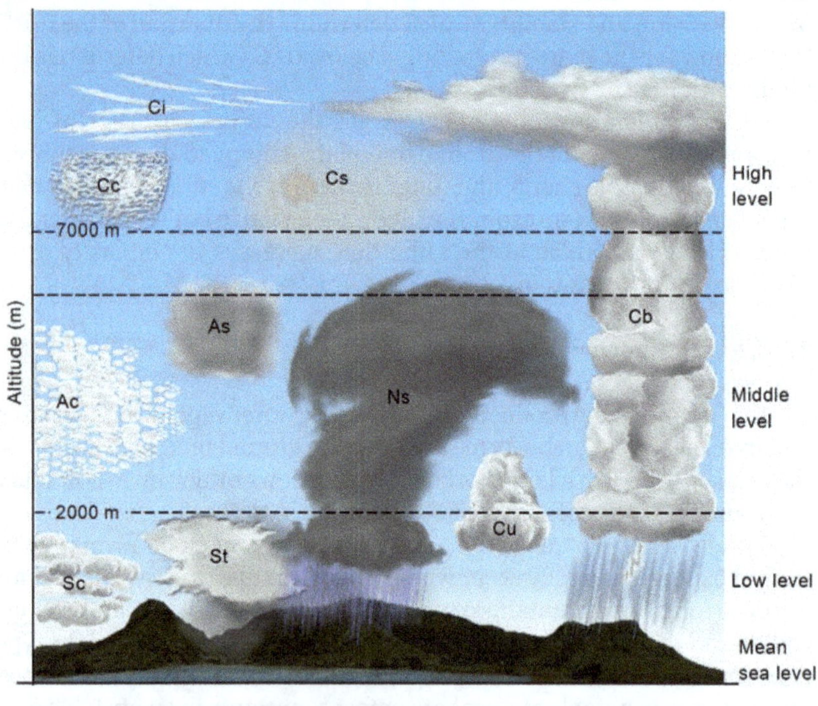

Figure 5.3: Classification of clouds: group, level, height, symbols.

Cloud names usually come from Latin words describing their appearance. In order to more accurately describe cloud characteristics, the prefixes *alto–*, for high clouds, and *nimbo–*, for rainy clouds, are added.

Cloud content depends on the height, place and mechanisms of its formation; it can be characterised by individual ice crystals, snow and drops of different sizes as well as different mixtures of all these forms.

According to their temperature and content, clouds can be identified as *warm* or *cold.* If a cloud's temperature over its whole profile is above *0 °C* and if it consists of liquid droplets, only then is it considered a warm cloud–typical for tropical regions. However, if the cloud top reaches heights with temperatures far below *0 °C, its content is determined by a mixture of* ice crystals and liquid droplets. This formation, typical for mid and high latitudes, is called a cold cloud.

5.2.1 Low clouds

If a cloud base lies below 2000 m, it is commonly classified as a low cloud. Even though mostly composed of water droplets, low clouds can also contain ice particles and snow when temperatures are low enough. Low clouds types are: stratus, cumulus, stratocumulus and cumulonimbus.

Figure 5.4: Low clouds: a) stratus, b) cumulus, c) stratocumulus and d) cumulonimbus. (Source: International Cloud Atlas, WMO, 2018, https://cloudatlas.wmo.int/home.html).

Stratus clouds are shapeless (Fig. 5.4a), low–level clouds with layers of grey to white and a uniform base. Depending on the surrounding air temperature, they can consist of water droplets, supercooled water droplets or ice crystals. Atmospheric conditions favourable for stratus cloud formation are associated with: a) radiative cooling of a clear, moist atmospheric layer; b) warm advection and c) forced rising of moist air in regions close to a front or to steep hills. In particular, advection of warm air over a colder surface can lead to fog formation. Wind can elevate the base of the fog layer, leading to the formation of stratus clouds. They often appear in fair weather but can occasionally produce drizzle or light snow.

Cumulus clouds are cotton–like (Fig. 5.4b) in appearance and are formed as a result of local convection of warm, moist parcels in a conditionally unstable atmosphere. The common ragged form of these clouds is usually underlined by a flat base at heights not much above 1000 m, while the tops of these clouds can reach or even shoot through the tropopause. They appear individually or in clusters and cloud lines. Depending on the temperature cloud top height, they can consist of water vapour, water droplets, supercooled water or ice crystals. Cumulus clouds are associated with fair weather and do not produce precipitation, but they can easily develop into cumulus congestus or even cumulonimbus, which produce heavy rain, strong winds and thunderstorms and sometimes hail.

Stratocumulus clouds are grey–to–dark clouds (Fig. 5.4c) whose layered structure is interrupted by cumulus elements, since they often appear in transitions from cumulus or stratus clouds. Conditions favourable for stratocumulus formation are associated with large–scale subsidence and strong static stability in the lower troposphere. The cloud dynamic, responsible for their vertical development, is primarily driven by convective instability caused by infrared radiative cooling from the cloud top. Stratocumulus clouds typically contain different forms of liquid condensate. Ice can also be present if the cloud top is cold enough. They commonly have a thickness of 200–400 m within an atmospheric layer of 500–2000 m. They are associated with fair weather, and precipitation is not common. Exceptionally, light rain or snow can appear. However, stratocumulus clouds often announce approaching storms.

Cumulonimbus clouds–"the king of clouds"–are one of the most impressive and easily observable weather phenomena (Fig. 5.4d). Grey or dark, dense with a pronounced tower structure, these clouds can exist alone or form the line of other clouds, with the potential for supercell development. Initially in the form of well–developed cumulus clouds and with an intensive vertical development reaching 93 km/h, these clouds extend over most of the troposphere. Their cloud base is commonly found at altitudes of between 200 m and 4000 m, while their cloud top appears at 7600 m and in some extreme cases at 12000 m. Conditions favourable for their formation are associated with strong atmospheric instability caused by great temperature decreases in the lower troposphere. Water vapour originates close to the cloud base and is then lifted by strong upward air currents. At higher altitudes, the cloud content is dominated by supercooled drops and the solid phase of water in the form of snow and individual ice crystals. Cumulonimbus clouds are a frequent cause of storms with strong winds, lightning, intensive precipitation, sometimes with hail. In some regions, such as the US Midwest, they spawn tornadoes. Cumulonimbus clouds usually appear on summer afternoons after well–developed atmospheric convection due to solar heating of the Earth's surface coupled with low–level convergence of moist air.

5.2.2 Medium clouds

Clouds are classified as medium or mid–level clouds if their bases are at altitudes of between 2000 to 6000 meters. Primarily, they are composed of water droplets but when temperature are low enough ice crystals can be present. Medium clouds types are: altocumulus, altostratus and nimbostratus.

Figure 5.5: Medium clouds: a) altocumulus, b) altostratus and c) nimbostratus. (Source: Source: International Cloud Atlas, WMO, 2018, https://cloudatlas.wmo.int/home.html).

Altocumulus clouds (Fig. 5.5a) are white and/or grey and appear in the form of layers, lines or patches of cloudlets. They form at altitudes of between 2000 m and 7000 m and consist of supercooled water droplets (even at -10 °C) and, sometimes, of ice crystals. They form by convection in an unstable, high–altitude atmospheric layer. They do not produce rain but can indicate weather changes. For example, on a summer morning they announce a high risk of thunderstorms in the afternoon. If associated with Foehn, they indicate weather deterioration within the next half day or the day after.

Altostratus clouds (Fig. 5.5b) are formless grey or blue–grey sheets of thin cloud that steadily thicken over a large horizontal extension. Since thickening is not uniform throughout the extension, in some parts altostratus clouds are thin enough to make the sun visible. Since altostratus form as a result of cirrostratus descending from higher levels, they contain a mixture of water droplets and ice crystals, commonly contributing to the formation of optical effects in the atmosphere (coronas). Altostratus are not rainy clouds but they often form in advance of a warm or occluded front. The passing of a front transforms altostratus into rainy nimbostratus. Their presence therefore often announces a change in weather.

Nimbostratus clouds (Fig. 5.5c) have a form and content that correspond to their name–nimbus, or "dark cloud." They are rainy, low–level layer clouds with a uniform base. They are formless, grey or dark grey and thick; they do not allow sunshine to pass through (hence their dark colour) and contain water droplets or snow. Nimbostratus clouds form in the lower or mid atmosphere through the thickening of altostratus clouds. They are often associated with a cloud system of mid–latitude cyclones which bring precipitation to the surface below.

5.2.3 High clouds

At heights above 6000 m, temperatures are so low that these clouds primarily consist of ice crystals. Therefore, high clouds are commonly thin and white but can affect energy balance. High clouds types are: cirrus, cirrocumulus and cirrostratus.

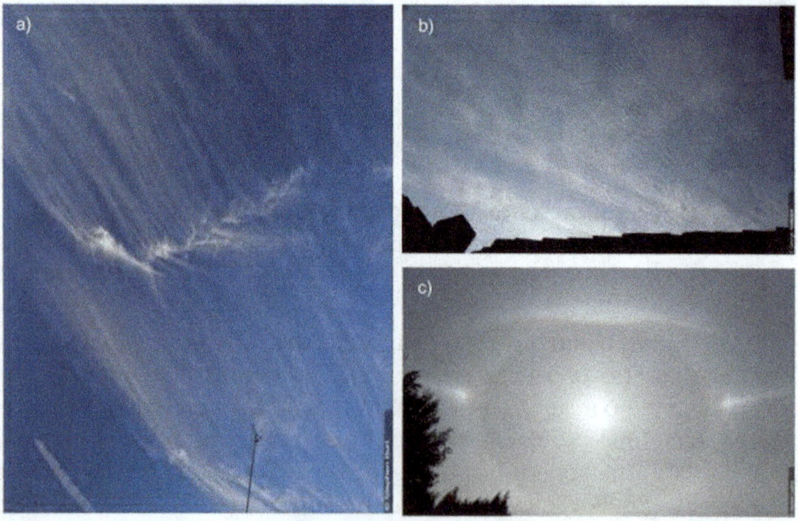

Figure 5.6: High clouds: a) cirrus, b) cirrocumulus and c) cirrostratus. (Source: International Cloud Atlas, WMO, 2018, https://cloudatlas.wmo.int/home.html).

Cirrus clouds (Fig. 5.6a) are high–level, typically white clouds with a thin and often wispy appearance. They consist of ice crystals formed by freezing droplets. Their presence usually announces nice weather, while their motion indicates wind direction at cloud heights. Even though they are very thin and high in the sky, cirrus clouds with their ice–crystal surfaces (ideal for solar radiation reflection) can affect surface energy balance; their passage and/or presence is clearly seen on pyranometers (see Chapter 13).

Cirrocumulus clouds (Fig. 5.6b) are usually white, high–level, thin clouds appearing in the form of long rows or layers of rounded puffs. They consist of supercooled water droplets appearing as a result of shallow convection rising within the thin and unstable high–atmospheric layer. At mid latitudes, they usually appear in the winter, indicating nice but cold weather, while in the tropics they often announce an approaching hurricane.

Cirrostratus clouds (Fig. 5.6c) are a very high, thin, transparent, layered clouds consisting of ice crystals, which typically form as a result of the rising of a broad layer of air. If a warm front is approaching, cirrostratus clouds become thicker and transform into a lower and much denser cloud type. They cover the sky either totally or partially and commonly produce a halo effect.

Cloud seeding is the dispersal of substances (AgI – silver iodide, KI – potassium iodide and dry ice) which can act as condensation or ice crystal nuclei in clouds, in order to change the appearance, type or intensity of precipitation. The aim behind cloud seeding is usually to reduce hail or to increase precipitation. Its efficacy in modifying weather has never been rigorously proven; indeed it has been a controversial topic in the scientific community for many years.

5.3 Growth of water droplets in clouds

The growth of water droplets in clouds starts from cloud condensation nuclei (typically 0.2 μm in diameter) and finishes as either a cloud droplet (typical = 20 μm, large = 100 μm), raindrop (2000 μm) or ice crystal (snowflake). The formation of droplets larger than 20 μm involves processes of diffusion and collision–coalescence as well as ice–crystal processes.

The rate of cloud droplet growth by condensation in a supersaturated environment is a slow process–it takes several days to produce one raindrop–which decreases with radius increase. Obviously, smaller drops grow faster. Therefore, condensation is a predominant mechanism of growth in the case of drops smaller than 0.1 mm, when the fall speed of the drop is very slow (due to almost balanced weight, friction and buoyancy force). The further growth of a cloud drop into a liquid form large enough to fall through the cloud bottom and reach the ground depends on cloud temperature: in warm clouds (Fig. 5.7), i.e. clouds with a maximum temperature above -15 °C, it is determined by collision and coalescence processes, while in cold clouds ice–crystal processes are additionally involved (Fig. 5.8).

The rate of cloud drop growth by collision and coalescence is enhanced by: high liquid water content (which increases the efficacy of collision), the range of droplet sizes (which increases the efficacy of collision and the chances that drops will coalesce after collision) and the high speed of the updraft and cloud thickness (which increase the length of the droplet collision path and the time spent in the cloud). Therefore, relatively thin stratus clouds with a weak updraft may produce only drizzle (very light rain), while strong cumulus clouds of intensive vertical development can easily produce 5 mm raindrops.

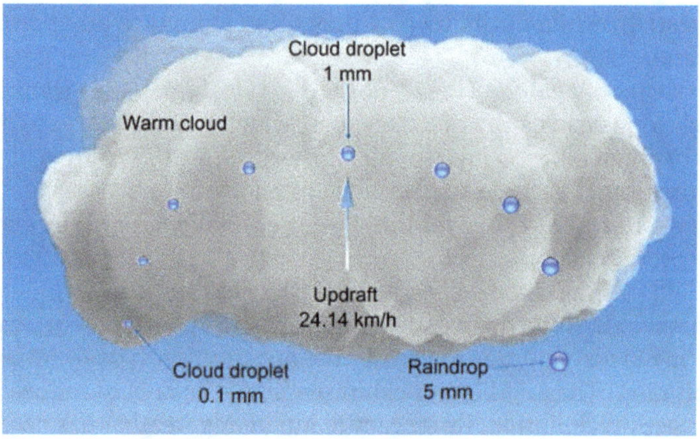

Figure 5.7: Growth of droplet in a warm cloud.

Ice–crystal processes are found in cold clouds where part of the cloud is at temperatures far below 0 °C. In the warm part of a cold cloud, collision and coalescence of cloud droplets is the dominant mechanism of droplet growth.

In air just above freezing level, supercooled cloud drops are still liquid. They predominate in the existing mixture with ice crystals, and are still present in large portions of a cloud with temperatures between freezing and -40 °C. The cause of this surprising finding is explained by the fact that the freezing temperature of water droplets decreases with size. Why does this happen? Condensation nuclei are usually hygroscopic and dissolve upon droplet formation. When that occurs, the smaller the droplet, the more concentrated the solution. Usually solutions have lower freezing point temperatures than pure water. Therefore, the smaller the droplet, the colder it can become before it freezes spontaneously at -48 °C.

Water droplets may also contain ice nuclei, substances with molecular structures similar to those of water. They may be part of composite condensation nuclei on which droplets form or they may be swept from the air as droplets fall. When this happens, droplets contain a sort of "ice seed" and can freeze rather quickly.

On the other hand, ice crystals can form at air temperatures below freezing if ice nuclei are present. Since the saturation vapour pressure just above a water surface is greater than that above an ice surface, the concentration of water molecules just above a droplet surface is higher than above one composed of ice. This concentration difference produces flux of water vapour molecules from the droplet to the ice surface, causing further evaporation of drops and the growth of ice crystals during the *ice–crystal* (Bergeron) process.

While falling, ice crystals collide with supercooled liquid droplets which freeze on their surface, enhancing ice–crystal growth in the process of *accretion*. Typical ice–crystal processes of raindrop formation can be found in cumulonimbus clouds.

Finally, let us note that the main source of ice in clouds results from the freezing of droplets on ice nuclei, and not the homogeneous freezing of drops or their direct sublimation on ice nuclei.

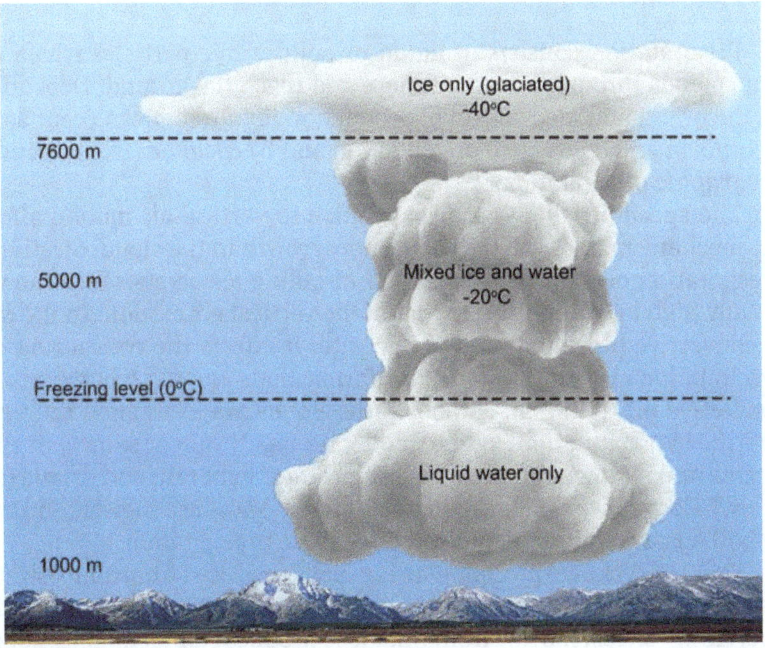

Figure 5.8: Growth of droplets in a cold cloud.

5.4 Hydrometeors

According to the Glossary of the American Meteorological Society, a *hydrometeor* is any product of condensation or deposition of atmospheric water vapor, whether formed in the free atmosphere or at the Earth's surface, as well as any water particle blown by the wind from the Earth's surface. Hydrometeors (meteors other than clouds formed by different forms of water) may consist of:

- the fall of an ensemble of particles, i.e. precipitation (rain, supercooled rain, drizzle, supercooled drizzle, snow, snow grains, snow pellets, diamond dust, hail, small hail, ice pellets),
- a deposit of particles (deposits of fog droplets, white dew, rime, glaze),
- an ensemble of particles raised by the wind (drifting and blowing snow, spray, snow devil, steam devil),
- a vortex of particles (spout) and
- a suspension of particles in the atmosphere (fog, ice fog, mist).

In particular, this chapter will discuss precipitation, deposits of particles and hydrometeors raised by the wind.

5.4.1 Precipitation

Precipitation consists of liquid or solid water particles which fall from clouds through the atmosphere and reach the ground. Depending on the origin and mechanism of production involved, it can occur as: a) *stratiform* and *convective* precipitation and b) *cyclonic* (frontal) and c) *orographic* precipitation.

The speed of falling particles relative to vertical air motion affects the mechanism of droplet, ice and snow growth in the cloud. Stratiform precipitation occurs when the speed of falling ice crystals and snow is usually high in comparison to that of the vertical air motion. In the case of convective precipitation it is opposite. It affects the typical time for precipitation growth: 1–3 h in stratiform clouds and 0.5 h in the case of convective precipitation. Nimbostratus clouds are a common source of stratiform precipitation (light to moderate steady rain or snow) but even stratus and stratocumulus clouds can produce light rain and drizzle (see below). Convective clouds (like cumulonimbus) produce showers and hail. In both classes, thicker clouds produce more precipitation.

Frontal (cyclonic) precipitation is usually produced from nimbostratus clouds in a frontal area between a warm, moist air mass and a cool, dry one. However, fast moving fronts may also produce cumulonimbus clouds.

When moist air is forced to rise, encountering mountains or hills, condensation starts in this air mass and orographic precipitation is commonly produced on the upwind mountain ridge.

Precipitation can also be classified according to its form, character, intensity or type. The *form* of precipitation is classified as liquid (rain, drizzle), freezing (supercooled rain, supercooled drizzle, mixed rain and snow–"snain") or frozen (snow, snow grains, snow pellets, diamond dust, hail, small hail, ice pellets).

The *character* of precipitation can be intermittent or continuous. It indicates not only the duration of precipitation but also the area covered.

The *intensity* of precipitation is the ratio of its amount (the volume of precipitation that turns to liquid per m^2) to its duration. Three terms are used to describe the intensity of precipitation: light (less than 2.5 mm h_{-1}), moderate (2.5–7.6 mm h^{-1}) and heavy (more than 7.6 cm h^{-1}) (AMS Glossary, 2018). A shower (of rain or snow) is precipitation of a specific character and intensity commonly produced by convective clouds. Typically, it starts and stops suddenly by rapidly increasing and decreasing intensity.

There are many different types of precipitation —drizzle, rain, snow, hail, and sleet among them. At this place, particular attention will be paid to rain, snow and hail.

Rain consists of water drops *(with a diameter larger than 0.5 mm) that fall from a cloud and whose* density and size (diameter) distribution vary considerably with the origin and intensity of the precipitation (light stratiform: 0.5–2.0 mm; moderate stratiform: 1.0–2.6 mm; heavy thundershower: 1.2–4.0 mm) (Fig. 5.9). Clouds which regularly produce rain are nimbostratus and cumulonimbus. The intensity of rain varies from 1 mm hour^{-1} for a light rain to 25 mm hour^{-1} in the case of heavy shower.

Snow consists of ice crystals (single or stuck together) that fall from a cloud and whose form, concentration and size distribution vary significantly with the temperature and supersaturation conditions at which they develop (Fig. 5.10). At temperatures above -5 °C, crystals, frozen drops and melted–frozen parts of crystals stick together and form typical snowflakes. At higher latitudes, snow falls in the form of ice crystals only. Higher stratus, i.e. nimbostratus, clouds with a prevailing portion above 0 °C in the atmosphere are responsible for snowfall (lower ones produce

Figure 5.9: Rain.

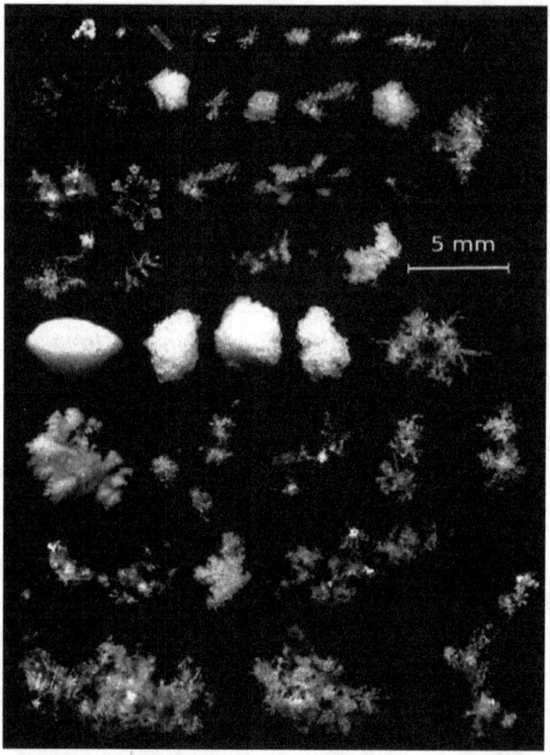

Figure 5.10: Snow (Garett *et al.*, 2012).

rain). The definition of precipitation intensity that we have already seen normally applies to snowfall; there is, however, an additional definition of snow intensity based on its impact on visibility. Hence the intensity of snow is described as light, moderate or heavy if visibility is 1 km or more, between 0.5 km and 1 km, and less than 0.5 km, respectively.

Hail are ice particles that are either transparent, partly opaque or completely opaque, with a diameter usually between 5 mm and 50 mm (Fig. 5.11). Each hailstone consists of a central nucleus (an ice particle up to 1 cm in diameter) usually surrounded by several layers of opaque and/or transparent ice. The environment required for hail formation is cumulonimbus clouds with intensive vertical development and high liquid water content: this implies a high vertical drift within the cloud (35 m s^{-1}) and a significant part of the cloud far above the 0 °C isotherm (Fig. 5.12). Captured by an upward drift, the nucleus rises, growing by condensation on its surface and by collision with cloud and raindrops (below the 0 °C isotherm) and snow and ice (above the 0 °C isotherm), as far as its weight overcomes the upward drift force. Afterwards, the resulting hailstone moves downward and starts melting when crossing the 0 °C isotherm. At the cloud bottom, formed hail can drop off the cloud if its weight is above

Figure 5.11: Hail (Source: International Cloud Atlas, WMO, 2018, https://cloudatlas.wmo.int/home.html).

the upstream drift force. Otherwise, it will be moved by inertia and captured for the next round of upstream growth and downstream reduction. During each passage, hail reaches the lower parts of the cloud and forms one more ice layer. After its critical weight is reached, hail falls through the cloud base. Hail is associated with high intensities of precipitation and if it occurs, it is always during thunderstorm showers. The spatial distribution of hail is often in the form of a elongated belt. Since it is associated with cumulus convection, hail usually appears in the afternoon (in 20% of cases, 12–3 pm; in 50% of cases, 3-8 pm), while it is less frequently observed in the morning and in the evening (in 13% of cases, 8 pm -1 am).

Cloud particles that fall from the cloud base in the form of a streak or shaft of precipitation without reaching the ground surface is called a *virga*. Virga is common on summer days, when raindrops falling through warm and dry environments evaporate quickly and disappear in the atmosphere.

5.4.2 Hydrometeors consisting of a deposit of particles

Deposition of liquid or solid water particles on an underlying surface can affect its humidity and friction. In some regions, such as South America and some countries of the Middle East, the harvesting of dew and fog deposition can be an important source of water.

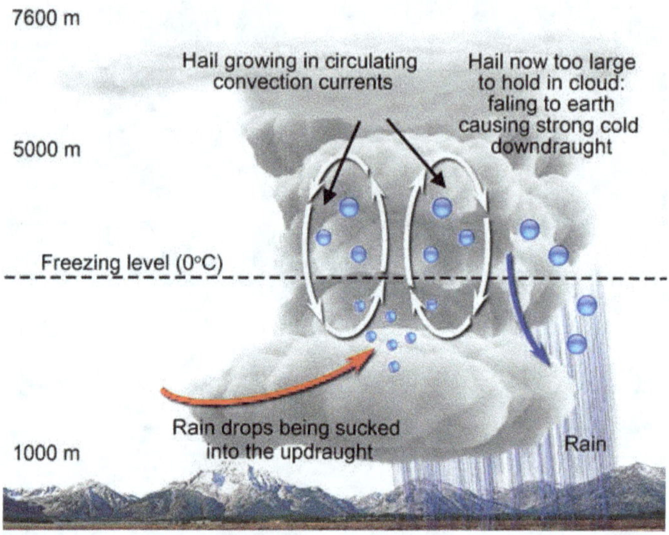

Figure 5.12: Hail formation.

Deposits of fog droplets are the result of fog droplet deposition on objects whose temperature is above 0 °C.

Dew is the result of direct condensation of atmospheric water vapor on underlying surfaces and objects whose temperature is above 0 °C (Fig. 5.13). Even a collected amount of dew rarely exceeds 0.5 mm, though theoretically it can measure up to 0.8 mm. At low temperatures, dew freezes, producing window frost.

Figure 5.13: Dew.

Hoarfrost (white frost) is the result of the sublimation of atmospheric water vapor on underlying surfaces and objects in the form of crystals (Fig. 5.14).

Figure 5.14: Hoarfrost (Source: International Cloud Atlas, WMO, 2018, https://cloudatlas.wmo.int/home.html).

Rime is the result of the deposition of ice from the surrounding air, which is the product of freezing of supercooled fog or cloud droplets on surfaces and objects whose temperature is slightly above 0 °C.

Glaze is the result of freezing of supercooled drizzle or rain on surfaces whose temperature is below or slightly above 0 °C.

5.4.3 Ensemble of particles raised by wind

Liquid or solid water particles raised by the wind, often on a small horizontal and vertical scale, can significantly affect air and soil humidity, snow cover presence and depth, and visibility.

Drifting and blowing snow is the result of snow carried up into the atmosphere by the *wind*. The distance and height of its transfer depend on the speed, direction and gustiness of the wind and the density and age of the snow. For many days after a snowfall, strong wind can cause serious problems and significant snow cover on roads, particularly on plains (Fig. 5.15).

Figure 5.15: Drifting snow (Source: International Cloud Atlas, WMO, 2018, https://cloudatlas.wmo.int/home.html).

5.5 Impact of clouds and precipitation on plants

Clouds and precipitation strongly affect the energy and water balance of soil, air and plant canopy. Stratocumulus clouds cover 20% of the Earth's surface in the annual mean, more above oceans (23%) than above land (12%). In the case of stratus clouds, the annual mean is about 10%. Even if the horizontal scales of cumulus clouds are typically much smaller than those of stratus clouds, they can affect energy and water balance locally, impacting sites directly below clouds through intensive absorption (over their significant vertical scale) and high reflectivity of solar radiation.

Clouds vs. energy balance. The impact of clouds on energy balance is twofold: a) they decrease incoming solar radiation by absorbing and reflecting it while passing through the atmosphere and b) they increase terrestrial radiation by increasing the longwave radiation emitted by the atmosphere and reflected by the Earth. Even the impact of thin cirrus and cirrostratus clouds on ground temperature is noticeable. In addition, a significant change in soil and air temperature can be expected in the presence of stratus and stratocumulus clouds, because of their horizontal extension, depth and average retention times. In summer days, the presence of these clouds will decrease temperature, since a decrease in incoming solar radiation prevails over the effects of increasing terrestrial radiation. In winter and particularly in early spring, when the intensity of incoming solar radiation is reduced, an increase in terrestrial radiation can significantly increase the energy budget of the Earth's surface. It is well known that the presence of clouds during a frosty night can increase air temperature by a few degrees and mitigate losses caused by cold.

The presence of cumulus clouds, particularly stratocumulus and nimbocumulus, can produce significant drops in soil and air temperatures. The effects of the high reflectivity of these clouds, particularly in the case of snow–white patchy cumulus distributed all over the sky, should not be underestimated. Functioning like perfect reflective surfaces, these clouds can cause a significant increase in on–spot solar radiation intensity, which can exceed the solar constant (the intensity of solar radiation at the atmosphere top). This short–term, high–intensity "spotlight" can produce sunburn and severe damage to plants.

Clouds vs. water balance. Even when precipitation does not take place, the presence of clouds can affect air and soil humidity. This impact is seen through their effect on evaporation and evapotranspiration, which occurs by changing soil and air temperature, producing ground surface shade and increasing the humidity of the surface atmospheric layer, which is caused by the reduction of the vertical transfer of water vapour from the ground to the free atmosphere.

Precipitation vs. energy balance. The impact of precipitation on energy balance is manifested through: a) changes in soil thermal characteristics, b) an increase in evaporation and evapotranspiration from the ground and canopy and c) its role of thermal isolator when covering soil and plants.

Increased soil moisture, caused by precipitation, increases the heat capacity and conductivity of soil, thereby affecting energy balance. A temperature decrease after a rain event (Fig. 5.16, top) is more significant and pronounced for air than soil. This effect is particularly important in autumn when moist soil accumulates more energy than dry soil, conducting it deeper into the ground. During winter and early spring, when soil temperature increases with depth, increased soil conductivity boosts energy transfer towards the top of the soil, thereby increasing surface soil and air temperatures and reducing plant damage from low temperatures.

Retention of precipitation, primarily liquid, on a plant surface causes the same degree of evaporation as from open water body. Energy invested in evaporation is taken from the plant, which as a result decreases its temperature. An increase in soil moisture usually increases evaporation from bare soil and evapotranspiration from a vegetated surface, decreasing the temperature of both.

When present on soil or on a plant surface, snow can play the role of insulator during frost days, particularly in the early spring, preventing plant cooling below 0 °C. When during winter soil is bare, snow cover reduces radiation losses and daily soil temperature variation (Fig. 5.16, bottom). If soil is not frozen before a snowfall, soil temperature near the surface under a closed snow cover stays at approximately 0 °C, which is important for the protection of winter crops from frost (Fig. 5.17).

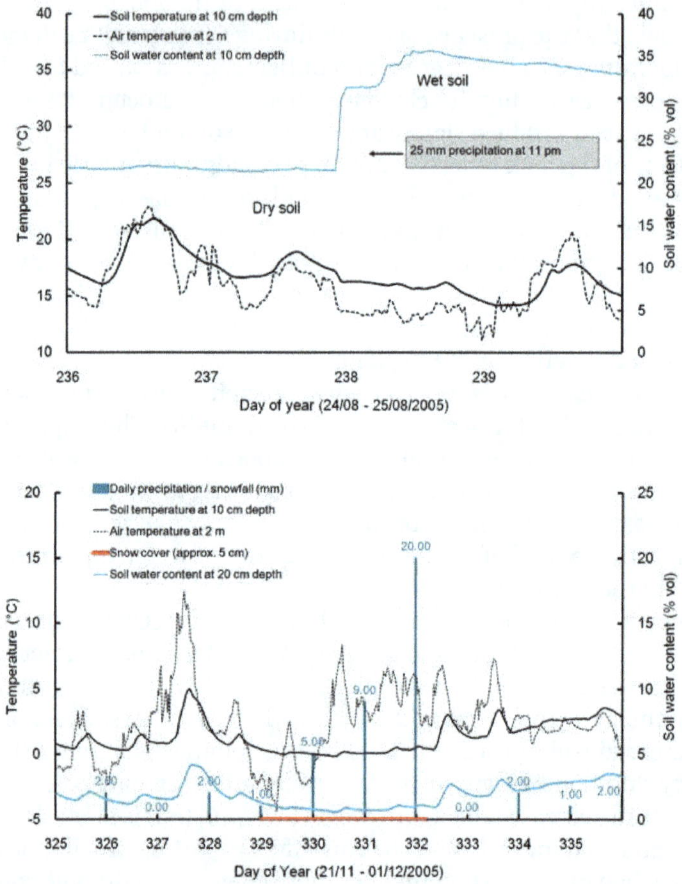

Figure 5.16: Daily variation of soil temperature before and after rain (top) and snow (bottom) event.

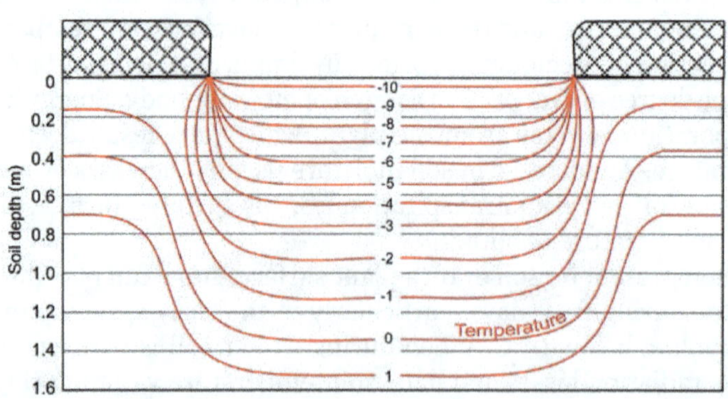

Figure 5.17: Variation of soil temperature over the soil column, with and without snow cover.

Precipitation vs. water balance. Liquid precipitation directly entering the soil increases soil moisture, evaporation, drainage and run–off. However, the increase in soil moisture is not always proportional to the amount of water reaching the ground surface. In the case of bare soil, the soil type and structure as well as the intensity of precipitation can significantly affect soil moisture increase. Even in the case of soil with high water–holding capacity, such as chernozem, high–intensity precipitation will only very moderately increase soil moisture when the soil is highly compressed and has been dry for a long time, with crust and cracks on the surface. It appears that due to water loss from surface run off and drainage to deeper soil layers, only a part of the incoming water flux is actually used for moisturizing the soil. In the case of optimally wet and structured soil, the type, structure and growing period of vegetation that is present can also affect precipitation absorption. For example, in the case of small grains, after tillering, light rain of 3.8 mm h^{-1} will be absorbed in the amount of 1.8 mm h^{-1}, while high intensity rain of 24.3 mm h^{-1} will be absorbed in the amount of 10.4 mm h^{-1}. A few months later, after flowering, the same intensities of precipitation will be absorbed in the amounts of 0 mm h^{-1} and 16.6 mm h^{-1}, respectively.

Snow cover melts slowly, often leaking moisture to the ground surface and providing prolonged moisturizing of the soil.

The presence of vegetation can disturb the effects of precipitation on the soil water balance by shading the vegetated surface to the extent that no rain can reach the ground and by retaining water on plant surfaces: this is known as *interception*. The effect of shade and the consequent irregular distribution of soil moisture in the top soil layer depends on vegetation type, growth phase and the structure and density of the canopy (Fig. 5.18). Interception additionally depends on the intensity of precipitation. In the case of 1 mm of precipitation, all rainwater will be intercepted. As intensity increases, the amount of water reaching the ground increases in proportion. In general, the maximum interception rate is up to 50%. The leaves of deciduous trees, for example, can retain from 20% to 30% of falling rain. Still, an important effect of interception is not only the reduction of water reaching the ground during a precipitation event, but leaf dripping which occurs long after precipitation stops, which sometimes confounds precipitation measurements within the canopy. A part of the intercepted water will evaporate, but a significant amount will still reach the ground; particularly in the case of high intensity of precipitation and dry soil, the water will produce an even more effective moisturizing of soil than in the case of bare soil.

Figure 5.18: Impact of vegetation on precipitation distribution over the soil surface.

5.6 Impact of plants on clouds and precipitation

Whether as crop land or Amazon forest, the presence of vegetation plays an important role in the energy and water balance of the Earth's surface, on scales ranging from a field to the entire planet. In parts of the world like Africa's Sahel region, the latest research shows that the presence of vegetation can change annual amounts of precipitation by up to 30%. Since more than 20% of Earth's surface is covered by plants, its role as climate modifier is clear, even if its direct impact on weather is perhaps less so.

The presence of plants changes surface albedo as well as the thermal and hydrological characteristics of soil. Plants in fact affect: a) energy exchange between the Earth's surface and the atmosphere, which is an important governing force of atmospheric circulation; b) humidity exchange between atmosphere and surface; c) changes in circulation patterns of air masses, and d) turbulent transfer between vegetated surface and atmosphere. Because of these changes in the temperature and humidity of soil and air, in the intensity of evaporation and transpiration, in wind speed and in the efficacy of humidity transfer from the surface to the atmosphere, vegetation significantly affects condensation processes in the atmosphere and therefore cloud formation (Fig. 5.19) and precipitation (Fig. 5.20). The impact of vegetation on these meteorological elements has been discussed in previous chapters, while in the next chapter we will consider atmospheric circulation and the impact of vegetation on it.

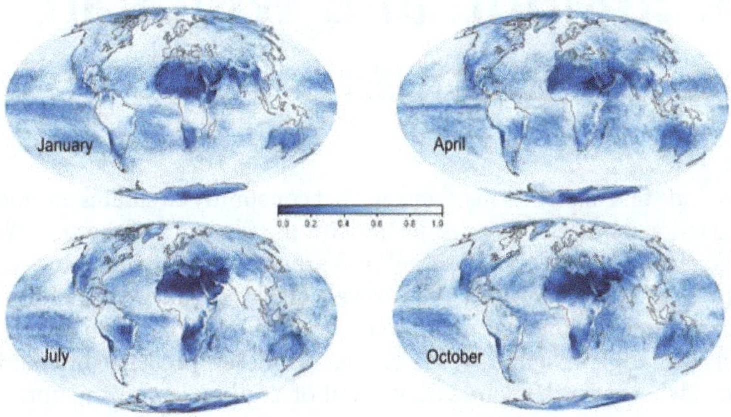

Figure 5.19: Global distributions of cloud fraction for 2001 (Source: NASA Earth Observatory, https://earthobservatory.nasa.gov/GlobalMaps/view.php?d1=MODAL2_M_CLD_FR).

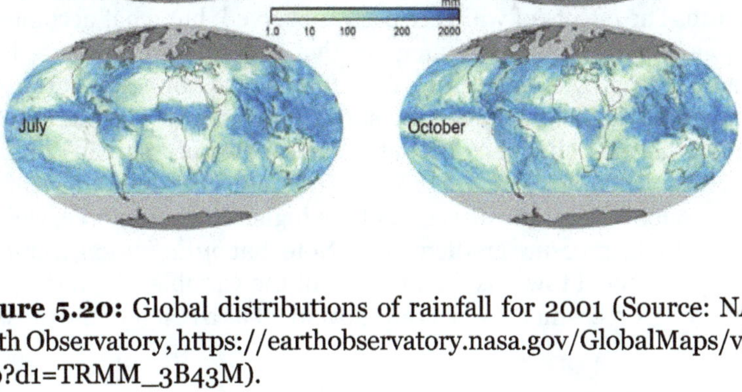

Figure 5.20: Global distributions of rainfall for 2001 (Source: NASA Earth Observatory, https://earthobservatory.nasa.gov/GlobalMaps/view.php?d1=TRMM_3B43M).

6

Circulation and Winds in the Atmosphere

Winds take many forms, from air that constantly circulates around the globe to local breezes that disappear with the first rays of the sun. Winds are governed by "forces of nature" and are modified by the nature of the Earth's surface over which they travel, bringing energy, humidity and momentum from one continent to another or from one crop field to another. Although invisible, we are reminded of their presence by countless effects, whether inthe gentle movement of a falling leaf or the uprooting of ancient trees from the ground.

Air is forced to move by the difference in atmospheric pressure between two places. The movement is further affected by centrifugal and Coriolis forces caused by the Earth's rotation, which produces atmospheric circulation. This chapter introduces the horizontal motion of air (wind), the governing force and modifier of all atmospheric processes.

6.1 Forces governing atmospheric motion

Pressure gradient force. Wind is a current of air moving almost horizontally above the Earth's surface, from higher to lower atmospheric pressure. The effects of wind are easily observed, but what accounts for this phenomenon? Let us remember that atmospheric pressure is the force produced by a certain quantity of air on surrounding surfaces. If this force on an imaginary vertical surface is higher at one place than at another, there is a change of force and of associated pressure in a horizontal direction, in other words, a pressure gradient. Net force, which affects air movement, keeps the direction of the higher intensity force, which is opposite to the pressure gradient (∇p) [Note that *by definition*, a gradient is always directed towards the increase of the variable of interest]. The intensity of force which governs air movement as a result of the presence of a pressure gradient–*pressure gradient force* per mass unit of air–can be expressed in the form

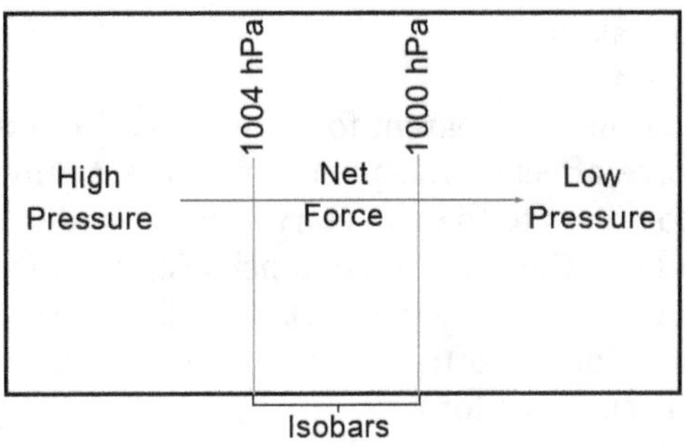

Figure 6.1: Pressure gradient force.

$$\overrightarrow{F_p} = \frac{1}{\rho}\nabla p \qquad (6.1)$$

where ρ is the air density. No matter what the cause of the pressure difference (net radiation imbalance, ascending or descending air), this is the main governing force which makes the wind blow.

Friction force. In the atmosphere, friction force resists the relative motion of air against solid surfaces and atmospheric layers. It always acts in the direction opposite to that of the motion in order to slow down the air. Therefore it can be calculated as

$$\overrightarrow{F_{tr}} = -k\vec{v} \qquad (6.2)$$

where k is a friction coefficient and is the air velocity. Since air is not very viscous, friction associated with molecular motion is important only in the very thin layer of air close to a solid surface, commonly called the viscous layer.

Centrifugal force. The circular motion of either the Earth or a small air particle is always affected by two forces: centripetal–directed towards the centre around which the body is moving–and centrifugal–resulting from the body's inertia and directed away from the centre of rotation. All objects on the Earth's surface, including the atmosphere, are under the influence of the centrifugal force that results from planet rotation. Its intensity depends on the speed of the object and its distance from the axis of rotation (r); it is commonly expressed in the form

$$\left|\overrightarrow{F_{cf}}\right| = \frac{v^2}{r}. \qquad (6.3)$$

In the atmosphere, however, circular depressions of low pressure associated with the circular motion of air parcels (as in case of tornadoes) occur quite often, which are the result of pressure gradient force and centrifugal force balance.

Coriolis force. As an apparent force resulting from the Earth's rotation, Coriolis force affects atmospheric movement only on scales from hundreds of kilometers to the planetary scale. In order to introduce the impact of latitude on Coriolis force intensity (and here we will leave out some details concerning its derivation), we will discuss the zonal (west–east: F_{cx}) and meridional (north–south: F_{cy}) components of Coriolis force; these can be written in the form

$$F_{cx} = 2\,\Omega\, v \sin\varphi \text{ and, } F_{cy} = -\,2\,\Omega\, u\, \sin\varphi \tag{6.4}$$

where: u and v are the zonal and meridional components of wind speed, respectively, Ω is the Earth's angular velocity ($7.27 \cdot 10^{-5}$ rad sec^{-1}) and φ is the latitude angle with respect to the Equator. From the definition of Coriolis force and Eq. (6.4), we can conclude that it has maximum values at the poles and diminishes towards the Equator (φ = 0 rad), therefore diverting moving air particles to the right in the Northern Hemisphere and to the left in Southern.

6.2 Scales of atmospheric motion

Pressure gradient is the "working force" of weather. It governs the circulation of air over the hemispheres, moves air masses from one continent to another, or simply provides light breezes from a mountain after a hot summer day. The hierarchy of air motion from global to local scales and their apparent life spans are called scales of atmospheric motion (Fig. 6.2). Additionally, some atmospheric motions are periodic in their appearance on different spatial and temporal levels, affecting weather and climate on even continental scales.

In the continuation of this chapter, the focus will be on global and partially synoptic scales of atmospheric motion. Readers should note that the term "local" in reference to winds denotes regional typicality as opposed to limited scale.

6.2.1 Global winds

Global winds patterns extends over the entire planet with a lifespan that exceeds days and even weeks. This pattern is associated with the global horizontal distribution of atmospheric pressure, which is the result of an imbalance in the Earth's radiation heating and cooling. On the other hand,

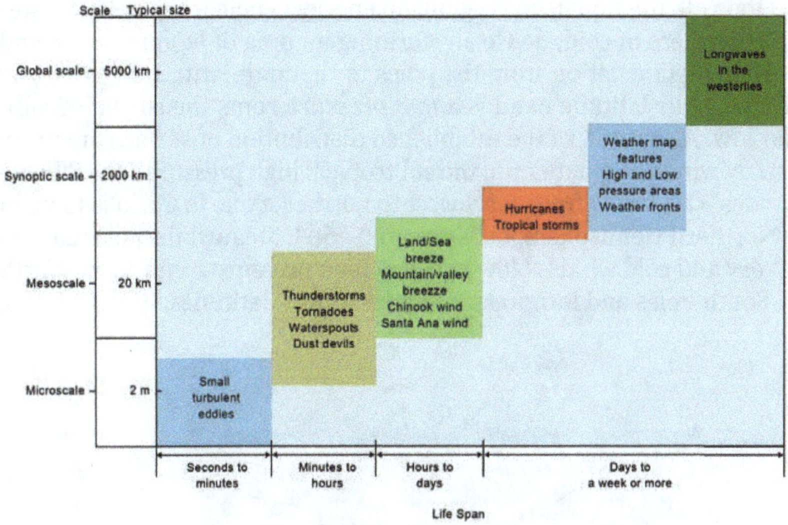

Figure 6.2: Hierarchy of atmospheric motion.

the produced global circulation reduces the effects of energy imbalance, transferring cool air towards the Equator and hot air towards the poles.

In order to better understand the horizontal distribution of atmospheric pressure and the resulting global circulation, an idealized *distribution of surface pressure* as well as the phenomenon of permanent winds will be examined (Fig. 6.3). However, it is important to keep in mind that on small scales prevailing winds can be dramatically changed by local effects.

Over the year, the Equatorial region receives an excessive amount of solar radiation. The warmth of the underlying surface and of the air produces a great quantity of thermally–driven convection, which rises up to the tropopause leaving a region of *low pressure* at the surface. The tropopause acts as a glass ceiling, forcing air to move laterally toward the poles up to 30° latitude. This produces an increase in the air mass above the surface and a *high pressure* area in the subtropical region.

What are the weather consequences of an established circulation? The rising of warm and moist air in the Equatorial region leads to condensation of water vapor and the formation of cumulus clouds, while liberated latent heat enhances the established circulation, forming a zone of intensive convective activity followed by thunderstorms in the vicinity of the Equator–the inter tropical zone of convergence (ITZ). The established pressure gradient between 30° and 0° latitude causes wind to blow towards the Equator, but deviating westward in the Northern Hemisphere and eastward in the Southern, due to Coriolis force. Winds are called trade winds since, according to tradition, trade boats are carried by these winds towards the east coast of South America.

Opposite the Equatorial region, an energy deficit at the poles creates an atmosphere of cold, dense air, forming an area of *high pressure* and establishing circulation from the poles to the mid–latitudes (50° – 70° latitude). This latitude band is a *low pressure* zone, the so–called subpolar low. As a result of the established distribution of surface pressure, westerly winds originate from the subtropical high pressure zone ("horse latitudes") and blow from southwest to northeast due to Coriolis force in the Northern Hemisphere, between 30° – 60°. Meanwhile, polar easterlies–dry and cold winds–blow between high pressure zones at the North and South Poles and low pressure areas at mid–latitudes.

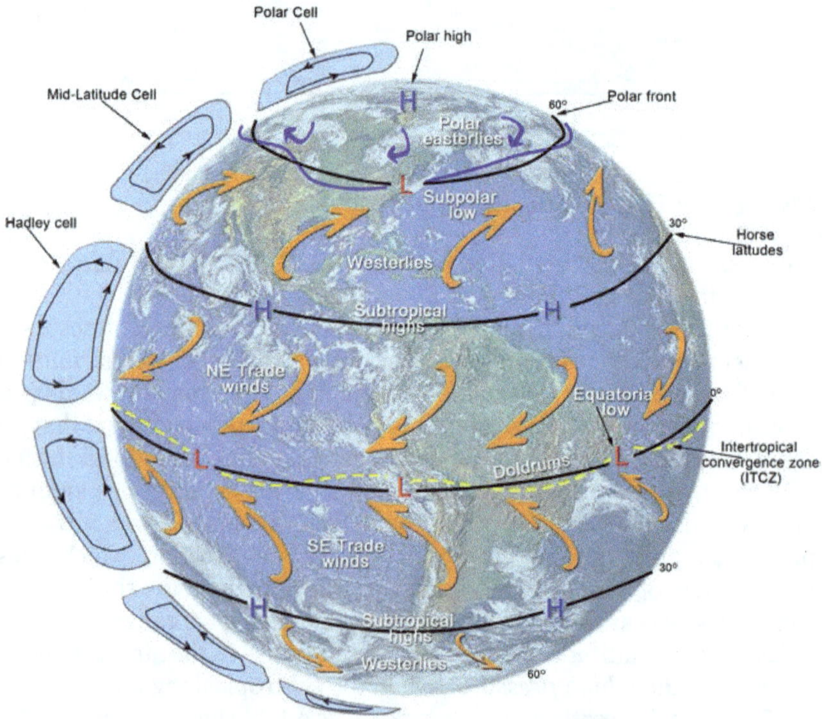

Figure 6.3: Idealized distribution of surface pressure and permanent winds.

6.2.2 Synoptic scale winds

Synoptic scale atmospheric circulation includes phenomena with a lifespan from one or several days up to one week. Spatial scales from several hundred to a thousand kilometers are associated with phenomena like high and low pressure areas and weather fronts. Associated wind systems are, for example, hurricanes and tropical storms, but some synoptic scale winds also appear as a result of a specific distribution of cyclonic and antyciclonic pressure distribution. An example is given in Fig. 6.4, which illustrates the synoptic situation over Mediterranean, southeastern and eastern Europe on February 1st, 2014. The top panel shows the atmospheric pressure field and the fronts present, while the bottom figure also indicates wind direction (blue arrows). A high pressure field in eastern Europe and a low pressure field in the western Mediterranean establish a strong pressure gradient force over the Balkan peninsula, producing wind from the high to the low pressure area. On its path over eastern Europe, the air flow encounters the Carpathian Mountains. Passing over the mountain ridge, it significantly increases in speed (above 11 m s^{-1}); when it reaches Serbia as a southeastern wind, it is known as *Kosava.*

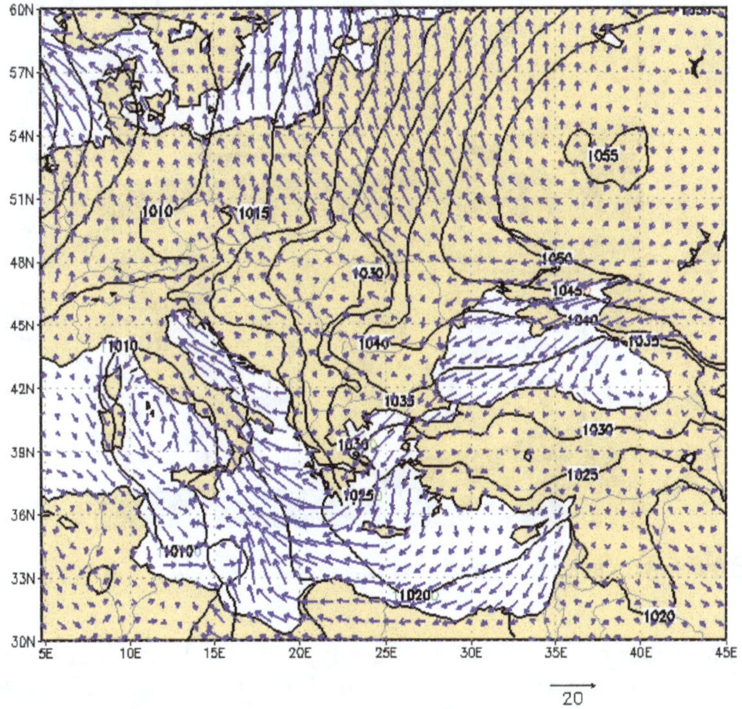

Figure 6.4: Synoptic situation over Europe at 1st February 2014.

6.2.3 Local winds

Local winds are small–to medium–scale winds (tens to hundreds of kilometers) with a duration of a few hours or days, caused mainly by regional topographic conditions. Depending on their origin, local winds are classified as: sea and land breezes, mountain and valley breezes, anabatic and katabatic winds, monsoons and Foehn.

Sea and land breezes (Fig. 6.5) are the result of heat capacity differences between land and water. They occur when wind blows from a large body of water to land and vice versa. With a much higher capacity, water has a slower rate of heating and cooling than does land for the same incoming radiation. Therefore, during summer water is cooler during the day and warmer at night than land. Above a cooler surface, cooler and denser air produces a higher atmospheric pressure. Pressure differences force winds to blow from the sea to the land (sea breeze) during the day and from the land to the sea (land breeze) at night.

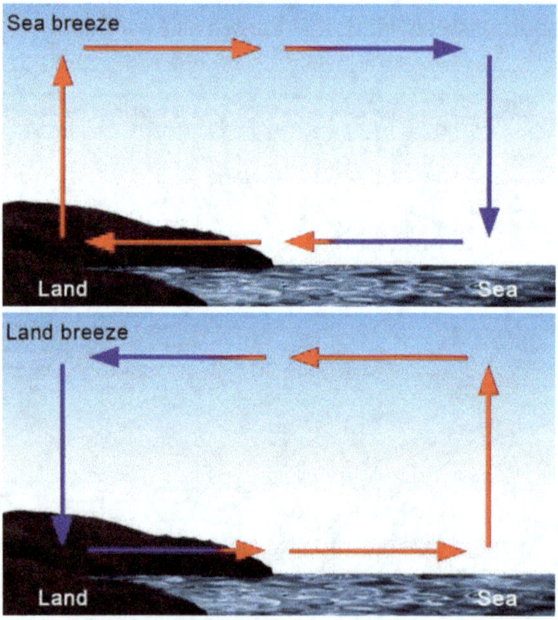

Figure 6.5: A sea and a land breeze.

Mountain and valley breezes are typical in hilly and mountainous regions. These breezes blow as a result of the different diurnal course of heating between the peak and slopes and the mountain valley. During the day, peak heating is more intensive, leaving colder air to remain longer in the valley. The resulting pressure difference produces upslope winds, which increase from sunrise until midday: these are *valley breezes*. By contrast, *mountain breezes* usually blow during the evening and night when the mountain slopes become colder: at that point the pressure gradient force supported by the force of gravity acts downward on the slope. This is also called drainage wind. Mountain and valley breezes are periodic within a daily cycle.

Anabatic and katabatic winds are likewise upslope and down slope winds, yet these terms are reserved for winds of much stronger intensity than mountain and valley breezes (Fig. 6.6). For example, katabatic winds can have hurricane speeds while blowing down a slope.

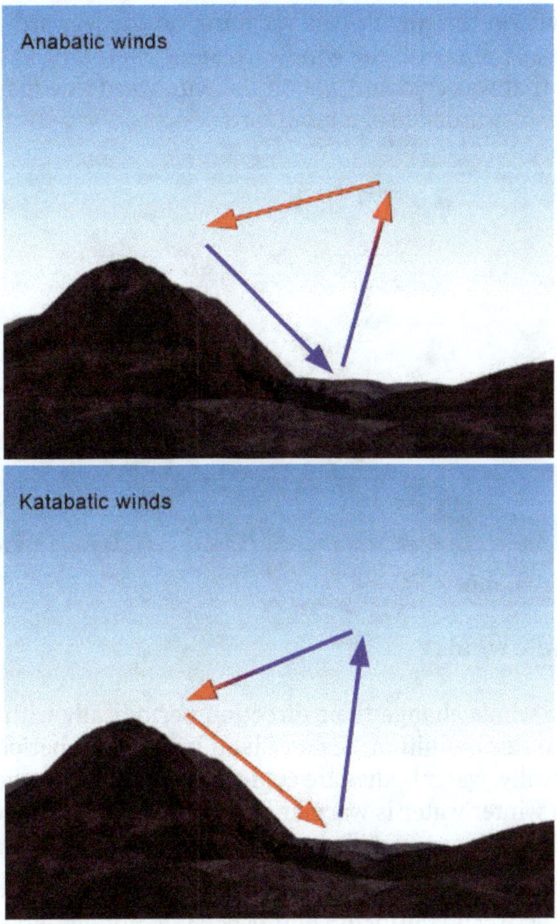

Figure 6.6: Anabatic and katabatic winds.

Foehn is the generic name initially used for dry, warm and often high–speed winds appearing on Alpine down slopes (Fig. 6.7). Now, the same term is employed for winds with the same mechanism of appearance. Foehn appears when air rises on high, steep mountains. At first, temperature decreases in dry adiabatic manner (~1 °C/100m). Soon enough, cooling will cause condensation and formation of cloud. Inside the cloud temperature follows wet adiabatic rate (~0.6 °C/100m). Along the way, some of the condensed water may precipitate. After passing the mountain peak, this air parcel descends, and its temperature increases along wet adiabat, until cloud evaporates. Afterwards, temperature will rise along dry adiabat. However, if precipitation on the windward side reduced moisture, cloud base on the leeward side is higher than on the windward side. As a result, dry adiabatic travel length on the way down is longer than on the way up, air warming on the leeward side is more intensive than cooling on the windward side. Therefore, at the same elevation, air that was cold and wet on the windward side may be dry and warm on the leeward side of a mountain

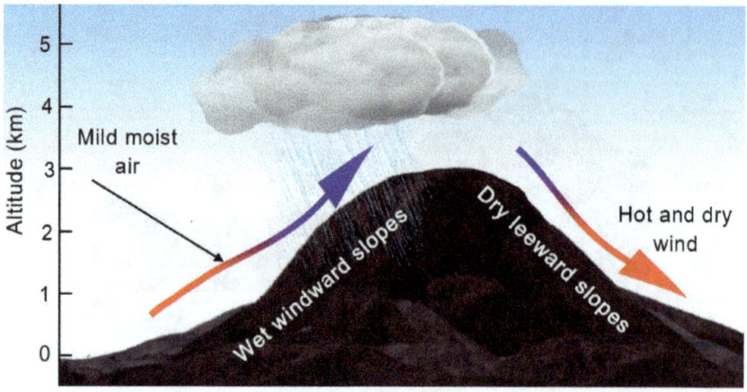

Figure 6.7: Foehn.

6.2.4 Periodic winds

Periodic winds change their direction periodically with the season. According to this definition, sea and land breezes are periodic winds as well. Specifically, water bodies are colder than land during the day only in summer. In winter water is warmer, meaning that pressure distribution and wind direction are opposite to what they are in summer.

The most prominent periodic wind is *monsoon*. Governed by pressure differences between continents and oceans in summer and winter, the monsoon life cycle looks like a large–scale sea or land breeze. It typically blows in eastern and southeastern Asia, but monsoon–like circulation can often be found in wider coastal areas of Australia, Africa and North and South America. The time of appearance, intensity and duration of monsoon and the weather associated with it–primarily precipitation–affect agriculture production and the existence of more than 900 million people.

The high pressure area that develops over the continent (Fig. 6.8) in winter brings fair weather to eastern and southern Asia. Such weather is characterized by clear skies (if there is no pollution) and wind–*winter monsoon*–which blows from the continent to the ocean. In summer, the high pressure area shifts to the air above the colder ocean, forcing circulation from the ocean to the continent. This circulation brings humidity to the continent, resulting in rainy weather followed by wind blowing from the ocean–*summer monsoon.* Summer monsoon season in eastern and southern Asia typically lasts from June till September, but only with 15 to 40 days of rain.

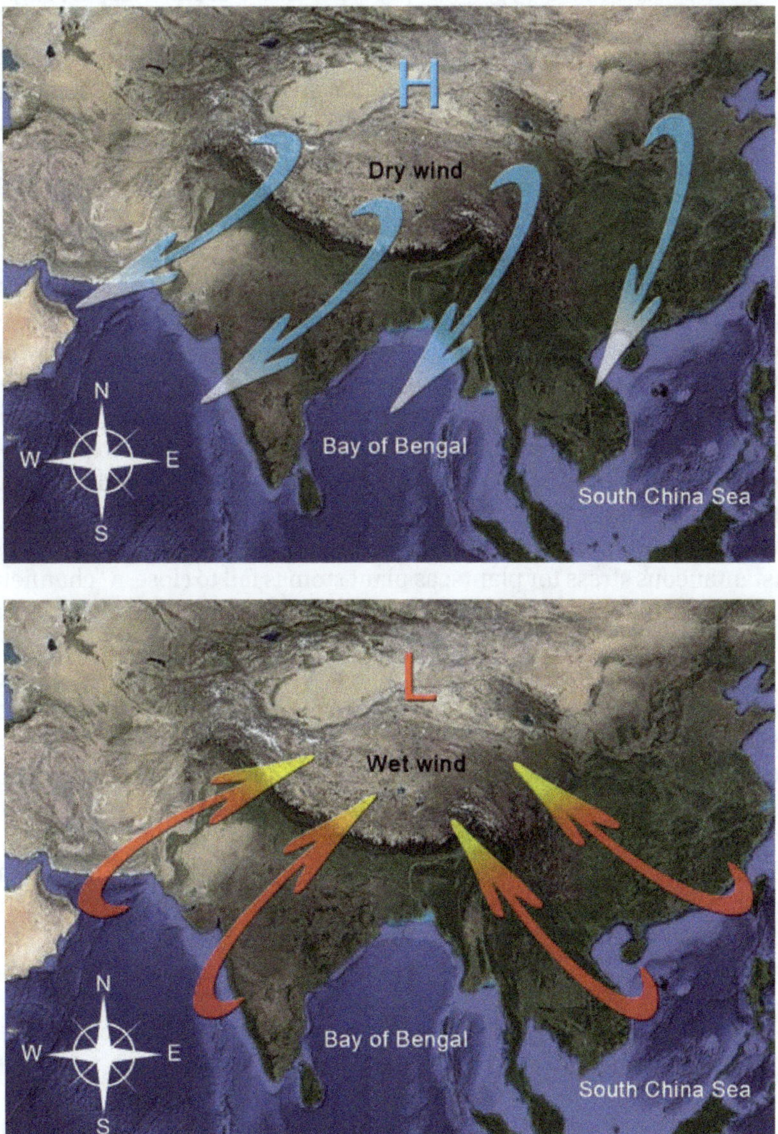

Figure 6.8: Wind flow associated with winter and summer monsoons.

6.3 Impact of wind on plants

Plant physiological and physical processes are the interface of plant–atmosphere interaction in which wind plays an important role, both positive and negative. The presence of wind enhances turbulent transfer between the atmosphere and the plant canopy and increases mixing of the atmospheric surface layer. Improved transfer of water vapor, carbon dioxide and other gasses between the atmosphere and plants increases, up to certain threshold, the rate of all processes involving these gasses, primarily evapotranspiration, respiration and photosynthesis. The direct physiological impact of wind on plants can be clearly seen in the cases of barley and rice, for example, where ethylene production and nitrogen concentration increase with wind speed. Additionally, wind can affect the balance of hormones in some plants.

High intensity winds and squalls lead to soil erosion and significant mechanical damage to plants by breaking branches and shoots, tearing flowers and uprooting shallow root crops. Blowing wind produces advection of warm or cold air, more or less humid than its surroundings, which can significantly affect the energy and water balance of soil, plants and canopy air. Hot and dry winds can produce intensive evaporation from soil and evapotranspiration from plants, causing high stress to plants. An extreme example is sukhovey, wind which is typical for Kazakhstan and the Caspian region. Appearing with speeds from 5 to 20 m s^{-1}, temperatures above 20 °C and low relative humidity (30% or less), they produce instantaneous stress for plants: as plant stomas fail to close, a "channel of communication" is left open which allows the surrounding air to absorb all plant water.

Light to moderate winds allow pollen to spread while decreasing soil moisture, which in early spring can help to remove excess water after snow melts and fields are prepared for sowing. If these winds bring additional humidity, they can enhance dew condensation, which is an important element of water balance in some regions.

During early spring, cold winds can produce frost damage, particularly when radiation frost protection is activated, by making heavy ice crystals on branches and leaves. In the case of atmospheric inversion associated with below–freezing temperatures, warmer winds can destroy inversion and allow mixing of the cold air overlying the surface with the warmer air above it.

6.3.1 Windbreaks and shelter belts

The negative effects of wind on plants can be significantly reduced by wind breaks, which are shelter belts of growing trees and tall crops placed perpendicular to the direction of the prevailing wind. The main aspects of air flow and windbreaks which affect the microclimate of a sheltered area are:

a. the turbulent characteristics and the wind velocity of air approaching the windbreak, which affect the flow dynamics and size of the sheltered zone, and in particular the spread of the mixing zone;
b. the height of the windbreak (H_{wb}), sometimes defined with respect to the average tree height, is actually the height at which the turbulent mixing layer is initiated; and
c. the porosity of the windbreak, which from an aerodynamic point of view is defined as the portion of wind, with respect to its speed, which passes through a windbreak. Since this is quite difficult to determine experimentally, the measure of optical porosity (the portion of the background visible through a windbreak) is often used instead.

On its leeward side, air flow passing over and through a windbreak creates zones of the specific microclimate, which include the approach flow zone, the bleed flow, the displaced profile, the quiet zone, the mixing zone and the equilibration zone (Fig. 6.9).

The *approach flow* zone is an upwind zone, at a distance of typically $5H_{wb}$ from the windbreak, where at heights below H_{wb} wind speed is reduced and diverges more intensively as it approaches the belt.

The *bleed flow zone* is located below the belt height immediately after the shelter. Wind speed in this zone depends on belt porosity and is greatly reduced because of leaf drag.

Approaching the barrier, flow streamlines converge in the area just above the shelter ($1.5H_{wb}$), increasing wind speed. Together with the flow in the bleed flow zone, they form, just after the shelter, a wind profile with extremely reduced speed; the profile has almost zero gradient up to the belt height, a high gradient at $z=H_{wb}$ and a quasi–logarithmic wind profile above this height: this is called the *displaced profile* zone.

Beyond the shelter, between the ground and the mixing zone, is located the *quiet zone*. This zone has the lowest wind speed and reaches distances from $5H_{wb}$ to $8H_{wb}$ downwind. The dimensions of the quiet zone depend on shelter porosity and the approaching flow, while the wind profile and turbulence transfer are mostly determined by the bleed zone flow.

The *mixing zone* is formed downward and downwind from the top of the shelter, reaching the ground at distances between $5H_{wb}$ and $8H_{wb}$. Within the mixing zone, there is a turbulent air region downwind of the quiet zone called the *wake zone*. At distances much greater than $10H_{wb}$, the mixing zone passes into an *equilibration zone*, where the logarithmic wind profile is restored.

In the sheltered area, the most pronounced differences in heat, water vapor and CO_2 profiles are between the quiet zone with lower and the wake zone with higher turbulence intensity. During the day, in the quiet zone the low intensity of sensible heat turbulent transfer from the ground to the atmosphere leads to higher near–surface temperatures with respect to those in the wake zone. The opposite occurs at night, when intensive surface cooling and low turbulence mixing produce lower temperatures in the quiet zone. Some experiments (e.g. Caborn, 1957) show that the presence of a shelter slightly reduces CO_2 concentration during the day and increases it at night in the quiet zone, but without significant impact on the photosynthesis rate.

Since windbreak reduces wind speed in the sheltered area, we might expect evapotranspiration and water loss to be generally reduced in the soil and canopy, particularly in the quiet zone. This line of reasoning is supported by experiments involving open water pans or well–watered plants. Although decreased wind speed reduces the intensity of soil evaporation and plant evapotranspiration, other factors affected by the presence of a windbreak play important roles as well: a) decreases in incoming solar radiation due to shading; b) increases in soil and air temperatures in the quiet zone, and c) changes in stomata resistance, which is affected by incoming solar radiation, air temperature and humidity and which will vary for different plants. Obviously, windbreak affects water loss, but the process is too complex to make straightforward conclusions.

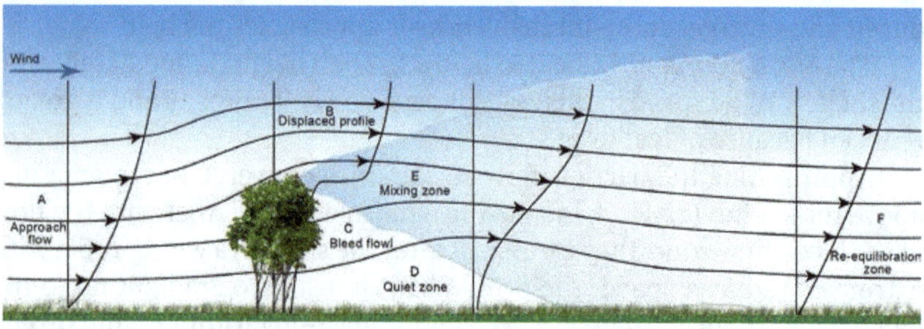

Figure 6.9: Wind flow in presence of a windbreak.

6.4 Impact of plants on wind

For wind flow, plants are often found on the underlying surface. Their presence affects flow directly by increasing friction between the air and the ground and indirectly by changing the energy balance of the Earth's surface.

Vegetation type, height and canopy density determine the impact of plants on flow. In the case of a fully covered surface, we might expect that taller vegetation with higher LAI produces greater roughness. During the vegetation period, plant growth and LAI development reduce the roughness of Earth's surface. When a surface is not uniformly covered by vegetation, the height and distance between plants can significantly affect flow (Fig. 6.10).

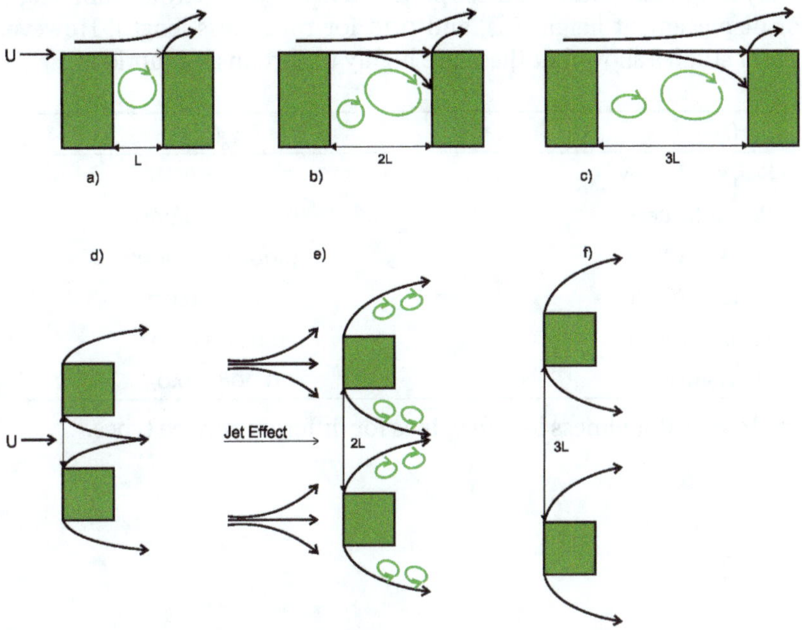

Figure 6.10: Flow over different canopy structures.

Wind speed above a horizontal surface changes with height. In the case of bare soil, we commonly use a logarithmic profile which has the form

$$u(z) = \frac{u_*}{k} ln \frac{z}{z_0} \tag{6.5}$$

where u(z) is wind speed at height z above the ground, u_* is friction velocity, z_0 is roughness length and k is von Karman's constant, normally 0.41.Roughness length refers to the height above the surface at which molecular transfer dominates exchange processes between the ground and the atmosphere (Tab. 6.1).

The presence of vegetation affects the vertical wind profile, transforming it into

$$u(z) = \frac{u_*}{k} ln \frac{z-d}{z_0} \tag{6.6}$$

where d is the displacement height–the level within the plant canopy at which turbulent transfer is considered the most intensive. Roughness length and displacement height are aerodynamic characteristics of the underflow surface, which in the case of vegetation can change during the vegetation period due to changes in plant height and LAI. In some studies, a very rough estimation of these parameters is 0.6–0.8 from plant height, for displacement height (d), and 0.1d for roughness length. However, recent studies show that these are highly uncertain assumptions.

Surface	z_0(m)
Sea, sand, snow	0.0002
Concrete, desert	0.0002 – 0.0005
Flat snow field	0.0001 – 0.0007
Rough ice field	0.001 – 0.012
Fallow ground	0.001 – 0.004
Short grass	0.008 – 0.03

Table 6.1: Roughness length values for different surface types.

7

Role of Water in Soil

7.1 The components of crop–soil water balance

The driving force of the water cycle on Earth is energy from the Sun. Solar radiation transfers energy through space and air by electromagnetic waves. Vegetation absorbs solar radiation to grow and generate biomass. The radiation balance at the Earth's surface determines the energy available for the energy balance components, where evapotranspiration (LE) is a direct link to the water balance of the earth's surface.

A common assumption is that the net radiation, R_{net} received by a canopy or any other surface is balanced by the turbulent fluxes of latent, LE and sensible heat, H together with flux into or out of the ground, Q_G (ground heat flux) and energy transfer in the biomass, Q_S through the photosynthesis process.

Q_S is often neglected in energy balance calculations due to its small amount of about 1 % of R_{net} over vegetation. Relation between energy balance components is given by Eq. (2.5).

Water balance is strongly driven by the energy balance for which processes radiation energy is used. Evapotranspiration (LE) term in water balance of any natural surfaces is the direct link to the energy balance. Water balance of a pre–defined soil volume in a basic form can be written as:

$$P = \mathrm{LE} + \mathrm{R} + \mathrm{D} - \mathrm{K} + Q_{SW} \tag{7.1}$$

where P – precipitation or any other water input into the system, such as irrigation, R – surface runoff, D – drainage out of the considered soil volume (e.g. into groundwater), K – capillary rise from ground water or lateral water flow input in the soil and Q_{SW} – water stored in the soil. Units of the different terms can be given as actual flux rates of water per unit of area (e.g. $l\ m^{-2}$) or as sum of water flux per unit of area over certain time period (e.g. $mm\ d^{-1}$), ($1\ l\ m^{-2} = 1\ mm$).

Precipitation provides the natural water supply from the atmosphere for plants and soils, at times supplemented with irrigation. In the presence of plant canopies, precipitation reaches the ground only partially, because it is partly retained by plant surfaces and evaporates from these surfaces again (interception). Like soil surface water runoff, this part of water will therefore not infiltrate into the soil. The amount of intercepted water depends largely on plant stage, canopy structure and precipitation characteristics. On average, fully developed crop canopies reach interception levels of about 20%–in forests up to 80%–of total precipitation under Central European precipitation regimes. Therefore, interception losses should not be neglected for soil–crop water balance calculations which use measured precipitation data as inputs.

Interception can be assessed as a result of leaf area index (LAI) development and precipitation amount (Fig. 7.1), but it also depends on canopy structure complexity (e.g. forests vs. homogeneous crop canopies vs. row crops). In general, the higher the amount of precipitation in one event, the lower the relative amount of intercepted to total precipitation. However, the absolute amount of intercepted water per rain event (including rewetting) is more or less constant for a specific vegetation type. This means that the precipitation pattern of rain periods, which are often characteristic for specific climates, determine to a great extent total interception losses over longer periods.

When water from precipitation reaches the soil surface, infiltration into the soil and runoff depends on soil properties (physical, biological and chemical soil characteristics). For soil water infiltration, pore size distribution at the soil surface as well as the soil surface structure play crucial roles, including aspects such as biotic pores (especially from rainworms), humus content and the stability of pores during rains. Furthermore, soil gaps, caused mostly by drying clay minerals, can significantly change the amount of water which can infiltrate the soil over a certain period of time (infiltration rate).

The water balance of a crop stand can also be related to the rooting volume (rooting zone) of crops, introducing additional components to the soil such as subsurface flows, which are often present in hilly terrains (Fig. 7.2).

Out of a certain soil volume (e.g. rooting zone), water can drain or percolate into deeper soil layers and reach the ground water level, or water can rise up by capillary forces into the rooting zone from wetter soil layers or from the groundwater level.

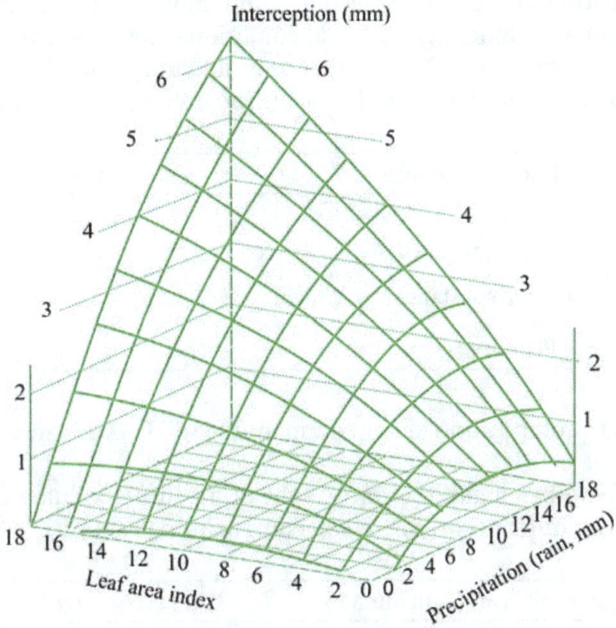

Figure 7.1: Interception as a function of precipitation and leaf area index (schematized, based on Hoyningen-Huene, 1980).

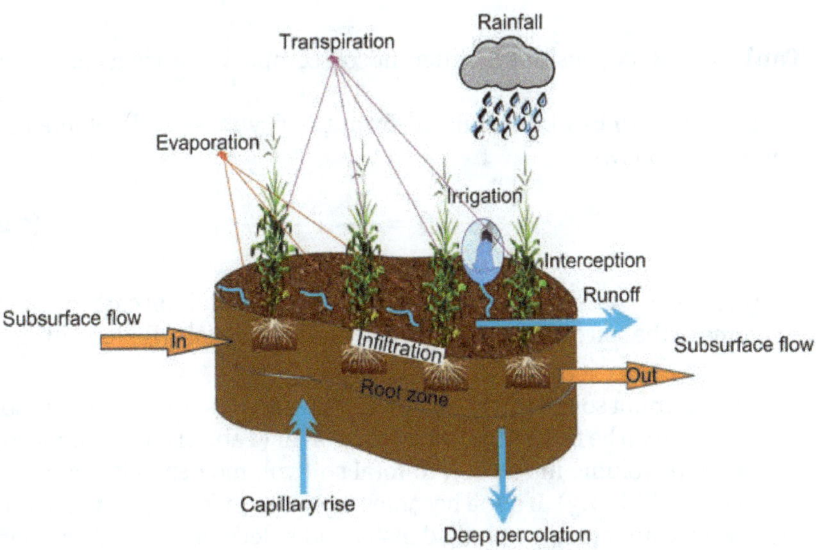

Figure 7.2: Crop–soil water balance components of crop stands (Source: BOKU–Met).

The **total soil pore** volume determines how much water the soil can take up in total. Under average field conditions, however, the soil is only partly filled with water (Tab. 7.1). For a medium mineral soil which is typical for agricultural use, the following mean composition can be given:

Mineral substance	45 %
Air volume	25 %
Water volume	23 %
Organ. substance – of which: • Humus • Plant roots	7 % 85 % 15 %

Table 7.1: Composition of a medium arable soil (in percent volume).

On the other hand, the dry weight of a soil per volume unit (bulk density) can differ according to soil composition (Tab. 7.2).

Upper soil layer (with humus)	0.9 – 1.3
Organic soils	0.2 ∸ 0.8
Mineral soils	1.4 – 1.5
Mineral soils compacted	>1.5

Table 7.2: Bulk densities of different soil compositions (in g cm^{-3}).

Bulk density can be calculated from the dry mass and volume of a soil core as follows:

$$BD = \frac{SM}{V} \qquad (7.2)$$

where BD – bulk density (g cm^{-3}), SM – Dry mass of soil core (g) and V – volume of the soil core (usually taken with small metal cylinders) in cm^3.

An important soil property that determines many soil functions are soil pores, which can be filled either with water or air (Tab. 7.1). The proportion of total pore volume in relation to total soil volume depends mainly on the soil type (Tab. 7.3). If a soil becomes more dense (e.g. by compaction), there will be an increase of bulk density and a decrease of total soil pore volume and soil pore size distribution. The relation between bulk density and soil pore volume depends, however, on the mineral composition of the soil and its pore size distribution.

Less organic, mineral soils (typical for arable soils)	50%
Soils with higher organic content (typical beneath forests, grassland)	40 – 45%
Soils with very high organic content (i.e. swamps)	10 – 35%
Compacted mineral soils	60 – 65%
Sand soil	44 – 64%
Silt soil	45 – 70%
Clay soil	30 – 65%

Table 7.3: Portion of solid soil particles from total soil volume (inverse to total pore volume).

Soil pore volume can be calculated as follows:

$$PV = 1 - \left(\frac{BD}{SD}\right) .100 \qquad (7.3)$$

where PV – soil pore volume (% vol.), BD – bulk density (g cm^{-3}), SD – specific mass of the solid soil particles (g cm^{-3}).

The total pore volume of soils determines how much water a soil can hold. However, this depends largely on the pores size distribution, which in turn determines the capillary forces and how strong the soil can hold the water, e.g. the soil water potential.

These factors, and especially soil pore size distribution, have direct consequences for the storage and movement of water in the soil (direction, speed and amount per time unit), for soil water uptake by roots, for drainage into deeper soil layers and for infiltration into the soil. Soil pore size distribution is more difficult to measure: it can be calculated, for example, by pressing out water from a wet soil core at different air pressure levels to determine the volume of pores in a certain pore size range.

According to the soil water potential, soil water can be primarily differentiated as follows:

- groundwater: a body of water in deeper soil layers, where all pores are filled with water;
- drainage water: water which flows top–down in the soil, driven by gravity;
- available water: the part of soil water that the soil can hold against gravity and is available for plants;
- dead water: water which is held firmly by the soil and can only leave the soil by vaporisation (e.g. through heating the soil in an oven).

7.2 Determining factors of water movement into and within the soil

The part of the water which does not run off and evaporate will infiltrate into the soil. The speed at which this process proceeds is the infiltration rate from the soil surface, which depends on the water flow within the soil medium itself. In a simple approach (one–dimensional), it can be calculated according to Darcy´s law:

$$Q_W = k_w \cdot \left(\frac{d\Phi}{dl}\right) \tag{7.4}$$

or

$$V_W = k_w \cdot \left(\frac{h_W}{dl}\right) \tag{7.5}$$

where Q_w – water flux amount (in $cm^3\ cm^{-2}\ s^{-1}$), k_w – empirical flux constant, V_w – speed of water flux (in $cm\ s^{-1}$), $d\Phi$ – water pressure difference (e.g. MPa), h_w – hydraulic pressure (e.g. MPa), dl – distance between two points in the soil (e.g. cm).

According to Darcy´s law, the large range of water flow and related infiltration capacities of soils impacts characteristics that are relevant for the whole water balance (Tab. 7.4).

Flux constant k	Water conductivity in soil (cm day^{-1})	Flux time needed for amount of 10 mm (= 100 $m^3\ ha^{-1}$)	Water conductivity (porosity)
10^{-1}	8640	10 s	High – fast flux
10^{-2}	864	1 min 40 s.	
10^{-3}	86	16 min 40 s	
10^{-4}	8.6	2 h 46 min	Medium
10^{-5}	0.86		
10^{-6}	0.086	11 days	Low – slow flux
10^{-7}	0.0086		
10^{-8}	0.00086	1 year	

Table 7.4: Range of soil water flow characteristics of soils.

7.2.1 Water–holding capacity of soils available to crops

Strong precipitation or the thawing of snow produces a quantity of water that is present for some time, until the excess water has infiltrated into the soil. The soil can take up and hold only as much water as it can hold against gravity. This level of soil water content at a specific soil water potential (pF 1.8 – 2) is called *field capacity* (FC) (Tab. 7.5).

Conversely, a plant can only take water from the soil as long as a water potential difference exists between the water taken up by the root cells and the surrounding water in the soil pores. If this is not the case, and if the soil water content has already decreased (because an increasing number of pores are filled by air), no water can move to the roots and be extracted from the soil. This state is called the *permanent wilting point* (PWP) (Tab. 7.5). Roots can also lose water to the soil if the soil water tension becomes very high. Plants have developed certain mechanisms to protect their roots against increased soil water tension, e.g. decreasing the root diameter and increasing resistance to loss of water.

Soil water tension (in cm water column)	pF– value	Value in bars	Classes
	At a soil water content specific to the soil type		
15 000 – 10 000	4.2 – 4.0	15 – 10	Permanent wilting point
1 000	3.0	1.0	Slow moving water
100	2.0	0.1	Moving water
	1.8 – 2.0		Field capacity
10	1.0	10^{-2}	Drainage water
1	0.0	10^{-3}	Percolation water

Table 7.5: Soil water tension characteristics (water potential) within soils.

Plants can take up water from the field capacity up to the wilting point. The amount of water between these two limits is called *plant available water* (Fig. 7.3). However, we must keep in mind that the soil water content of both the field capacity and the wilting point can vary significantly between soils, depending largely on soil pore size distribution.

Therefore there exists a relationship between *soil water tension* (water pressure or potential) and *soil water content*, which is expressed by the pF–curve (Fig. 7.3) and is specific to the soil type.

Soil water content can be given as a fraction of mass (e.g. g/g) or as a fraction of volume (e.g. cm^3/cm^3), depending on which measurement method is used:

- Gravimetric method (weighing a soil probe before and after drying in an oven at 100°C)

$$SWC_G = 100 . \left(\frac{Soil_{wet} - Soil_{dry}}{Soil_{dry}} \right) \qquad (7.6)$$

where SWC_G – soil water content (% mass), $Soil_{wet}$ – wet soil probe (as taken from the field) in g, $Soil_{dry}$ – oven dried soil probe in g.

• Volumetric method (measuring soil water content by electronic methods, such as time–domain reflectometry, to determine the volume of water in a soil core).

When the soil bulk density (BD, in g cm³) is known, the volumetric soil water content (can be calculated from the gravimetric unit (SWG, in g cm^{-3}) as follows (assuming that the density of water D_W = 1 g cm^{-3}):

$$SWC_V = SWC_G \cdot \left(\frac{BD}{D_W}\right) \tag{7.7}$$

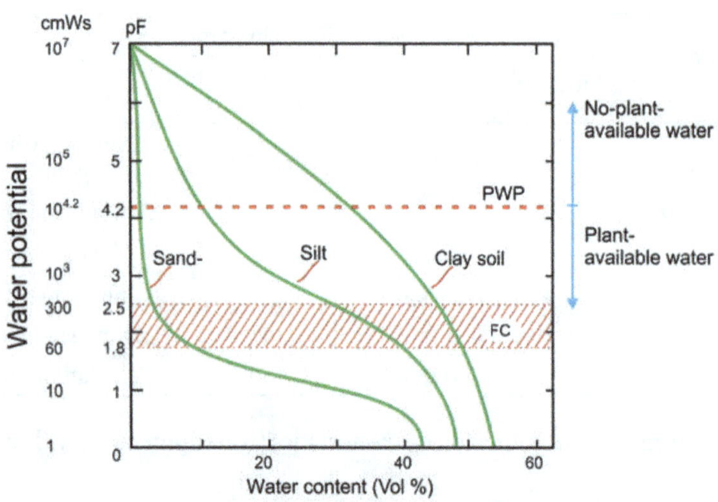

Figure 7.3: Relation between water potential and soil water content for different soil types; FC – field capacity, PWP – permanent wilting point; cmWS – centimeter of water column (unit of water pressure or potential); pF – logarithmic value of cmWS (schematic, based on Scheffer und Schachtschabel, 1982).

The water potential changes from soil to crop and atmosphere, and within the plant itself. For example, if the soil wilting point is at 1500 kPa and the water tension at 200,000 kPa with air at 20°C and 50% humidity, a huge potential difference from soil to atmosphere is created, such that the plant acts as the transfer medium of water in between.

This potential difference from root to leaves (Fig. 7.4) is also the main force of the sap flow against gravity in the xylem of the plant organs, in which water also transports nutrients and minerals to the growing parts of the plants.

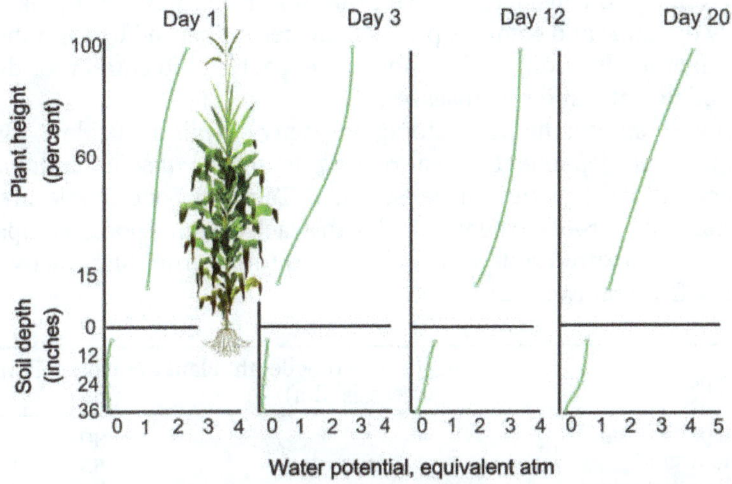

Figure 7.4: Development of vertical gradient of soil and plant water potential in maize under condition of maximum evapotranspiration during drought development after irrigation (modified, based on Shinn and Lemon, 1963).

7.3 Water extraction from the soil by plant roots

In view of the transport of water from the soil to the atmosphere, plants create a "short cut" whereby water is transported more efficiently to the atmosphere from the soil than from bare soil without any crops. In this sense the presence of growing plants significantly reduces the resistance of water transport from the soil into the atmosphere. This also leads to faster extraction of water from soils, which has practical relevance in crop farming in arid regions. For example, crop rotations in arid regions without irrigation options often require years in which fields lie fallow (bare soils) to recover soil water reserves.

Water (and nutrient) uptake by the roots mainly occurs at the root hair cells, which are located on growing roots.

In a generalized way, potential water extraction by roots can be calculated according to Gardner's formula, which accounts for the potential difference and water flux resistance between roots and soil as:

$$W_A = \mathrm{EA} \,/ \left(\frac{\Phi_{soil} - \Phi_{root}}{\Sigma r} \right) \tag{7.8}$$

where W_a – potential amount of water taken up by roots, EA – exchange area of roots (in $cm^2\ cm^{-3}$), Φ – hydraulic potential of roots / soil, r – water flux resistance in soil and between soil and roots.

We need to consider, however, that water and nutrient uptake is a highly dynamic and complex process, further influenced by a number of other factors, including soil temperature, root architecture, root depth and soil density, and root diseases.

If we combine the water storage capacity of soils available to plants with the soil–dependent typical rooting depth, the *absolute amount of water available to plants* can be estimated (Tab. 7.6). For example, in sand it is difficult for roots to penetrate due to a lack of stable pores, compared to better structured soils with stable pores (often from biotic activities, such as from rainworms).

Soil type	Mean effective root depth for cereals (dm)	Plant available soil water (mm)
Large particle sand	5	30
Medium sand	6	55
Small particle sand	7	80
Loamy sand	7	115
Silty sand	8	140
Loamy silt	11	220
Sandy loam	9	155
Silty loam	10	190
Clay loam	10	165
Loamy clay	10	140

Table 7.6: Effective root depth and plant–available water as a function of soil type (medium bulk density).

The water demand of growing crops is in the range of 200–600 kg water per 1 kg of dry biomass. Thus, for optimum crop yields a(more or less regularly distributed precipitation amount of about 300–500 mm is necessary during crop growing periods.

The water demand for biomass accumulation of a single plant is determined by its transpiration coefficient. In contrast, the total water consumption of a crop field (including unproductive evaporation) related to achieve crop yield is defined by the term crop water footprint or water productivity (where the green water footprint is water coming from precipitation and the blue water footprint that from irrigation). Further information on water footprints can be found in Chapter 12.7.2 on climate change adaptation.

As plants effectively extract stored water from the soil by their rooting system, soil water content within the main rooting depth will in general decrease during the main growing period in summer (Fig. 7.5). Depending on the soil water storage capacity available to plants, the soil water content will often approach the wilting point during dry periods and be replenished partly during rain episodes.

When the soil water content under a growing crop approaches the wilting point, the decrease in the unit of time slows down (e.g. summer 2017 in Fig. 7.5), as water extraction by the roots takes more time, due to increasing water flow resistances in the soil. In this case, plants are already under drought stress, showing typical reactions, such as stomata closing during the daytime or wilting of leaves.

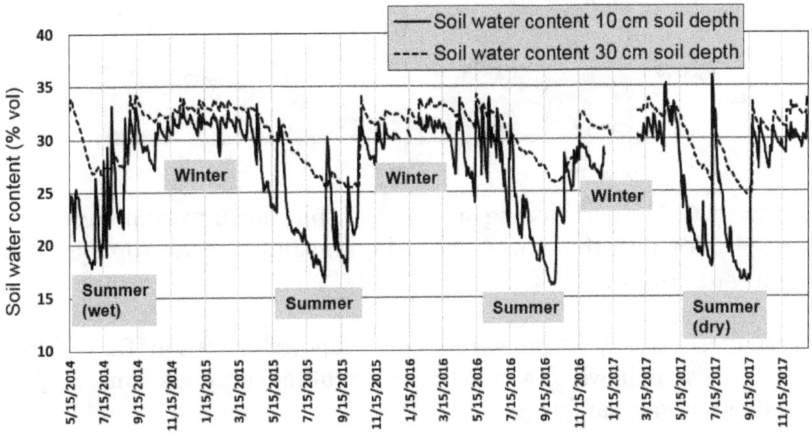

Figure 7.5: Measured soil water contents in two soil depths over four years (2014–2017) at a Central European site (Marchfeld, Eastern Austria) under a permanent crop mixture. Increasing soil water contents are caused by precipitation (Source: BOKU–Met).

7.4 Soil evaporation and transpiration (evapotranspiration)

Soil–water balance of a changing or growing crop canopy is determined by evaporation and transpiration, respectively. The share of soil evaporation in relation to the proportion of transpiration to total evapotranspiration of a crop field changes according to the transpiring canopy biomass (Fig. 7.6).

Simplified approach to calculating actual soil–crop water balance allowing agrometeorological applications (FAO Method).

The FAO (Allen *et al.*, 1998) developed an approach for soil–crop water balance simulations for practical applications, notably for crop irrigation planning. This approach is based on a simplified calculation of the actual evapotranspiration of crops, originally defined by Penman Monteith (see also “Numerical examples”).

The same calculation is used in the “AQUACROP” software, which is freely available from the FAO website. It has been tested for many crops and regions across the world. This approach starts with a calculation of

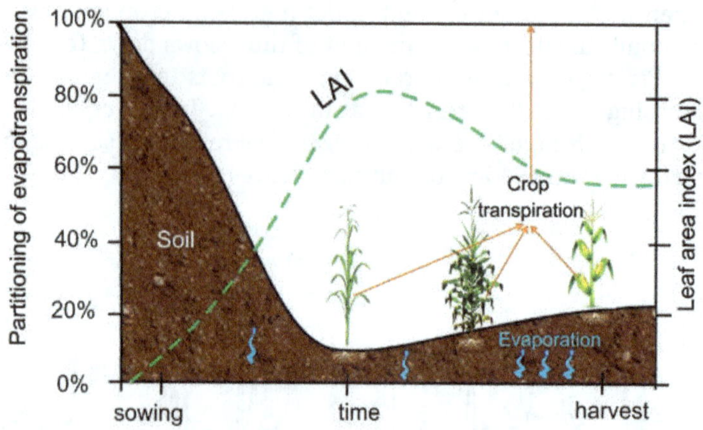

Figure 7.6: The partitioning of evapotranspiration into evaporation and transpiration over the growing period for an annual field crop (based on Allen *et al.*, 1998).

parameterized (Fig. 7.7) grass reference evapotranspiration (Eq. 7.9) for a daily time step, in which standard daily weather variables from standard weather stations can be used.

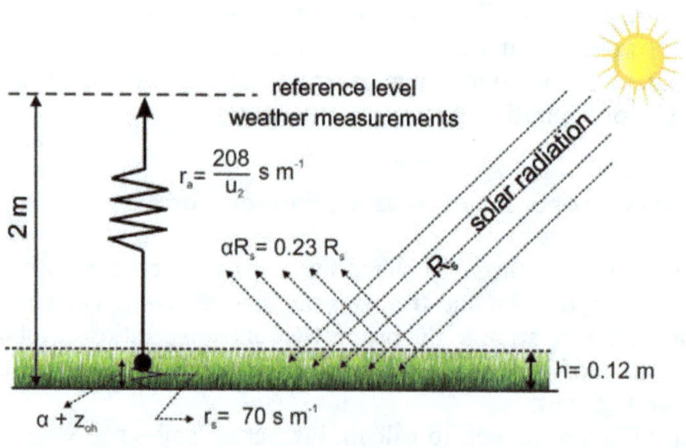

Figure 7.7: Scheme of parameterization for FAO Penman–Monteith grass reference evapotranspiration equation (based on Allen *et al.*, 1998).

A panel of experts recommended the adoption of the Penman–Monteith combination method as a new standard for reference evapotranspiration and advised on procedures for calculating the various parameters. The FAO Penman–Monteith method was developed by defining a hypothetical reference crop with an assumed height of 0.12 m, a surface resistance of 70 s m^{-1} and an albedo of 0.23; it therefore closely resembles the evaporation of an extended surface of green grass of uniform height which is actively growing and adequately watered. The method overcomes the shortcomings of the previous FAO Penman method and provides values which are more consistent with actual crop water–use data worldwide.

From the original Penman–Monteith equation and the equations of the aerodynamic and surface resistance, the FAO Penman–Monteith method to estimate grass reference ET_0 is written as:

$$ET_o = \frac{0.408\,\Delta\,(R_{net} - Q_G) + \gamma \frac{900}{T+273} u_2\,(e_s - e_a)}{\Delta + (1 + 0.34\,u_2)} \tag{7.9}$$

where ET_0 – reference evapotranspiration of short grass (mm day^{-1}), R_{net} – net radiation at crop surface (MJ m^{-2} day^{-1}), Q_G – soil heat flux density (MJ m^{-2} day^{-1}), T_a – mean daily air temperature at 2m height (°C), u_2 – wind speed at 2m height (m s^{-1}), e_s – saturation vapour pressure (kPa), e_a – actual vapour pressure (kPa), e_s-e_a – saturation vapour pressure deficit (kPa), – slope vapour pressure curve (kPa $°C^{-1}$), g – psychrometric constant (kPa $°C^{-1}$).

In the next step, a crop–specific K_c factor (Fig. 7.8), which describes the phenological development and canopy status of a crop during the growing period, is used to modify the grass reference value to receive the actual crop value (introduced also in Chapter 4). Both are calculated for conditions without any water stress. In the final step, water stress is additionally included as a reduction factor for actual evapotranspiration through stress coefficient, K_c. The daily actual water content by the calculated actual evapotranspiration is then updated to the actual day.

7.5 Approaches to basic soil–water balance calculations

Different approaches are used to calculate the soil water content in different soil layers (Fig. 7.9); these depend on the specific application as well as on available data for the physical soil.

The procedure that requires the least amount of data is the application of a simple, one–dimensional, layer–specific soil water balance approach, also called the cascade approach. It assumes only vertical input water flow, from the top soil layer to deeper soil layers. In this approach, artificially defined soil layers are used to fill up the layer–specific field capacity: if the layer in question is "saturated", water goes to the next deepest layer to be filled up to the field capacity. This approach requires us to know "only" the soil water holding capacity up to the field capacity (or saturation) in order to calculate vertical water movement into the soil and the related change

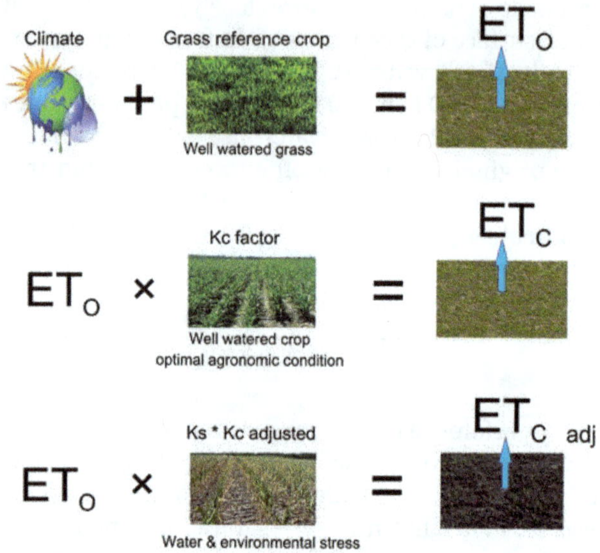

Figure 7.8: Reference (ET_o), crop evapotranspiration under standard (ET_c) and non-standard conditions ($ET_{c\ adj}$) for use in soil water balance calculations (Allen *et al.*, 1998).

of the soil water content of the soil layers after rain. If this approach is applied on a daily time step, it works relatively well in homogeneous soils with medium water conductivity (Tab. 7.5).

In another, more complex approach, we can apply the water conductivity (according to Darcy's approach) from one layer to the next. However, the smaller the soil layers are defined, the higher will be the data demand for an accurate calculation: such data, however, are normally not available for agrometeorological applications. Still, this approach is often used in agrometeorological, crop yield and irrigation models. Vertical capillary rise from groundwater levels is sometimes introduced into this calculation.

The most complex and physically sound approaches consider the three–dimensional water flow in the soil matrix. These take into account soil water potential differences in the three–dimensional matrix together with sinks and sources (based on Richard´s equation), and include the Van Genuchten method for calculating the soil water retention curve for the wetting and drying of soils. Only such complex approaches can well address soil physical heterogeneities, if the relevant soil physical data are known or have been estimated. The more complex approaches are particularly relevant for solute or contaminant transport in soil layers.

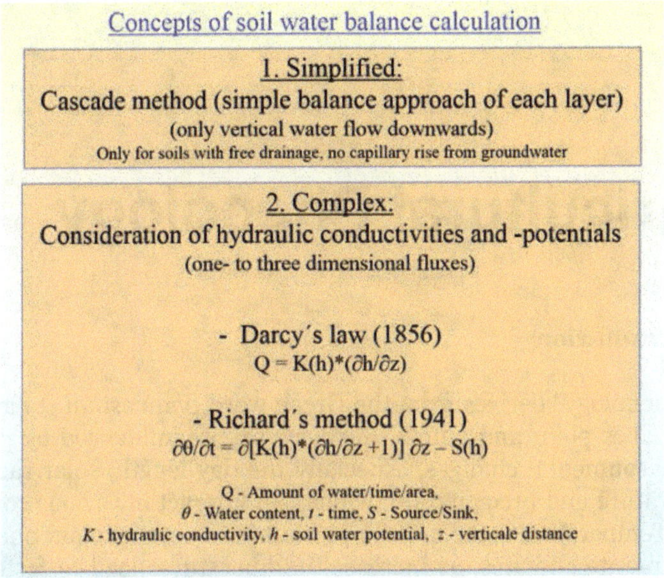

Figure 7.9: Main soil–water balance calculation approaches for agro-meteorological applications (Source: BOKU–Met).

8

Agricultural Phenology

8.1 Introduction

"Phenology" derives from the Greek word *phainesthai* (to appear). It describes plant and animal life cycle events influenced by periodical environmental changes, especially the day length, solar radiation, temperature and precipitation, which can interact in various combinations. Temperature is the most important parameter of plant phenology in temperate climates, as increased temperatures lead to faster crop development. A strong correlation between plant development and air temperature (probably also influenced by the strong correlation between temperature and solar radiation) was already reported by Reaumur (1735), who originally suggested the temperature sum method for calculating phenological development. Since then, this observation has formed the basis for all phenological models. However, phenological observations had already been made in ancient times, such as in China 2000 years ago, or in Japan, where the start of cherry flowering has traditionally been recorded for many centuries (Nejedlik and Orlandini, 2008).

Seasonally changing weather conditions are the primary triggers of the phenology of plants or animals (especially insects). Seasons more or less replicate annual climate variations, which are mainly caused by the latitudinal annual variations of solar energy absorbed by the Earth´s surface and the related influence on surface energy balance. Other factors, such as the distance to oceans, the influence from regional wind regimes and ocean currents, and the general topography also influence seasonal climates at a regional level.

Phenological events and their inter–seasonal and inter–annual variations have important impacts on terrestrial ecosystems and human societies by altering global carbon, water and nitrogen cycles, crop production, the duration of the pollination season or the occurrence of diseases and pests. Crop phenology is therefore a basis for understanding intra– and inter–annual variations of agroecosystem processes. Consequently, many model applications in agriculture, such as for crop risk warning, pest and disease warning, yield prediction and crop growth models, have to implement phenological models as a key element.

Plant and animal species have responded to climate conditions in their environment throughout their evolutionary history. The phenology of plants as a climatic indicator was therefore also used in climate mapping methods in the past (e.g. due to the coincidence of a climate class with a typical type of vegetation in a region). Nevertheless, in the past climatic changes were commonly considered as a relatively slow and gradual process, which allowed a smooth process of evolutionary adapting or replacement. Today, though, scientists are concerned about how different species and ecosystems will respond to the recent rapid rates of climate change. Many species of a particular ecosystem adapt to ecological niches with climatic conditions that have been prevalent over long periods of time and can therefore be quite vulnerable to even modest changes in climate. Not only do increasing temperatures combined with spatially different trends in precipitation influence natural growing plants and their environment, but they also affect crops of managed ecosystems in direct and indirect ways.

Increasing temperatures and other climate changes have already affected ecosystems on all continents. Observed changes include the increase of overall vegetation periods and changes in overwintering conditions in temperate climates. In combination with these temporal shifts of growing conditions, spatial shifts of climate and thus of ecological zones, such as latitudinal and altitudinal shifts, affect production conditions in agriculture.

Phenological observations of various plants, such as grapes (Fig. 8.1), confirm accelerated crop development (phenology) due to climate warming. Thus, monitoring of phenology serves as a life cycle indicator reflecting the impact of weather on vegetation, which is carried out in many countries by permanent observations in, for example, phenological gardens or in regular observations in nature. Climate warming has therefore stimulated increased interest in phenological processes as indicators for climate change impacts. Nevertheless, the timing of phenological events is also of importance for agricultural crop management, such as the scheduling of fertilization, harvest and plant protection measures.

On the other hand, the phenological development of crops at a specific location can be influenced by various factors such as sowing and cutting dates, irrigation, soil cultivation, fertilization and crop selection. Phenological data also play a key role in the calibration and validation of many ecosystem models.

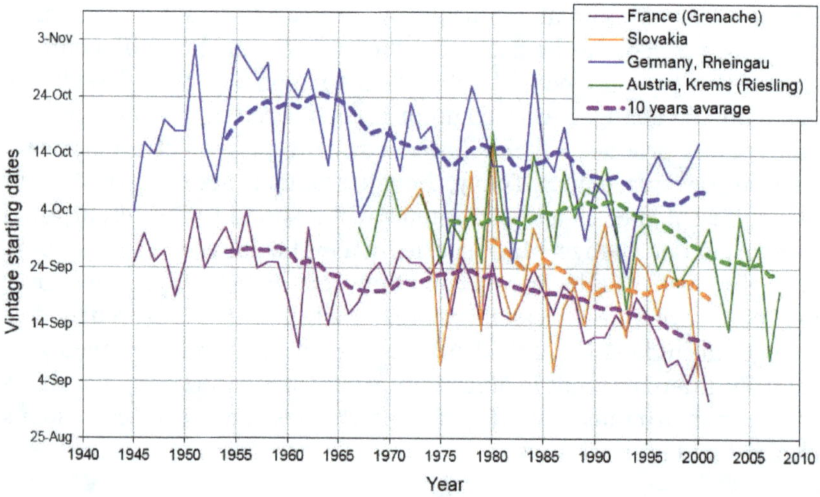

Figure 8.1: Tendencies in grape harvest dates in different wine–growing regions of Europe (France, Germany, Austria, Slovakia) confirm ongoing warming trends over the past decades (Eitzinger *et al.*, 2009).

8.2 Vegetation period and phenological phases

At different latitudes, the length of the vegetation period is related to temperature and solar radiation. When we relate mean monthly temperatures of > 10 °C as an indicator for the vegetation period duration to latitude, the strongest increase of day length during the growing period can be observed, especially between 40° and 60° latitude (Fig. 8.2). Accordingly, the vegetation period of 12 months at the equator decreases to only a few months as one moves towards the poles. Also, the mean temperature during the vegetation period decreases from the equator towards the higher latitudes.

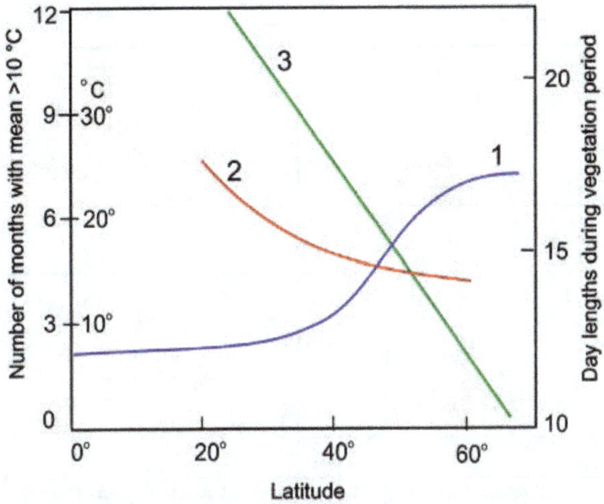

Figure 8.2: Length of vegetation period (3, defined as >10 °C mean monthly temperature), mean monthly temperature (2) and day length in hours (1) by latitude (schematic, based on different sources).

Natural plants are normally well adapted to local climates due to the evolutionary process, whereby their seasonal requirements for growing conditions fit well to the mean seasonal changes of the local climate (e.g. seasonal radiation and temperature). If plants are transferred to other climates and latitudes with other seasonal characteristics, this fit might be lost, meaning that local growing conditions can no longer be used in an optimum way. The plant will therefore lose productivity and will be less competitive in relation to other adapted local plant species.

For example, if a plant adapted to warm mid–latitudes is transferred to a cooler location at a higher latitude, the temperature limitation might cause it to start its development too late, and the shorter growing season will prevent it from completing its development (Fig. 8.3, top). Conversely, crops which are adapted to low temperatures start their growth prematurely under warmer conditions at lower latitudes, developing faster and therefore completing their growing period too early (Fig. 8.3, bottom), which results in lost crop productivity (because less time is available for biomass accumulation). This effect is well known in the cases of some cereals such as winter wheat, which has its highest productivity in northern European climates and decreasing productivity in lower latitudes where mean temperatures are above the optimum.

On the other hand, plants (and even plant cultivars) whose phenological development is additionally determine by day length (photoperiodic reaction, such as short– and long–day plants) might not be able to reach their generative phase under changed day lengths during the temperate growing period. Examples of agricultural crops are soybean (a typical

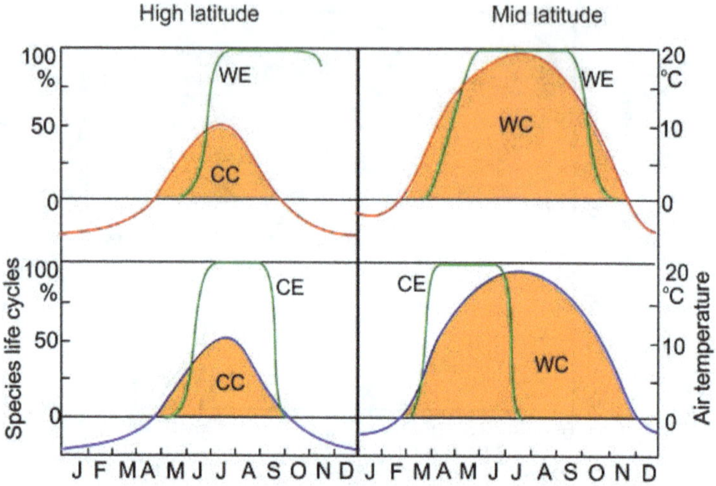

Figure 8.3: Crop ecotype seasonal temperature requirements (WE, CE) in comparison to seasonal temperatures of high and mid latidudes. Top left: Warm ecotype (WE) not adapted to high-latitude cold climate (CC); top right: WE well adapted to mid-latitude warm climate (WC); bottom left: Cold ecotype (CE) well adapted to high-latitude CC; bottom right: CE not adapted to mid-latitude WC (schematic, Bünning, 1953).

short–day plant that requires short days to reach the generative phase for flowering and producing seeds) and different vegetables such as lettuce (which needs long days to maintain the vegetative period in order to produce leaves as its harvested product). Other crops such as oak, sugar beet and potatoes also show similar reactions to day length; these are, however, cultivar specific and are optimized by crop breeding for relevant latitudes.

In summary, seasonal changes of phenology are mainly driven by the following environmental variables:

- variations in day length or duration of isolation;
- temperature (including extremes);
- precipitation amount and drought episodes (especially in regard to their distribution);
- other life–controlling factors (crop stresses of any kind).

8.2.1 Phenophase classification

In order to describe the phenological phases of plants, several classification schemes have been developed, which are used for phenological observations. Only with a phenological classification scheme can phenological observations from different years or sites be systematically compared and phenological data sets be then interpreted independently of any observer.

In recent decades, the use of the so–called extended *BBCH code* (Meier, 1997; 2003, 2009) has been recommended. Based on Zadok's cereal code (Zadok *et al.*, 1974), this system can be used for uniform coding of phenologically similar growth stages of all mono– and dicotyledonous plant species.

The BBCH code (Tab. 8.1) uses a general scale, such that it can also be applied to those plants for which no special scale is available. Clear and easily recognized external morphological characteristics are used to describe the main (longer–lasting) phenological development stages, called the principal growth stages. Secondary growth stages define a shorter step of development.

Principal Growth Stages	Description
0	Germination / sprouting / bud development
1	Leaf development (main shoot)
2	Formation of side shoots / tillering
3	Stem elongation or rosette growth / shoot development (main shoot)
4	Development of harvestable vegetative plant parts or vegetative propagated organs / booting (main shoot)
5	Inflorescence emergence (main shoot) / heading
6	Flowering (main shoot)
7	Development of fruit
8	Ripening or maturity of fruit and seed
9	Senescence / beginning of dormancy

Table 8.1: BBCH code for cereals (Meier, 1997).

Agroclimatic indices (see also Chapter 11) are used to quantitatively assess available resources for different agricultural needs and related risk assessments or warning tools. These needs are determined by the physiological processes of plants and animals, which again strongly depend on their phenological or development status.

As we have already seen, the major drivers of plant development (phenology) are temperature and day length, while the availability of photosynthetic active radiation, water and nutrients limits plant biomass accumulation (growth). Indices characterizing plant development can provide indirect information about the impact of climate variability and also about the conditions for further plant growing. Beyond the standard indices, specific indices focus on crop responses to weather obtained either from statistical or simulation models. All basic principles of phenological indices and models are based on the close relation between temperature and plant development and growth.

Phenological development of plants can be compared on both spatial and temporal scales. Comparisons on spatial scales during the same period can help to identify variations in regional climates over large or small areas (see Chapter 12). Comparisons on temporal scales provide important information on crop responses, which help, for example, plant breeders to identify abiotic stress resistances. Comparisons between single years or multi–year period can provide insight into important crop responses such as yield and yield quality (e.g., the quantity and quality of proteins in cereals, or the annually changing quality of wine). They can also supply information on trends in phenological phases, with important implications for crop management options (e.g., the timing of fertilization, pest management or harvesting).

Inter–annual variations in a local climate and related crop phenology can be caused by several factors, including human–induced ones, such as microclimatic manipulation by human measures (irrigation, establishing wind breaks or cold air lakes etc.). Furthermore, monitoring the length of the growing season by recording phenological changes helps scientists better document and understand our changing weather from year to year or climate change over many decades.

8.3 Pest and disease phenology

Diseases, such as those caused by fungi, have life cycles which are often coordinated to the phenological phases of their hosts (e.g., crops). However, they also respond in their development to meteorological and climatic factors such as wetness, light and temperature. We are thus able to create various algorithms and models for pest and disease development, which are widely used for pest warning and management in agricultural practice (Fig.8.4)

Population dynamics of pests and insects are influenced by a wide range of biotic and abiotic factors. Thermophile insects are in particular sensitive to temperature during their life cycles. However, compared to diseases and plants they can move and find more suitable environments for their development: larvae of soil–born pests, for example, can move to deeper soil layers over winter or during cool weather periods during the vegetation period. Insects such as mosquitos can survive winter in warm places, such as in buildings. Due to climate change, climate niches have shifted to allow new pests with a higher temperature demand to establish populations in higher latitudes.

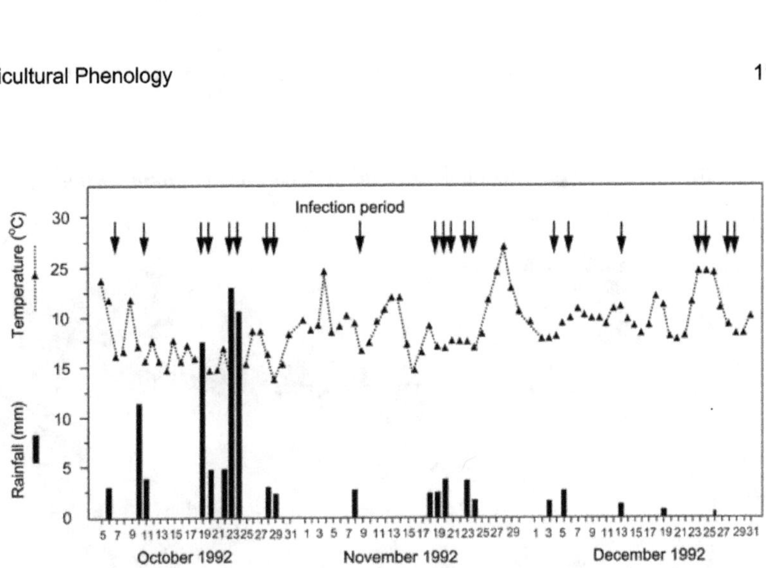

Figure 8.4: Monitoring and forecasting system of powdery mildew based on disease phenology algorithm using air temperature and precipitation as the driving parameters (Source: BOKU-Met).

Many pests causing significant damage to agriculture respond to changing temperature conditions, thus affecting risks associated with agricultural production. For example, the western corn rootworm (WCR, *Diabrotica virgifera virgifera)* is one of the most destructive maize pests worldwide. It was introduced accidentally into Europe in 1993 and was found in Austria for the first time in 2002. Since the development of eggs, larvae and pupae takes place completely below ground, this pest is very sensitive to soil temperature. It is therefore an ideal candidate for applying mathematical soil temperature models and temperature sums based on phenological models (Fig. 8.5).

8.3.1 Calculating crop/pest development by phenological models

The main approach used for calculating crop development (phenology) is based on the air temperature sum method (also called growing degree days). In this approach, the mean daily temperature above a certain base temperature (at which the growth activity of the relevant plants starts) is added up over the growing period. For most crops, the base temperature is defined at between +5 °C and +10 °C. The occurrence of specific phenological phases (such as the start of flowering or physiological maturity) are mainly determined by temperature sums (but still other factors can contribute, such as day length or environmental stresses). The basic equation of the temperature sum calculation is as follows:

$$GDD \sum (T_a - T_b) \quad (8.1)$$

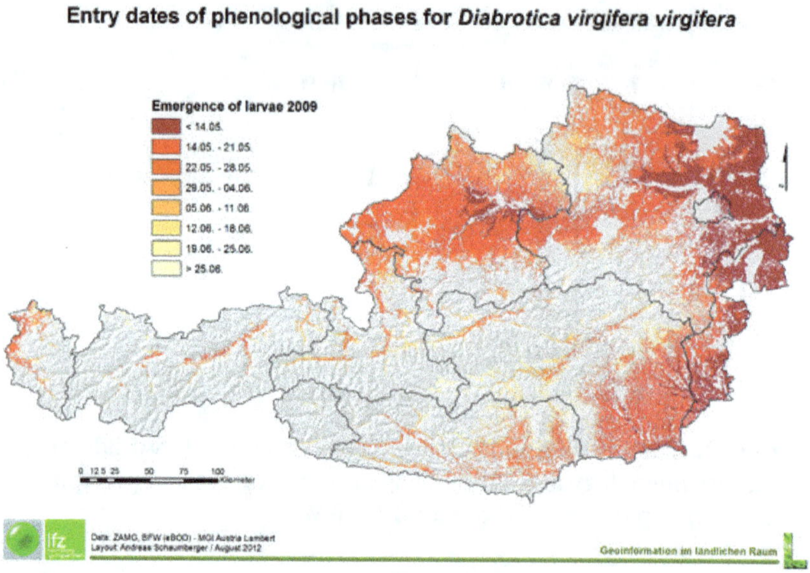

Figure 8.5: Entry dates for emergence of larvae of western corn rootworm based on soil temperature sum model, calibrated for Austrian conditions (base temperature of 11.7 °C at 6 cm soil depth, and temperature sum of 280 °C), (Source BOKU-Met).

where GDD – temperature sum (or growing degree days, see also Chapter 3), T_a – mean daily air temperature and T_b – base temperature.

We should consider that air temperature can have different effects on phenological development in different phenological phases, especially between the vegetative and generative stages. There are also differences between plant ecotypes and crop cultivars, which means that phenological models have to be calibrated using phenological observations and air temperature measurements over several (>10) years before they can be applied.

The simple form of the temperature sum method as written above therefore has some disadvantages:

- T_b has to be re–calibrated for different plants, cultivars, environments;
- the impact of high temperatures is often overestimated (gradual decrease of response with higher temperature).
- It does not consider:
    a. potential changing temperature responses of phenological development over the growing period;
    b. the effects of daily maximum and minimum temperatures and daily temperature amplitude;

c. seasonal effects such as the impact of cold– and heatwaves and various combinations of cold/warm seasons;
d. other effects on crop phenology such as drought or nutrient stress.

Given the high degree of simplification of the basic equation, several more complex temperature sum equations (such as those which include extreme temperature effects) and crop–specific equations (such as for grapes) have been developed. An example of a well–known crop–specific index is the HUGLIN index (Huglin, 1986), which considers the temperature sum of daily mean and maximum temperatures over the growing period of grapes, using a base temperature of 10°C. It can be, for example, easily adapted to the demands of different grape varieties. This index does not, however, consider overwintering conditions or frost risk for grapes. The Huglin index is calculated as follows:

$$IH = \sum_{September\ 30th}^{April\ 1st} KL \frac{(Ta-10)+(Tmax-10)}{2} \quad (8.2)$$

Where IH – Huglin index value calculated from April 1 to September 30 of a specific year, KL – latitude factor (i. e. day length effects) (K (40°) = 1.02; K(60°) = 1.06), Ta – daily mean temperature (°C) and Tmax – daily maximum temperature (°C).

8.4 Low temperature effects on phenological phases

In temperate regions, many overwintering plants or crops have a low temperature demand in correspondence to complex genetic–based physiological mechanisms for dormancy regulation during the cold winter period, when growth is not possible and frost resistance is necessary. The dormancy period is a necessity to survive episodes of cold as crops and plants overwinter.

Depending on plant type, low temperature demand is called either the chilling requirement (used for permanent plants such as orchards) or vernalisation/jarowisation (used for annual overwintering crops such cereals). These low temperature demands are especially necessary for the timing of the dormancy break in spring, when active growth is initiated–whether in the form of stem elongation in annual crops (cereals) or the sprouting of seeds or bud breaks in orchards–as well as for initiating flowering in the coming growing period. Day length (through photoperiodic effects) can also influence these processes.

The low temperature demand of crops during the winter period lies mainly between 0 °C and 10 °C over a certain period of time. Large differences in temperature requirements during the growing period exist between plant types, ecotypes and cultivars. Low temperature demand is also calculated by temperature sum methods, for orchards in hours (chilling hours) and for other crops such as cereals (vernalisation) in days (using daily mean temperatures below, e.g., 10 °C).

Dormancy can be broken by warm periods during the winter and re–initiated upon the return of lower temperatures. However, once bud break has occurred (in orchards and grapevines) or active growth has started in spring (in winter crops), dormancy cannot be re–initiated. From that point on, the temperature sum method (or forcing method) can be applied to calculate the timing of the subsequent phenological phases, such as flowering time.

Phenological models for orchards combine "chilling and forcing" algorithms with different parameterizations of the required sums of chilling hour units (CU, Tab. 8.2) and growing degree hours (GDH).

The chilling requirement is estimated by the sum of chilling hour units (CUs) during winter, starting from September 1st (Northern hemisphere) and based on hourly temperature data (UTAH method, see Tab. 8.2). The date of the release of endodormancy (t1) is estimated by using crop specific calibrated CU.

In the next step, Growing Degree Hours (GDH) are accumulated starting from (t1) to calculate the start of flowering (t2), (Anderson *et al.*, 1986; Luedeling *et al.*, 2009).

Growing degree hours for day k are calculated as:

$$GDH(k) = \sum_{i=1}^{i=24} \mathrm{Max}[0, (\mathrm{Ta(i)} - 4.5)] \qquad (8.3)$$

where $T_a(i)$ is air temperature at hour *i*.

The GDHs accumulated from t(1) to t(2) and named as GDH_f are calculated as below:

$$GDH_f = \sum_{k=t1}^{k=t2} \mathrm{GDH(k)} \qquad (8.4)$$

where GDH_f is the forcing requirement of flowering, and k is the day in the interval from t_1 to t_2.

Temperature (°C)	CU
T < 1.4	0.0
1.5 < T < 2.4	0.5
2.5 < T < 9.1	1.0
9.2 < T < 12.4	0.5
12.5 < T < 15.9	0.0
16 < T < 18	-0.5
T > 18	-1.0

Table 8.2: CU values as a function of hourly temperature (Richardson *et al.*1974).

To receive suitable results, both CU and GDH values need to be calibrated on observed flowering dates and measured temperatures for specific sites and cultivars. For example, CUs and GDHs were used for estimating blossom dates (t2) of apple trees in Austria (Fig. 8.6).

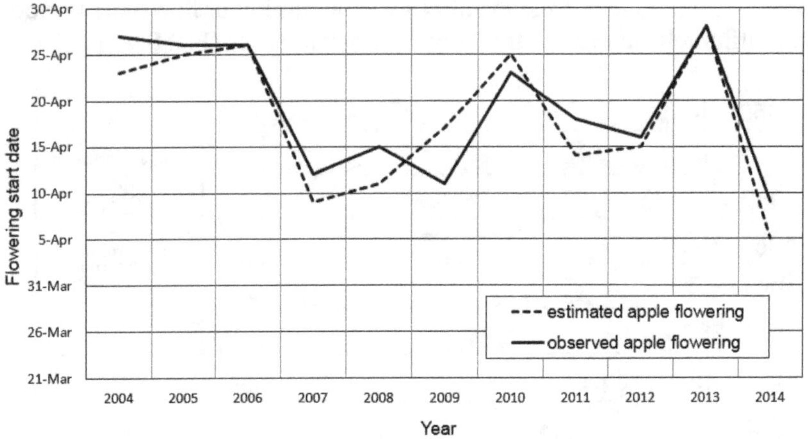

Figure 8.6: Observed and simulated apple flowering dates in Austria using combined chilling–forcing UTAH model with calibrated values of CU = 1000 and GDH = 6250 (Source: BOKU–Met).

Further to the above described basic phenology model concepts, different variations of considering driving factors of phenological development were developed. These various approaches try to consider different plant characteristics and their complex responses to phenology driving factors.

Such examples can be found in the Appendix 2.

8.5 Crop phenology as an important information source for agricultural management

The phenological phases of crops, especially the start and duration of important phases such as dormancy, germination, stem elongation, flowering, corn filling and physiological maturity, accurately determine the optimum timing of crop management options (for soil cultivation, disease and pest management, crop protection measures against abiotic extremes, fertilization and harvesting) in order to receive maximum crop yield quantity and quality. Thus crop phenology in the context of seasonal weather conditions has important economic implications for farming. For example, the timing and duration of various phenological phases of crops as well as pests and diseases usually differ from year to year due to short–term weather conditions. To adapt crop management options to these highly variable conditions every year in order to reduce cropping risks and ensure stable and high crop yields is a great challenge in agricultural and food production worldwide.

One example is the coincidence of late frost events (see Chapter 9) with flowering periods of orchards, which both show high variability from year to year (frequency of frost events as well as timing of flowering period) and ultimately determine frost damage risk to crops (Fig. 8.7).

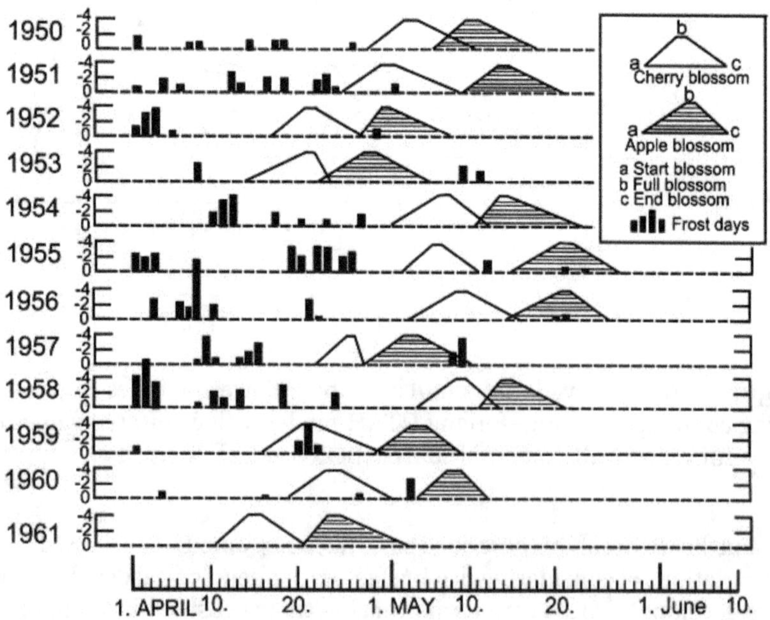

Figure 8.7: Coincidence of late frost events and apple and cherry flowering periods from 1936–1961 at a German site, determining late frost risk damage (based on Schnelle, 1963).

The forecasting of main phenological events affects the determination of such field operations as irrigation schemes, sowing time, fertilization and pest management planning, and harvesting period, all with important economic implications for farmers in light of resource optimization options (labour) and machinery costs. Phenological models with acceptable accuracy (+/-3 days) are therefore required to forecast many practical field operations.

Phenological events or phases can be affected also by extremes and crop stresses. Therefore phenological models are also employed for specific applications considering these effects in combination with crop physiological processes.

For example, low nocturnal temperatures reduce dark respiration and affect factors related to crop yield (e.g., the sugar and starch content of harvested products), especially for tomato, potato, green paprika, tobacco, sugar beets and others.

Daytime temperatures determine photosynthesis activity and biomass accumulation in relation to the relevant temperature optima of the physiological processes of a specific crop. The fertility of crops during flowering/blossom can also be sensitive to day temperatures and especially to extremely hot conditions: this is the case, for example, for peas, strawberries, maize and cereals, as high temperatures adversely affect pollination. On the other hand, low day temperatures can have a beneficial impact on aromatic constituents, positively affecting the taste of strawberries at < 10 °C, for example.

Finally, soil temperature itself also has an influence on crop physiological processes, although in this case there is always a certain correlation with air temperatures. For example, experiments have revealed a correlation of soil temperature and dry matter accumulation (growth) in maize. Studies have also shown that soil temperature has an direct effect on the phenology on crops, such as in the case of grapes. Experiments have demonstrated that the effects of soil temperature on crop phenology and growth can significantly differ between cultivars, as in the case sweet corn (e.g. Chang, 1968 and others).

Including soil temperature in phenological models could further increase the accuracy of phenological predictions; this step, however, is still being investigated.

Linking the climatology of day and night temperature combinations at a specific site with optimum plant response can identify the best growing sites (in sense of maximum productivity) for a specific plant or crop cultivar (Fig. 8.8).

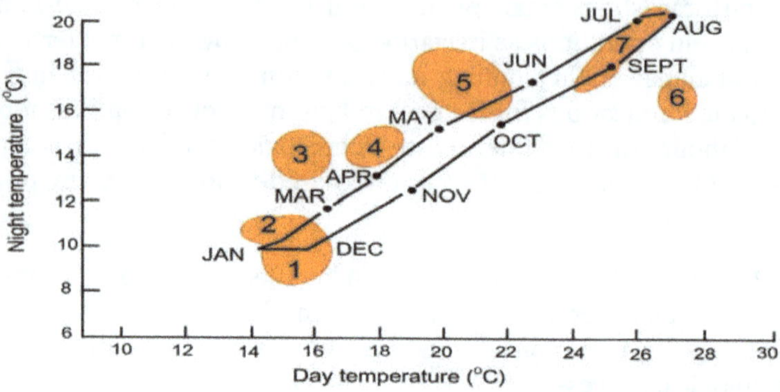

Figure 8.8: Scheme of potential coincidence of monthly day and night temperature combinations at a specific site vs. optimum plant requirements (various plant types indicated by numbers) (schematic, based on Went. IN: Chang, 1968).

By combining all main climatic effects – day length, day and night temperature requirements, etc. – on crop phenology and related growth conditions on a global scale, we can sketch a latitudinal classification of main growing areas of main crops (Fig. 8.9). Similar classifications can be drawn also by altitude, however, only in relation to the regional type of climate (e.g. classification of crop growing conditions by altitudes in the tropics; Manshard, 1968).

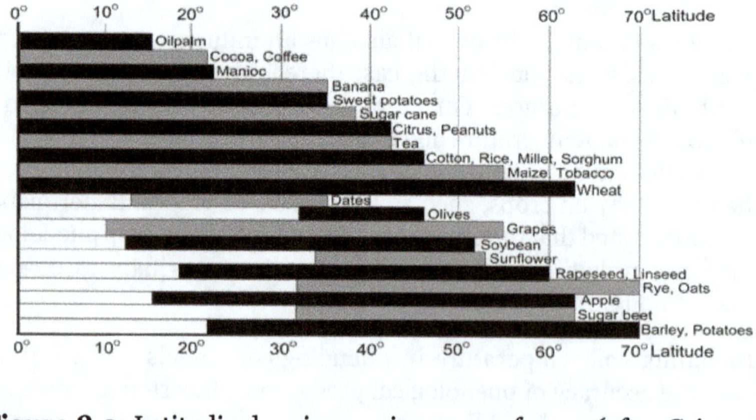

Figure 8.9: Latitudinal main growing areas of crops (after Grigg, 1984 and other sources).

Related to crop phenology and mean latitude–dependent temperatures there are interesting basic relationships between annual and single crop yields. For example, due to the longer growing period in a cooler environment (because of lower temperatures), mean rice yields increase, for example, with increasing latitude (Fig. 8.10). Similar behavior can be observed for winter wheat between mid and high latitudes. At low latitudes, however, rice can be grown 2–4 times a year (because of the longer vegetation period and the shorter rice growing cycle), thus producing higher total annual yields. However, this basic relationship can be varied by many other site specific growing factors, such as genetic potential of cultivars, cultivar response to abiotic stress, crop management, soil conditions, and others.

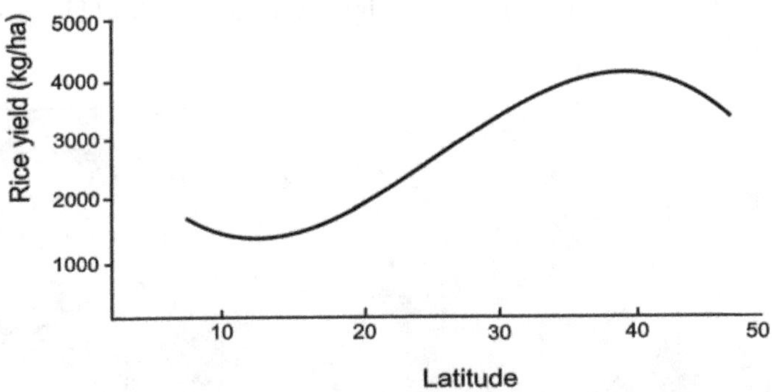

Figure 8.10: Approximated rice yield change from low to mid latitudes (schematic, based on v. Uexküll, 1965).

8.6 Phenology as an agroclimatic indicator

If phenological phases of the same plant cultivar (ecotype) are observed spatially, small–scale climatic variations can be estimated, such as in regions with complex topographies. Small–scale changes in local climates due to the effects of altitude, slope inclination, cold air lakes, etc., can therefore be described using systematic and multi–year phenological observations. In mountainous regions, for example, specific phenomena can be observed. At higher altitudes, temperature normally decreases (by approx. 0.5 °C/100m): decreasing temperature sums delay plant phenological phases within the same time period. However, the phenomenon of temperature decrease with higher altitude can be reversed during inversions: within an inversion layer (see Chapter 3) at the valley bottom (e.g. in a cold air lake), temperatures will rise as altitude increases. Under such conditions, especially in spring, this effect can also influence plant phenological phases. For example, plants will flower earlier at higher elevations than at the valley bottom, even though it is here that higher annual average temperatures (and temperature sums) are normally observed.

Over a large spatial scale, observed phenology of the same plant type can also indicate spatial climate variations, such as delayed phenology at higher latitudes and altitudes. This phenomenon is also monitored via satellite (observation of the greening of the main vegetation over large areas in spring).

Mapping climate suitability for crops by phenological models

Once phenological models have been established or calibrated for specific plants or crops, phenological mapping can be carried out by using weather data on various spatial scales. For example, crop suitability maps can be drawn based on the temperature sum method, such as for grapes using the HUGLIN index for spatial high resolution scales (Fig. 8.11) as well as for larger scales.

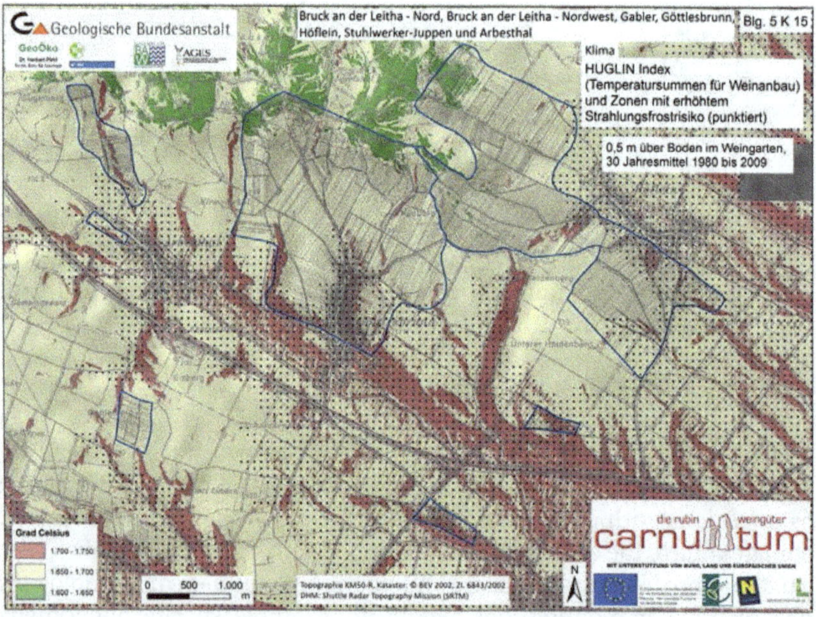

Figure 8.11: HUGLIN Index and frost risk areas (scattered area) for an Austrian grape growing region over the climatic period 1980–2009 (Source: BOKU–Met; Interactive web map: http://www.arcgis.com/home/webmap/viewer.html?webmap=5fbb85508f7d49fdab74c8e255ef40).

9

Significant Meteorological Events

9.1 Drought

Drought is a phenomenon that originates from a deficiency of precipitation over an extended period of time, usually a season or more, resulting in water shortage that can affect human activities, in particular agriculture (see Chapter 4.4). This phenomenon has to be contextualised to be explained: it is changeable in type, intensity, duration, timing and spatial extent. It is important not to confuse drought with aridity, which is a permanent feature of climate and is restricted to low rainfall areas and heat waves. A typical time scale associated with a heat wave is a week, while a drought may persist for months or even years. Drought effects may be also different based on the period when its occur. Intense damages are related to severe drought events that occur when water demand from crops is high.

Drought impact does not involve structural damages, in contrast to floods, hurricanes, and most other natural hazards. Its effects are diffused over larger geographical areas than those involved in damage from other natural hazards. In addition, unlike other disasters, it is unclear when a drought period actually begins. For example, no one knows for sure how severe a drought will be – until the rains return. However, droughts can be costlier than other natural disasters. For these reasons, the quantification of impact and the provision of disaster relief are more difficult tasks for drought than they are for other natural hazards (Wilhite, 2005). Moreover, several types of drought exist, and the factors or parameters that define drought will differ from one type to another, as is explained in Chapter 4.4.

Drought must be viewed as an integral part of a natural climatic cycle, even though extreme droughts can have disastrous consequences, causing such disasters such as desertification, crop failure, food shortages, malnutrition and famine, and epidemics. Climate change is gradually aggravating drought, making it increasingly frequent and severe in many parts of the world. In many cases, then, it is important to develop adequate risk management for this adversity.

9.1.1 Agricultural drought

For each drought occurrence, we can determine its spatial extent and its duration and development over time. For example, it is important to understand changes in inter–annual variability in connection with each region, or trends in the average amount of rainfall, or a different distribution of rainfall during the year. Moreover, it is crucial to consider the increase in water use for economic sectors like agriculture, industry and tourism as well as the possible effects of this increase on water balance and water quality. At the same, we have to examine the possibility of developing adequate agronomic technologies to optimize water use. While for a single farmer the temporal development of a drought episode might be the most important factor, for regional water management and policy both the spatial extent and evolution of the drought are the crucial considerations.

Many indices have been developed to monitor drought and rainfall conditions for agricultural purposes; some of these are explained below:

The Standardized Precipitation Index (SPI) is a meteorological drought index that shows the level of drought due to rainfall deficit compared to normal climate conditions. Mathematically speaking, the SPI is based on the cumulative probability of a given rainfall event occurring at a given site. Therefore, based on historic rainfall data, an analyst can determine the probability of rainfall being less than or equal to a certain amount (WMO, 2012) (Tab. 9.1).

The SPI calculation (Eq. 9.1 a, b) for any location is based on the long–term precipitation record for a certain period. This long–term record is fitted to a probability distribution, which is then transformed into a normal distribution so that the mean SPI for the location and desired period is zero (Van Loona and Laahab, 2015).

$$SPI = \left(t - \frac{C0+C1t+C2t^2}{1+d1t+d2t^2+d3t^3}\right);\ t = \sqrt{\ln\left(\frac{1}{H(P)^2}\right)} \quad \text{if } 0 < \mathrm{H(P)} < 0.5 \quad (9.1a)$$

$$SPI = \left(t - \frac{C0+C1t+C2t^2}{1+d1t+d2t^2+d3t^3}\right);\ t = \sqrt{\ln\left(\frac{1}{(1-H(P)^2)}\right)} \text{ if } 0.5 < \mathrm{H(P)} < 1 \quad (9.1b)$$

where: P is the cumulated precipitation for the given time–scale, H(P) is the cumulative probability of the observed precipitation, and C_0, C_1, C_2, d_0, d_1, d_2, and d_3 are constants.

The calculation can be based on different lead times, from one to six months, for example. For an area with historical data on precipitation, an analyst can then show that the probability of rainfall being less than or equal to average rainfall for that area will be about 0.5 (with the gamma distribution). Therefore, if a particular rain amount gives a minimum probability on the cumulative probability function, it is an extreme, such as too dry or too wet. Alternatively, a rainy event that gives a high probability on the cumulative probability function is close to the normal condition.

SPI	Classification	ΔP (%)
2.00 >	extremely wet	2.3
1.50 to 1.99	very wet	4.4
1.00 to 1.49	moderately wet	9.2
0.00 to 0.99	mildly wet	34.1
0.00 to -0.99	mild drought	34.1
-1.00 to -1.49	moderate drought	9.2
-1.50 to -1.99	severe drought	4.4
-2.00 <	extreme drought	2.3

Table 9.1: Classification according to the SPI index and probabilities (ΔP) that the index lies within each class.

The Palmer Drought Severity Index (PDSI) is a long–term index used as an indicator of drought severity derived from temperature and precipitation data. It is an agrometeorological index in which a simple soil water balance model is incorporated into the algorithm. This model measures, on a scale from −10 (maximum drought) to +10 (maximum humidity), the water needs in the soil. A particular index value is often the signal to begin or discontinue elements of a drought contingency plan.

The index and its variations have been used extensively for monitoring drought and for making operational water management decisions. Hydrological accounting over a period of time is summarised to obtain coefficients which are dependent upon the climate of the area being analysed. The data series is then reanalysed using the derived coefficients to determine the amount of moisture required for normal weather during each month. Monthly departures from normal conditions are converted to indices of moisture anomaly. Finally, these indices are converted to the drought index (Guttman, 1998) (Tab. 9.2).

PDSI value	Classification
4.00 more	extremely wet
3.00 to 3.99	very wet
2.00 to 2.99	moderately wet
1.00 to 1.99	slightly wet
0.50 to 0.99	incipient wet spell
0.49 to -0.49	near normal
-0.50 to -0.99	incipient dry spell
-1.00 to -1.99	mild drought
-2.00 to -2.99	moderate drought
-3.00 to -3.99	severe drought
-4.00 or less	extreme drought

Table 9.2: Classification according to the SPI index.

The Surface Water Supply Index (SWSI) "was designed so that it would be an indicator of basinwide water availability for the MWD (mountain water dependent) sector, be predictive, and permit comparison of water supply conditions between basins to assess relative drought severity" (Shafer and Dezman, 1982).

This index is used in cases in which the Palmer Drought Index does not adequately reflect conditions in snow–dominated regions. It is based on probability distributions of monthly time series of individual component indexes. The original SWSI was formulated as a rescaled weighted sum of non–exceedance probabilities of four hydrologic components: snowpack, precipitation, streamflow, and reservoir storage. The index is calculated starting from equation 9.2:

$$SWSI = \frac{aPsnow+bPpree+cPstrm+dPresv-50}{12} \quad (9.2)$$

where: a, b, c, and d are the weights for each hydrologic component; a + b + c + d = 1; P = the probability of non–exceedance (in percent) for component i; and snow, prec, strm, and resv = the snowpack, precipitation, streamflow, and reservoir storage hydrologic components, respectively. By subtracting 50, the SWSI values are centered around zero, and dividing by 12 compresses the range of values between -4.17 and + 4.17.

The following indices used for drought monitoring are based on remote sensing data:

The Vegetation Condition Index focuses on the impact of drought on vegetation and can provide information about the onset, duration and severity of drought impact on vegetation by observing vegetation changes and comparing them with historical values. The "proxy" indicator of the vegetation humidity conditions (Kogan, 1995) of the last 16 days, with respect to the minimum and maximum ecosystem limits, is defined by the NDVI (*Normalised Difference Vegetation Index*). It indicates weather stress that is due to dryness or excessive wetness.

$$VCI_i = ((NDVI_i - NDVI_{min}) / (NDVI_{max} - NDVI_{min})) \cdot 100 \qquad (9.3)$$

Other indicators of drought often used to monitor conditions are soil moisture, snowpack, streamflow, groundwater levels, reservoir and lake levels, vegetation health, and short, medium, and long–range forecasts. Remote sensing offers innovative opportunities to monitor drought conditions because of higher resolution. These techniques are especially advantageous in regions lacking adequate weather station networks.

A survey of operationally used indices for drought or wetness prediction (applied to past– now– and forecasting) in agriculture in Europe was carried out by EU Project COST734 Action and is presented in Table 9.3.

Used index/model	Type of use	Time step	Spatial realization	Aim of use	Country of use
		Drought			
Soil water content for top 10 cm	F	5 day	regional	estimating drought–affected regions	CH
Index of hydrometeorological drought	N	weekly	national	estimating drought–affected regions	CZ
Precipitation –potential evapotranspiration (NPET)	N, P	daily	national	estimating drought–affected crops and regions	AT
Water balance components	N, P	daily, weekly monthly, yearly	from site–specific to national	estimating drought–affected crops and regions	CZ, FR, IT, NO, PL, SI, SK, SR
Soil moisture content	P	daily, weekly, monthly	10x10 km to national	estimating drought affected areas	DE, FI, IT, RO, SI, SR
Vegetation Health Index (VHI)	P, N	monthly	national	estimating drought–affected crops and regions	GR

Palmer Drought Severity Index (PDSI)	P	upon user request	national	estimating drought–affected regions	HR, SR, GR
Water table depth levels	P	monthly	regional	estimating drought–affected regions	IT
Usable water supply	P	weekly	national	water supply	CZ
Palfai Aridity Index	P	Year	national	estimating drought–affected regions	HU, SI
Precipitation totals and anomalies	P	weekly, monthly, yearly	regional, national	estimating drought–affected areas	AT, IT, PL, SI, SR
Rainfall percentile	P	monthly	national	estimating monthly rainfall towards the normal	HR
Standard Precipitation Index (SPI)	P	monthly	regional	precipitation deficit	ES, IT, SI, SR, GR
WOFOST crop model	P	daily	regional	generating dry days	NL, SK
Excess rain					
Rainfall amount	Table 9.3 Reported operational use of agroclimatic indices including statistical models estimating water availability and drought status (Orlandini *et al.*, 2008). Legend: CH Switzerland, CZ Czech Republic, AT Austria, FR France, IT Italy, NO Norway, PL Poland, SI Slovenia, SK Slovakia, SR Serbia, DE Germany, Fi Finland, RO Romania, GR Greece, HU Hungary, ES Spain, NL Netherland, HR Croatia	hourly, daily	national, regional, local	estimating affected regions	All countries
Rainfall intensity	F	hourly	national–district scale	estimating affected regions	AT, DE, PL, SK
5–day probability forecast of 1.0 mm+	F	daily	regional	estimating affected regions	CH

Daily forecast of 25.0 mm+ precipitation	F		national	estimating affected regions	GR
Rainfall regime classes	F		regional	estimating affected regions?	RO
SPI maps	P	daily calculation on monthly, thirty–, sixty- and ninety–day basis	regional, national	estimating affected regions	IT, SI, SR, HR
Precipitation total anomalies	P	monthly	regional	estimating affected regions	AT, IT, SI
Palmer`s Z index	P	decade	national station network	estimating affected regions	SR

(F) forecasting, (N) nowcasting, (P) pastcasting

Table 9.3: Reported operational use of agroclimatic indices including statistical models estimating water availability and drought status (Orlandini *et al.*, 2008). Legend: CH Switzerland, CZ Czech Republic, AT Austria, FR France, IT Italy, NO Norway, PL Poland, SI Slovenia, SK Slovakia, SR Serbia, DE Germany, Fi Finland, RO Romania, GR Greece, HU Hungary, ES Spain, NL Netherland, HR Croatia.

Looking at the future, climate change can increase the probability of worsening drought in many countries, causing greater stress on water supplies and agriculture (see Chapter 12). For example, drought globally struck several major breadbasket regions simultaneously in 2012, adding to food price instability. In countries already coping with food insecurity, cost spikes can lead to social unrest, migration and dearth.

Once established, drought can continue thanks to a cycle of "positive feedback", in which very dry soils and diminished plant cover can further suppress rainfall in an already dry area (through less evapotranspiration and thus water vapor transport into the atmosphere). In drought situation, possible solutions include more efficient use of water and the development of more drought–resistant crops (see Chapter 10.4), but also the deployment of green infrastructures for storm water management or for increasing energy efficiency in buildings ("Center for Climate and Energy Solution").

9.2 Heat

Temperature is a primary factor affecting plant life and growth (Hatfield and Prueger, 2015). In particular, plant growth can be expressed as the amount of cell differentiation, biomass increase and reproductive phase achievement. In each of these, temperature plays a key role during the entire growing period and influences various aspects of plant growth according to season and phenological phase. Temperature trends during the year are fundamental considerations for sowing and the organization of other agricultural practices. In this regard, plants can be distinguished between cool–season and warm–season plants, based on their thermal requirements.

To avoid crop damage and yield losses, specific temperatures ranges are defined:

a. optimum temperatures represent the range in which plant growth is greatest;
b. cardinal minimum and maximum temperatures are those above and below which plant growth stops and then starts again when temperatures reach better levels;
c. critical maximum and minimum temperatures represent the limit range for plant survival.

In general, high temperatures increase the rate of plant development. Vegetative development usually has a higher optimum temperature than does reproductive development. However, from a physiological point of view, an increase of temperatures beyond critical levels can lead to reversible or irreversible alterations in plant metabolism (Blumenthal *et al.* 1991) that concern:

- plant growth and photosynthesis inhibition;
- uncontrolled increase in cell respiration;
- toxin accumulation;
- inhibition of protein production;
- proteolysis;
- cell membrane degradation;
- physiological alteration of plants;
- leaf wilting.

It is, however, quite difficult to distinguish heat damage from other types caused by water shortage or excess of light. Nevertheless, heat stress symptomatology includes uncontrolled leaf fall, leaf scorch, necrosis and metabolic imbalances (Baisi *et al.*, 1999; Ceccon *et al.*, 2017). Leaves and sprouts are the most susceptible parts of plants (both for herbaceous and perennial plants) due to their high transpiration and water content. Intense drying of those parts, from heat stress, reduced metabolic activity of plants and–in severe cases–apoplexy, can lead to plant death. Leaves change in color and become chlorotic. If stress persists, necrosis may occur on the entire leaves. In extreme stress conditions, scorch can appear in correspondence of necrosis and leaves may fall after a few days. Scorch on leaves is normally caused by the interaction of high transpiration, low water provision from the root system and heat. Fruits are also sensitive to heat and solar radiation action. In this case, fruits show sunburn on peel, especially on those species with thin skin (tomatoes, grapes, apples, pears and peaches).

If plants are subjected to heat during the first growing stages, additional delayed damage can occur. In these cases, stems show uncontrolled elongation and early and low flowering, while fruit produces a tough skin with decreased yields.

In this regard, extreme events occurring during the summer period have the most dramatic impact on plant productivity. This is mainly due to damage which is caused during the reproductive stage which affects pollen production, fertilization and grain or fruit formation.

In particular, *pollination is one of the phenological stages that is most sensitive* to temperature extremes at a lower critical level when compared to other physiological processes, across all species. During this developmental stage, temperature extremes greatly affect production. Chronic exposure to extreme temperatures during the pollination stage of the initial grain or fruit set will reduce yield potential.

In the context of climate change, which is generally increasing global temperature levels, extreme high–temperature events (heatwaves) are projected to become more intense, more frequent and to last longer with respect to what has been observed over the last decades (see Chapter 12). Extreme temperatures are more dangerous for annual crops, which have a shorter growing season and faster vegetative development, than for perennial crops. Extreme weather conditions that occur during summer, characterized by temperatures higher than usual values which persist for days or weeks, are defined as heatwaves.

The World Meteorological Organization (WMO) has not formulated a standard definition for heatwave. In several countries, the definition is based on exceeding defined temperature threshold values through the identification of the highest observed values in historical series of data recorded in a specific area (Tab. 9.4). Heatwave is defined according to the climatic conditions of a specific area; for this reason, it is not possible to propose an absolute threshold temperature valid for all latitudes.

Used index/model	Type of use	Time step	Spatial realization	Aim of use	Country of use
Heat stress					
Heat Index	F	Daily	regional	days with heat stress	HU
Maximum temp. of soil surface, upper canopy and under plastic cover	P	monthly, growing period	national station network	heat stress indication for different crops and periods	AT, DE
Temperature sum	P	if needed	national	estimating affected regions	HR, SI
Temperature percentile	P	monthly	national	estimating monthly air temperature towards normal	HR
Heat Units Index /sum of daily maximum above 32°C/	P		national	days with heat stress	RO
Number of days with maximum daily air temperatures ≥ 30 °C and 35 °C	P	monthly	national station network	number of days with heat stress	AT, HR, SI, SR
Deviation of mean maximum air temperature from long–term average	P	daily, decade and monthly	national station network	number of days with heat stress	AT, IT, SR

Table 9.4: Reported operational use of high temperature–related indices for agricultural purposes in Europe (Orlandini *et al.*, 2008). Legend: HU Hungary, AT Austria, DE Germany, HR Croatia, SI Slovenia, RO Romania, SR Serbia, IT Italy.

One aspect of extreme high temperatures that is often overlooked is their effect on atmospheric water vapor demand. In particular, a positive correlation exists between air temperature and saturation vapor pressure (see Chapter 4). The increase in water vapor demand causes more water to be transpired by plants until water supply becomes limited and stomatal conductance decreases. This condition leads to an increase in leaf temperature and a reduction in the photosynthesis process. If crops are exposed to extreme temperatures, water stress could occur quickly because of high water atmospheric demand and the limited capacity of crops to extract water from the soil. If such water deficit lasts for a limited period, crops are able to restore normal conditions. In the case of chronic water stress conditions, crops can be significantly damaged (IPCC 2007).

Finally, at high temperatures drought will develop faster and become more severe, due to high evaporative demand and the related rates of high actual evapotranspiration.

9.2.1 Protection methods against heat

Short periods of high temperatures may have a detrimental impact on crop yields (lower grain weights, sterility, reduction in grain quality, etc.). For this reason, changes in agronomic practices are fundamental to ensure crop adaptation and preservation of yields. In particular, protection methods can be defined as indirect and direct.

Indirect methods:

- use of species with high morphological and physiological tolerance to high temperatures;
- early seeding and short–growing season varieties to avoid critical summer temperatures.

Direct methods:

- field selection based on exposure and slope;
- intercropping between cash crops and shading crops (widely used in tropical and sub–tropical areas);
- agroforestry systems (see Chapter 10.4);
- use of specific coverage (i.e., plastic or wooden shading nets) to reduce solar radiation;
- windbreaks to reduce dominant winds;
- crop and soil surface cooling with irrigation (evaporative cooling);
- greenhouse cooling systems (natural and artificial ventilation, shading with nets or paints, reflective screens, artificial humidification – fog systems).

The use of *shading coverages* allows reduction of direct solar radiation at different levels in open fields and greenhouses. It reduces energy absorption by leaves and canopy surfaces as well as their heating. Based on crop requirements and climatic conditions, various covering strategies can be employed. Shade nets are the most commonly used coverages: according to the dimensions of the net holes, different shading levels can be achieved: 30, 50, 70 or 90% coverage (Fig. 9.1). Generally, shade nets are black or green, though white nets might also be used to improve solar radiation reflectance. Obviously, in an open field a pole and wire structure is needed to support the net. Hail nets have a similar structure and can be used as a combined measure.

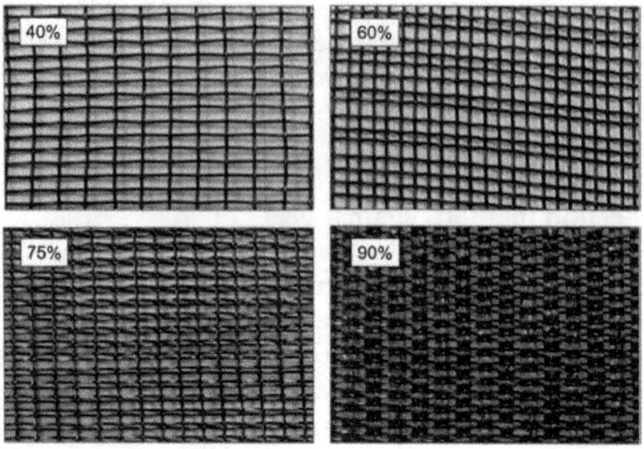

Figure 9.1: Shading nets with different degree (%) of shading.

Temperature reduction can also be achieved through the use of a *cooling irrigation system* (ideally combined with irrigation itself). Sprinkler irrigation represents the most effective strategy to reduce temperature and is used during heatwaves or the more susceptible phenological phases (before flowering, flowering, fruit ripening). At high temperatures and high evapotranspiration levels, irrigation provides water that increases air humidity and slows down transpiration. On the other hand, the adoption of this strategy increases total evapotranspiration and irrigation water demand. For that reason, the use of a cooling irrigation system is advisable only when extreme heat events pose the risk of great damage to crops.

This strategy is widely used on sports ground turf and ornamental turf, which goes by the name of syringing. This approach uses fine nebulized water during the hottest hours of the day to reduce plant temperature and to clean leaves of exudates (amino acids and carbohydrates) produced by plants during stress. In addition, the reduction of the amount of exudates on leaves decreases the risk of attack by fungi (*Pythium* spp.), which proliferate on those substances.

Plants that are subjected to extended periods of heat stress may develop some heat resistance characteristics. These processes are commonly due to eco–physiological adaptation of the species involving cell morphological structure and other physiological modifications such as:

- increase in cytoplasm concentration with a consequent raising of the boiling temperature;
- water chemical bond modification in cells;
- leaf size reduction;
- production of specific organs for storing water.

Those adaptation strategies allow the reduction of heat stress on crops. However, some detrimental processes may still occur, based on crop type, when temperatures exceed specific levels. Table 9.5 summarises the critical maximum temperatures of the main agricultural crops.

Crop	T_{max_l} (°C)
Wheat	47.5
Barley	35.0
Rice	40.0
Maize	35.0
Sorghum	44.0
Tomato	35.0
Soybean	39.0
Cotton	35.0

Table 9.5: Critical maximum air temperatures (T_{max_l}) for main agricultural crops (Luo, 2011).

9.3 Frost

The word "frost" in agriculture is commonly used to describe a meteorological event when crops and other plants experience freezing injury. In general, the term refers to the formation of ice crystals on plants or object surfaces through a phase change from vapor to ice (sublimation of water vapor). A freeze is defined by air remaining below freezing temperature over a widespread region (0 °C) for a sufficient period of time (at least one or two days) measured at a height of between 1.25 and 2.00 m above soil level, inside an appropriate weather shelter. Many definitions of frost exist:

- incidence of temperature less than or equal to 0 °C that is measured in a "Stevenson–screen" thermometer at a height of between 1.25 and 2.00 m;
- occurrence of air temperature less than 0 °C, without characterisation of the shelter type and height;
- absence of reference to ice formation with surface temperature dropping below 0 °C.

This weather hazard is subdivided into white and black frost. The former occurs when atmospheric moisture freezes in small crystals on a solid surface, while the latter involves the formation of a few ice crystals because the air is too dry. The consequences of both these types of frost for vegetation are the same.

In addition, frosts are usually categorised into two further types: radiation and advection frost (or their combination). Radiation frosts develop at night and result from intense, long–wave radiation cooling under calm, clear and dry atmospheric conditions. Strong surface inversions develop in the settled atmosphere associated with radiation frosts. Advection frosts develop as a result of a large–scale incursion of cold air, with temperatures below 0° C, moderate–to–strong winds and a well–mixed atmosphere. Sometimes these types of frost are mixed: for example, a region might experience a cold air mass (advection frost) and then several days of calm conditions and clear skies (radiation frost).

Radiation frost

Radiation frost is caused by an overnight heat loss from the earth's surface by radiation. Its duration is generally a few hours and it is followed the next morning by sunshine and clear skies. While air temperatures are above 0 °C during the day, they generally fall below 0 °C at night, starting from when the Sun goes down.

These frosts can appear in winter or spring. Radiation frost often follows the passage of a dry–air cold front, preceding a mass of cool air. In these conditions, air temperature increases with height in the micro layer close to the soil. Regarding frost protection for crops, temperature inversion is considered to be strong when the temperature at 10 – 15 m above the ground is -10 °C. Otherwise when the temperature difference is less than -15 °C, inversion is considered weak. Very common in the case of radiation frost is that the greatest temperature drop occurs around sunset. There is a rapid shift in net radiation from a positive to a negative level, which is caused by solar radiation decreasing from its highest value at midday to zero at sunset, with net long–wave radiation that is always negative.

Schematically, this phenomenon can be explained as follows: The leaves are in direct contact with the surrounding air. As leaves are warmer, they radiates heat to cooler air. If the surrounding air is drier, they will lose heat faster. As this cold air penetrates fast, it can be risky for plants, though not lethal: leaf edges turn brown and crinkly, as if they had been subject to herbicide damage (FAO, 2005) (Fig. 9.2a).

Advection frost

This type of frost is characterized by cold air that replaces warm air in a certain place. It is associated with clear night conditions, moderate–to–strong winds, no temperature inversion and low humidity. Since there is no inversion it is difficult to protect the crop from this type of frost, compared to the others. These frosts can occur in Mediterranean climates, but they tend to be more common in the eastern part of continents (such as Florida) where there are usually cold continental air masses that occasionally come from arctic regions into subtropical areas (FAO, 2005) (Fig. 9.2b).

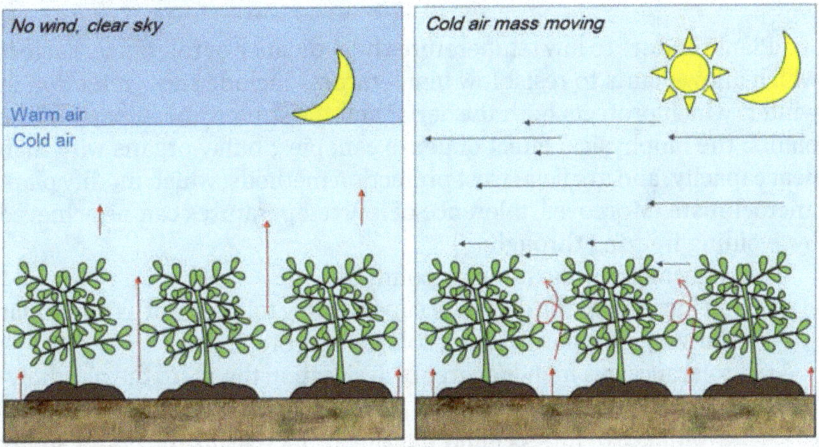

Figure 9.2: Radiation (a) and advection (b) frost.

Frost damage

Frost damage is usually related to different situations and plant types. We can think, for example, of tropical plants that are often subject to serious frost damage when exposed to temperatures just below zero. Different species or varieties experience different degrees of frost damage at the same temperature and phenological stage, depending on previous weather conditions as well as on "hardening", which is their adaptation to low temperatures prior to a frosty night. Frost damage occurs when ice forms inside plant tissue and injures plant cells. We can distinguish between two different types of damage, called direct (when ice crystals form inside cell protoplasm) and indirect (when ice forms inside the plants but outside of the cells) (FAO, 2005). The problem is that the formation of ice causes a "mechanical rupture of the protoplasmic structure" (Levitt, 1980). Moreover, we differentiate between cold damage, which occurs when plants are exposed to temperatures lower than 5 °C and as low as -2 °C; desiccation damage, which follows when ice forms on the outside of leaves at temperatures from 0 °C to -2 °C, and freezing damage, which happens at temperatures below -2 °C when there is rapid ice nucleation and ice crystals form within tissues.

The entity of the damage depends on the rapidity at which the temperature drops, and not so much on the duration of the low temperature. It seems that the speed of defrosting after freezing is also partially related to the quantity of damage.

The temperature at which a given level of freeze damage is expected is called the "critical temperature", which differs according to species and variety, phenological stage and many hardening factors (FAO, 2005).

Plants endure to low temperatures by avoidance or tolerance. Factors which allow plants to resist low temperatures include snow retention in winter, which protects both the aerial and the subterranean parts of the plants; the biophysical effect of dense canopies; bulky organs with high heat capacity, and artificial frost protection methods, which modify plant microclimate. Moreover, tolerance of low temperatures can be achieved by avoiding freezing through:

- a decrease of the freezing point;
- an increase in the degree of supercooling tolerance of extracellular freezing by concentrating solutes in protoplasm;
- tolerance of a higher degree of desiccation thanks to the plasmolysis of the protoplasm; and,
- avoidance of intracellular freezing by increasing the permeability of the plasma membrane.

Rainfall can also determinate a differentiation in frost damage. In fact, a slightly wet canopy may experience freezing damage at warmer temperatures compared to a dry canopy. This is because water contains ice nucleators that raise the freezing point of water.

Frost indices

Many indices have been used to describe the impact of frost on natural and managed ecosystems. Such indices include the timing of the last frost day in spring and first frost day in fall of each year, the number of consecutive frost days, the duration of frost–free days, and length of the growing season. Many of these indices are calculated using daily minimum air temperatures. Below are some examples:

- a frost or freeze day is defined as one with a minimum temperature less than a base temperature;
- the number of frost days (nFDs) is the number of days with frost. We determine the nFDs on monthly, seasonal, and annual time scales;
- the number of frost–free days (nFFD) is defined as the number of days without frost. We total the nFFDs on monthly, seasonal, and annual time scales;
- the last spring freeze (LSF) is the last day in March through May with minimum temperature below 0 °C for the last time until fall;
- the first fall freeze (FFF) is the day in September through November with minimum temperature below 0 °C for the first time since spring;

- the growing season length (GSL) extends from the beginning of spring to fall. If our interest is in vegetation growth rather than resilience, higher temperature thresholds (of 10 °C and 6.1 °C, respectively) are taken for the onset of spring and the end of autumn. The number of days between the LSF and the FFF of the same year is used to determine the GSL.

These indices are very important climate indicator in particular regions. The number of days without frost in a year, for example, reflects an overall warming trend in the climate system. As can be seen from the graph (Fig. 9.3), over the past 30 years there has been an increase in the length of the ice–free season over the contiguous United States and Alaska, compared to the 1979–2014 average. Such information can support decisions related to the management of agricultural and natural resources, including crop planning and forest fire risk management (www.globalchange.gov).

9.4 Storm, hail and wind

9.4.1 Hail

When cloud temperatures reach from -5 to -20 °C, an additional detrimental phenomenon can occur – hailstorms. Hail effects on crops aggravate rainfall damage. Specifically, hail causes stems and fruit skins to rupture, flowers and fruit to fall and the herbaceous and woody part of plants to break. As a secondary impact, fungi diseases can use damaged tissues as invasion paths and cause considerable yield losses in terms of both quality and quantity. Intense hailstorms may affect crop production with significant and often total direct yield losses. In these circumstances, crop protection can be achieved by active and passive strategies.

Active strategies aim to prevent hail with the use of chemical compounds. Such methods use silver iodide or hygroscopic compounds to reduce drop size and encourage early precipitation to prevent hailstorms. Nowadays, three methods are mainly used to spread those chemical compounds:

- directly from the soil (cheap but inefficient);
- anti–hail rockets that shoot chemical compounds from the soil inside the clouds (most widely used);
- planes that fly inside clouds and spread chemical compounds (most effective but expensive).

Hail canon are among the active strategies. These work with shock waves that reach clouds and break water drops, though cavitation, with a consequent production of rainfall. However, in general, hail canons are inefficient for crop protection compared to other active strategies.

Figure 9.3: Hail cannon (Source: Massimo Telò – Own work, CC BY–SA 3.0, https://commons.wikimedia.org/w/index.php?curid=19250054).

Passive strategies are represented by anti–hail nets that are widely used in fruit production. Hail interception is ensured by the passive protection of nets placed above plants. Normally nets are made of polyethylene with square or rectangular holes of 2 to 11 mm. Nets are supported by pole and wire systems with specific slopes to ensure that hail can be removed from nets.

As a very important measure to protect against economic losses, *hail insurance* has become quite usual for farmers in dangerous regions, such as near the Alps. In Austria, for example, more than 40% of farmers have hail insurance for all types of crops, and especially for orchards. Hail insurance fees can also be reduced if farmers install hail nets, for example.

9.4.2 Wind and storms

Wind plays a role in several fundamental aspects of agricultural productive systems, such as gas exchange at different atmospheric levels, wind erosion and pollen transport. However, when winds reach excessive speeds (storms), their action may produce negative effects in a short time,

Figure 9.4: Anti-hail nets on an orchard.

including crop lodging, branch breaking, fall of flowers and fruits, eradication and soil erosion. In particular, soil wind erosion is a detrimental phenomenon that affects a wide range of areas in the world and reduces year by year global agricultural surface by intensifying the desertification process. Wind soil erosion occurs in a specific climate (determined by wind, precipitation and temperatures) with a specific soil wetness and texture and specific soil surface and vegetation conditions. Wind speed, direction and turbulence are able to remove significant amounts of soil particles.

Turbulence is the most challenging factor to measure, as it decreases in relation to wind speed reduction and increases with soil surface roughness. A wind speed of 15 km h^{-1} at 30 cm above the soil surface is the critical level for the onset of soil erosion. Precipitation re-duces wind soil erosion by improving soil particle cohesion and favoring plant proliferation.

Wind soil erosion is more likely to occur in the presence of:

- dry and sandy soil with small particle size;
- flat soil without vegetation or soil cover material;
- large fields without any obstacles (stones, crop residues, cover crops) that act as windbreaks.

Wind erosion can be estimated by, for example, the Woodruff and Siddoway (1965) equation:

$$E = f\ (I + K + C + L + V) \qquad (9.4)$$

where E is the annual soil loss (t ha^{-1} $year^{-1}$); I is the erosion index (t ha^{-1} $year^{-1}$) in an unprotected open flat field without vegetation; K is the roughness of a harrowed soil, taking into account the irregularity of soil surface; C is a factor that accounts for the climate in question; L is the maximum length of the field exposed to dominant wind without protection, and V indicates the effectiveness of vegetation in countering wind soil erosion. Wind erosion evaluation requires the cumulative addition of previous factors and not simple multiplication.

Several strategies are available to protect crops from negative wind effects. The most effective is the use of windbreaks or shelterbelts (see Chapter 6.3.1). Windbreaks are one to several rows of trees or shrubs based on crop requirement and climatic conditions. They are commonly planted in hedgerows around the edges of fields perpendicular to the direction of the main winds.

However, windbreaks include artificial solutions such as drywalls, wood or plastic fences, and polyethylene nets supported by pole and wire structures. If dominant winds come from one direction, windbreaks should be placed perpendicularly to that direction. Windbreaks can be organized in a single row or in several parallel rows according to wind speed. On the other hand, in the case of dominant wind blowing from different directions, windbreaks are organized in an "L" structure, a semicircular structure or–in the case of turbulence of the air–a net structure, which protect a field from each side; these are often employed in agroforestry systems.

The action of windbreaks consists in reducing wind speed and deviating winds. When wind hits a windbreak, a part crosses it with a significant reduction in speed, while the main thrust is deviated above the windbreak. Based on its characteristics, the length of the area protected by a windbreak can be longer or shorter. In the area protected by a windbreak, different microclimatic and environmental conditions occur:

- reduction of wind speed;
- decrease in soil erosion caused by winds;
- increase in nocturnal temperatures;
- increase in amount and duration of dew;
- higher soil temperatures, which ensure early seeding and early crop germination;
- atmospheric moisture increase of 2–4 % (though an overabundance of windbreak density can cause higher leaf and soil wetness and increase the risk of disease);
- lower evaporative losses and proportionately less drought stress for crops;
- more snow accumulation during winter, providing additional water to spring crops.

The height (H_{wb}) of a windbreak is the most important factor for its impact. Generally, fields are protected by wind within a distance from the windbreak of five to eight times H_{wb}, depending on its structure. Wind recovers its initial speed at a distance of roughly 10–20 times H_{wb}. The density of the windbreak is an additional factor that strongly affects its effectiveness. In particular, density affects air flow and turbulence. A high–density windbreak strongly reduces air flow that is deviated above the windbreak. Behind the windbreak, a low pressure zone occurs, and wind is quickly directed to the ground, which can cause crop lodging.

On the other hand, a windbreak of medium density reduces wind speed, but its deviation action is low and turbulence is reduced. In this case, atmospheric pressure is roughly constant behind the windbreak, and the area protected by the windbreak is longer leeward.

Artificial windbreaks have different densities. Drywall, for example, diverts wind above the windbreak and creates a strong turbulent wind, in this case reducing the protected area. Plastic nets are the most effective solution among artificial windbreaks. Nets stop 50% of wind, greatly reducing wind speed such that only weak turbulence occurs. Wind flow is thus deviated above windbreaks; because of low turbulence, however, the protected area increases. Natural windbreaks can have different densities depending on planting distance and canopy density. *A canopy density of 40–60% produces the best results*, while a density higher than 80% creates too much turbulence and a density lower than 20% has no effect. In addition, windbreak effectiveness is closely related to thickness. For a natural windbreak, ten rows are commonly used on external boundaries and one to three rows between each field (Fig. 9.5).

Windbreaks, however, occupy fertile areas of farm fields and have negative effects on the growth conditions of crops that grow within a few meters from them (shading, competition for soil water and nutrients). Nonetheless, their positive economic effects outweigh these problems, especially in drought–prone regions and hot climates. Furthermore, they have positive effects on biodiversity, water retention and several other ecosystem services, which may not be apparent in the calculations of short–term economic impact made by farmers.

Windbreak effectiveness, then, depends on the interaction between structure, composition and thickness. Thus windbreaks with different structures and thicknesses may offer the same degree of protection.

Certain measures need to be taken to ensure the high protection potential of a natural windbreak:

- use of evergreen species to ensure field protection throughout the year (but not necessarily in temperate climates with a winter crop dormancy season);
- use of species with high tolerance to pests, diseases and severe weather conditions;

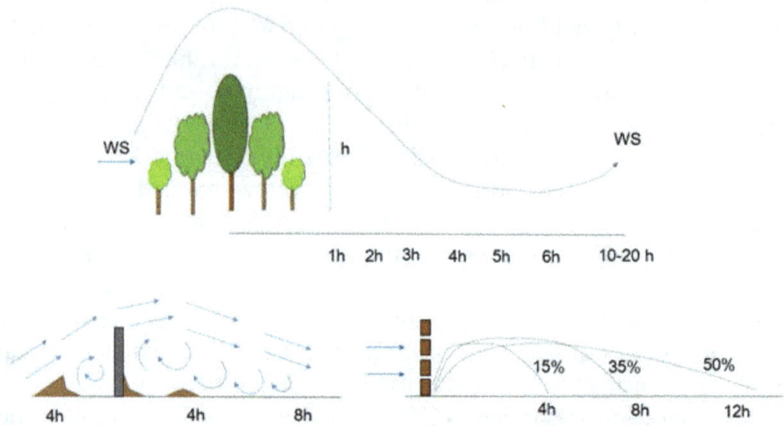

Figure 9.5: Effect of natural (a) and artificial (b, c) windbreaks. Artificial windbreaks have a different effect on wind direction, speed and turbulence (Ceccon *et al.* 2017).

- species with upward canopy development;
- high cutting tolerance (wood production may represent an additional economic activity);
- species with taproot systems to reduce competition with cash crops and increase resistance to eradication;
- consideration of biodiversity effects (e.g., flowering species for bees).

Generally, in warm–winter areas (e.g., Mediterranean climates) the most widely used species are Eucalyptus spp., *Cupressus arizonica, Pinus radiata, Acacia melanoxylon, Thuja gigantea, Chamaecyparis lawsoniana, Opuntia* spp., *Juniperus* spp., *Tamarix* spp. and *Ligustrum japonicum*. On the other hand, in continental areas with cold winters the most frequently used plants are Cedrus spp., *Platanus orientalis, Alnus* spp., *Ulmus pumilia, Robinia pseudoacacia* and *Populus spp.* In temperate climates such as those of Central Europe, local winter–hardened species are used in combinations of bushes and trees (mostly poplars). Furthermore, in areas where strong winds predominate, several rows of windbreaks are required. As such an intensive use of windbreaks can cause a reduction in agricultural soils, in many cases they are placed at the edge of fields or on unproductive strips within them.

Storms are combinations of extreme severe weather conditions such as strong wind, heavy precipitation, thunder, lightning and hail, as well tropical storms (hurricanes and typhoons). Depending on the climatic area, storm effects can be more or less intense. In more susceptible zones, such as Central America, they may cause serious disasters for human populations and agriculture, including storm surge, heavy rain, flooding and soil erosion. Effects of storms on crops and on crop recovery depend on several factors, such as crop type, stage of growth, weather conditions immediately after storms and prevalence of disease organisms. The strong winds that occur during storms may cause plant lodging, or in extreme cases eradicated plants and falling trees. In addition, heavy precipitation during storms may cause severe soil erosion or, worse, landslides with significant soil and crop losses. In general, storms include and exacerbate the negative effects of wind and heavy precipitation on both crops and humans. In addition, the detrimental effect of storms can be aggravated by the persistence of warm and wet conditions, leading to the incidence of disease (Kovats *et al.*, 2014). Storms frequently occur in the tropics. Due to climate change, however, more and more often storms events are occurring in temperate or continental areas.

9.5 Floods and heavy precipitation

The meteorological factor of precipitation is characterised by very high variability in all climates of the world, causing problems for agricultural management in general. Due to climate change, wet areas have become wetter, and dry areas have become drier in the last decades. In Northern Europe, North and South America and North Asia, precipitation has increased. On the other hand, in the tropics and sub–tropics outside the monsoon trough (the Mediterranean, Southern Africa and Southern Asia) precipitation has been decreasing. In addition, increased precipitation intensities have been observed in many regions: even in areas where total precipitation has decreased, significant increases in heavy precipitation have been recorded.

Changes in precipitation dynamics are also affected by temperature variations on a global scale. Temperature increases cause an improvement in the water holding capacity of the atmosphere and drive the shift to heavier and less frequent precipitation. In particular, the increased capacity of the atmosphere to store moisture raises the chances of more intense but less frequent rainfall, as more time is required to recharge the atmosphere with water vapor.

The increase in precipitation intensity represents a serious concern for agriculture. In particular, heavy precipitation increases crop and soil damage and especially soil water erosion. The weight of raindrops falling on plants increases the risk of physical damage in the form of lodging, the fall of leaves, flowers and fruits, and the breaking of branches. Such damage is more serious during the vegetative phase of plants and is exacerbated by strong winds.

The impact of heavy rain can take these forms:

Plant lodging: if heavy rains occur in the first phases of growth, crop stems are bent, so decreasing the negative impact of lodging. Plant can easily recover the normal position. Indeed, plant lodging poses a serious problem that can dramatically reduce yields. In particular, the persistence of crops in very moist soil creates the conditions for the occurrence of fungal attack.

Damage to the vegetative parts of plants: this normally occurs in the first phases of plant growth and may reduce plant development and yields. On trees the breaking of vegetative parts compromises annual production; in sensitive species, this may result in alternate bearing. Intense leaf fall strongly reduces photosynthetic activity and physiological processes.

Broken tree branches: damage in the form of broken branches affects annual production and the physiological activities of plants, which may increase their susceptibility to fungi attacks. In this case as well alternate bearing may occur.

Flower fall: when flowers fall, fruit differentiation and thus yields are greatly reduced.

Wounds to fruits: skin breaks render fruit more susceptible to pests and diseases. Extreme weather conditions can further cause fruit to fall, thus compromising yields. Even the persistence of fruit with broken skins on plants reduces production, as yield quality is decreased.

Even more than crops, agricultural soils are adversely affected by heavy precipitation. *The main effect of heavy precipitation on soil is erosion.* Water erosion affects soil directly, as splash erosion, and indirectly, in the form of rills and gully erosion. Splash erosion is on–site erosion caused by raindrops falling on soil. Splash erosion occurs when the kinetic energy of raindrops is able to detach and move soil particles. Yet even if splash erosion is able to move considerable quantities of soil, soil movement may be limited to only a few centimeters, such that soil soon resettles over the surface. For this reason, *precipitation intensity and duration, soil cover, soil texture and the slope of the field strongly affect the degree of erosion.*

Indirect water soil erosion is represented by runoff in rills, small channels that can be removed by tillage, and gullies, large channels that are too wide to be removed by tillage. Precipitation amounts that do not infiltrate into the soil flow downhill along slopes under the action of gravity. Runoff may occur for two reasons: (i) precipitation intensity is greater than soil infiltration potential (infiltration excess runoff), and (ii) the soil has already absorbed all the water it can hold (saturation excess runoff).

At first, runoff is incapable of moving soil particles, as water film protects the soil from splash erosion. In addition, water infiltrated into the soil improves cohesion between soil particles and reduces erosion. However, as water flows downhill and rain continues, it gains kinetic energy and is concentrated into morphological depressions of the soil. If precipitation proceeds for only a short time, soil particle deposition prevents erosion from occurring. However, with heavy and intense precipitation, the water film increases in depth, such that water may begin transporting and even detaching soil particles that are not able to deposit. Small rills increase in flow and converge to form bigger rills with higher erosion potential. As flows continue, soil slope increases and the soil's surface is lowered. Lower areas provide dominant flow paths for subsequent flow, and these flow paths are in turn eroded further. Between the dominant erosion channels, inter–rill areas form where water does not flow and soil is susceptible to raindrop erosion action. The amount of soil that may be transported by rills is considerable and is deposited in the lowest part of field where the slope ends.

Protecting the soil from rill erosion can be achieved through tillage and soil cover management. Normally several small channels perpendicular to the slope can be created (with a harrow or small plough). In this way, water flow that occurs during heavy precipitation is collected in several defined spaces and erosion potential is reduced. In sensitive areas, however, periodic management of water channels is recommended. The removal of deposited soil maintains water channel efficiency and reduces additional flooding risks.

The estimation of annual soil erosion is a fundamental factor to evaluate risk and to organize protection strategies. Water soil erosion estimation is commonly performed through the Revised Universal Soil Loss Equation (RUSLE) model:

$$A = R \cdot K \cdot L \cdot S \cdot C \cdot P \tag{9.5}$$

where A represents annual soil loss (t ha-1 year-1); R is the rainfall factor and represents the erosion potential of rainfall (MJ mm ha-1 h-1 year-1); K represents soil losses in standard plots of 22.13 m in length with 9% slope, harrowed in the direction of the slope and cultivated by the fallow system; L is the field length factor and represents the ratio between soil losses in studied fields and in the standard plot (22.13 m); S is the slope factor that considers soil losses in studied fields and those in the standard plot (9%); C is the crop factor and considers soil losses between the studied field and the standard plot cultivated by the fallow system; P is the factor of soil conservation practices that indicates the ratio between soil losses in agricultural systems with conservation practices (agricultural hydraulic systems) and those produced by fields cultivated in the direction of the slope without agricultural hydraulic systems. Considering L, S, C and P as dimensionless factors, soil loss (A) units will be the product of the R and K factors. In any case, the most effective strategy to reduce soil erosion is water management through hydraulic systems and tillage.

Figure 9.6: Different types of soil water erosion: rill and sheet (a), rill and gully (b), hydrogeological instability (c) and landslide (d).

Furthermore, the detrimental effects of heavy rains on soil include soil structure degradation and compaction due to falling raindrops. In this regard, the most sensitive areas are flat land, coastal areas and clay soils. *Soil compaction reduces the infiltration potential* of water that persists on the surface, with a reduction in the amount of oxygen. Oxygen abatement leads to an anaerobic condition that in turn reduces the population of soil aerobic microorganisms.

In these circumstances, soil fertility is greatly compromised due to following factors:

- nitrogen consumption by anaerobic bacteria (denitrification). The denitrification process transforms nitrogen from compounds that are available to plants (nitrates) to those which are lost by volatilization. Volatile compounds produced by anaerobic bacteria are elemental nitrogen (N_2) and nitrous oxide (N_2O). More than fertility reduction, N_2O emissions represent an environmental concern due to their high Global Warning Potential (GWP), which is 265–298 times that of CO_2 for a 100–year timescale. Moreover, anaerobic conditions strongly affect crop growth and productivity due to a reduction of root system respiration;
- reduced respiration. In the absence of respiration, carbon dioxide and other compounds such as ethanol and butyric acid accumulate near roots. In this way, plant roots become nonfunctional and ultimately die. Without functioning roots, plants are not able to

take up water–even when they are submerged in it–or nutrients. In a wet soil system, the incidence of disease in roots and crowns, such as *Pythium* spp., increases. Moreover, surface and deep water flows cause nutrient leaching, which increases nutrient losses (especially cations). Nutrient removal (nitrates and phosphorus) from soils is one of the main causes of aquifer and river pollution and in extreme cases can lead to eutrophication. Eutrophication induces an uncontrolled increase of plants and algae in a water ecosystem with rapid oxygen depletion and biodiversity reduction.

To reduce anaerobic stress conditions, crops produce more superficial roots or additional roots from stems. Superficial root systems, however, reduce plant stability and the absorption of water and nutrients. Those conditions negatively affect crop growth, encouraging the proliferation of weeds with higher anaerobic tolerance.

To reduce soil saturation in flat areas, two main solutions are normally implemented:

- trough ploughing. A slight slope from the central part of field to the boundaries, together with water channels, can be ploughed. In the presence of such troughs, water that arrives on soil slowly flows, following gravity, to the surrounding water channels, with negligible erosion potential. In this regard, it is fundamental to recreate the field slope every year before the cultivation season. Water channel management is also important to ensure the best water management efficiency;
- subsoil drainage. This strategy is, however, effective only in areas where soil saturation is caused by surface groundwater. The concept is to place a perforated pipe system under the ground that collects water and releases it into external water channels.

9.6 Early warning systems for agriculture

An early warning system is commonly defined as one that allows for directing and forecasting impending extreme events. This is achieved by formulating warnings on the basis of scientific knowledge, monitoring and consideration of the factors that affect disaster severity and frequency. The objective of warning systems is to prepare people for the risk that an injurious event can bring. Warnings need to inform people and make them ready to react in case of damage. A number of considerations are involved in creating and perfecting warning systems: these include extending the lead time of warnings; improving the accuracy of warnings; increasing demand for probabilistic forecasts; improving the communication and dissemination of warnings; using new technologies to alert the public, and targeting warning services and messages to relevant and specific users. Today there exist many types of warning systems for industrial, geophysical and biological hazards as well as for personal health risks. Several alert systems have evolved to manage the risks imposed by a wide range of natural processes (WMO, 2010). Early warning systems must include four identifying elements: risk knowledge, monitoring and warning service, dissemination and communication and response capability.

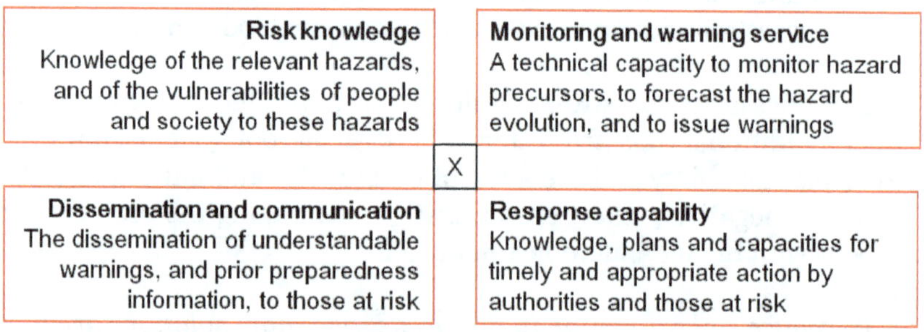

Figure 9.7: Elements of systematic people–centred early warning systems. Source: UN/ISDR Platform for the Promotion of Early Warning.

A particular risk takes the form of a series of connected factors. These are generally of an environmental nature, but may also involve social, political and other considerations. For this reason, a disastrous event should be monitored not only in terms of its short–term risks but also in light of its potential long–term effects, such as political changes or disruptions to social communication. Management of natural risks must therefore be based on an integrated approach (or "multi–hazard approach") (Basher, 2006).

The most common method of early warning systems is a warning chain, which starts from an analysis of the risk and proceeds to informing users.

Early warning systems can involve different aspects of human activity, beginning with meteorology. An example of an early warning system that begins with meteorology is provided by the Regional Technical Preventive Service Administration, which has established a 24/7 operations centre, the Natural Hazard Situation Room (SSRN), to forecast dangerous floods in Piedmont (Italy). The SSRN produces a daily report of the observed and expected meteorological situation, paying particular attention to the forecast of precipitation. The community participates in creating a local flood emergency plan, studying local flood and hill slope dynamics and exchanging updates and advice with national and regional authorities. Once a flood has started and is judged be imminently dangerous, an alert is transmitted to the public. Information systems in use at SSRN include an automatic observation network for meteorological and hydrometric monitoring, a meteorological radar, automatic upper air soundings of the atmosphere, performed twice daily, numerical modelling for meteorological forecasting on global and local scales, and numerical modelling for flood forecasting.

There are two different methods for assessing risk level: one compares the quantitative precipitation forecast with predefined rainfall thresholds, while the other uses real–time numerical simulations. These simulations are carried out via FloodWatch, a decision support system for real–time flow forecasting. ArcView's graphs automatically display and continually update status and forecast conditions. Finally, a team of experts produces forecast text bulletins of the expected hydrologic hazard risk (A, B or C) and the corresponding danger level (1, 2 or 3) for each alert zone. Another example is the warning system of Tuscany Region (Italy), which delivers information for citizens on the potential risks related to meteorological phenomena and inform about the behavior to be adopted in critical situations. The risks that are taken into account by the warning system are related to flooding (hydraulic risk), landslides and flooding caused by minor watercourses (hydrogeological risk), strong thunderstorms, strong winds, storms, snow, ice (Fig. 9.9).

Warning services are important for predicting hazard parameters. Risk assessments must be coordinated, where possible, to obtain the benefits of institutional, procedural and shared communication networks. Risks must be communicated effectively through a network that is developed at regional, national, and global levels. The population must be informed in advance of the risk in order to defend itself and develop disaster management plans. The goal of institutional arrangement is the development of national, institutional, legislative and policy frameworks that support the implementation and maintenance of effective early warning systems.

The main point of early warning systems, as we have seen, is a model that reproduces the danger characteristics of the event. These can be models of a biophysical process (such as a drought model in which the loss of soil moisture may occur over months) or of a response process (such as the dynamics of evacuation processes). In the absence of data, models are based on probabilistic values of the occurrence of a harmful event, such that they can give rise to false alarms. We can distinguish between four types of early warning systems: pre–science early warning systems, *ad hoc* science–based early warning systems, systematic end–to–end early warning systems and integrated early warning systems.

Early warning systems are widely used also for the protection of citizens in urban areas when meteorological conditions represent serious issues for human health. In this regards, an example of the early warning systems is represented by the Heat Shield Project (Horizon 2020) aims to ensure worker's health preventing heat–related risks through the

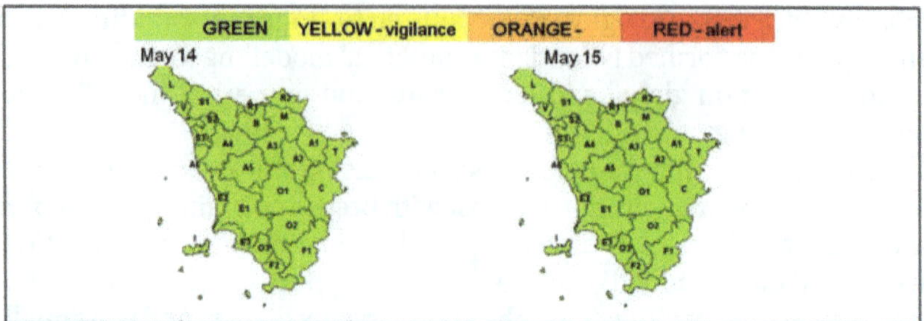

Type of criticality	Level of warning	Impact scenario	Individual behaviour rules
- 0 - No criticalities	Normality	None	To be informed
- 1 - Low impact	Vigilance	Circumscribed events, even intense, difficult to locate and forecast	To be aware
- 2 - Medium impact	Alert	Widespread weather phenomena with possible effects on slope stability and on watercourses	To be prepared
- 3 - High impact	High alert	Widespread weather phenomena of particular intensity and persistence, with consequent relevant ground effects, such as widespread landslides and flooding of major waterways	Ready to act

Figure 9.8: Weather warning system of Lamma – Tuscany Region.

evaluation of the effects of heat on workers during the summer period. In particular, based on ECMWF probability forecasting, the project is able to produce prediction on the probability of heat risks on 45 days period and specific information about the strategy to prevent heat risks on a period of 5 days. Information are mainly related to the level of hydration and rest period for each category of workers. In particular, using specific information about the subject as age, height, weight, kind of job, personal protective equipment (PPE), the system is able to provide personalized information (Watts *et al.*, 2017).

Another example of warning system is Meteoalarm, which provides the most relevant information needed to prepare for extreme weather, expected to occur somewhere over Europe (http://www.meteoalarm.eu/?lang=en_UK). The system website integrates all important severe weather information originating from the official National Public Weather Services across a large number of European countries. This information is presented consistently to ensure coherent interpretation as widely as possible throughout Europe. Meteoalarm allows to have information at European scale but detail on national warning texts are also available through links to the relevant National Weather Services.

Early warning systems are widespread in the agricultural sector. This is because it is important to know plant diseases in advance, before they affect crops. The first agricultural task is to monitor environmental parameters that can induce disease in crops. An interesting example is represented by an early warning system developed in the frame of a transnational project for grapevine downy mildew (*Plasmopara viticola*) monitoring in Friuli Venezia Giulia plain (Italy) and part of the Slovenian territory. Meteorological and epidemiological information are produced at a local not punctual level through the integration of meteorological ground measurements with radar data. Continuous meteorological information layers are obtained (500m x500m pixels) and specific models are applied to each pixel for the simulation of grape downy mildew. Daily local weather and epidemiological outputs (maps) are available on a dedicated website (Fig. 9.9).

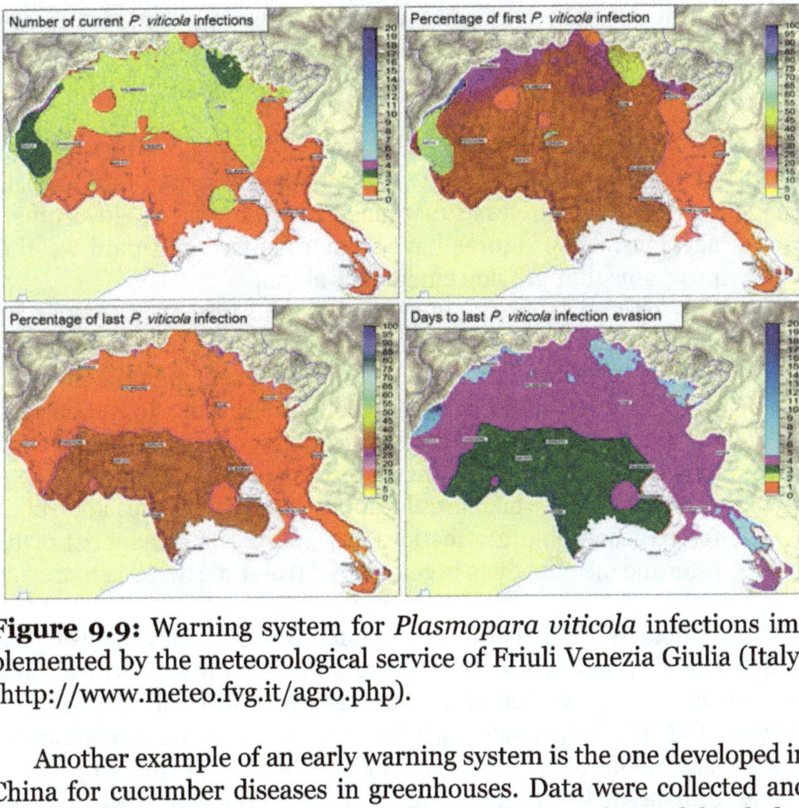

Figure 9.9: Warning system for *Plasmopara viticola* infections implemented by the meteorological service of Friuli Venezia Giulia (Italy) (http://www.meteo.fvg.it/agro.php).

Another example of an early warning system is the one developed in China for cucumber diseases in greenhouses. Data were collected and transferred to develop a model to predict the disease. The model included the five following steps. The first is a pre–alarm that is based on the degree of disease expansion. In this case two indices were developed: a danger index and a disease index. The second step concerns an analysis of the indicators that warn of the probability of illness, such as relative humidity (RH), the duration of leaf moisture and the air temperature. The third step is the search for sources that cause or increase the risk of disease, such as excessive irrigation times or a daily temperature range which is too high. After the onset of disease, alerts are issued via a sound alarm and caution light. The light shows different colors that correspond to five classes of illness severity. Finally, users can enter information in real time, which is useful for decision making. This early warning system uses techniques of inference machines from artificial intelligence and expert systems to provide more informed support for decision–makers (Li *et al.*, 2008).

The available technologies for monitoring and warning are:

- forecasting technology, which begin with temperature and rain data to develop a forecast model;
- remote sensing and geographic information system (GIS) applications, whose advantages are a systematic approach to data collection, reduction of collection costs, accessibility of data to a wider range of decision makers, and the possibility of spatial data analysis;
- satellite communication technology;
- mobile phone technology, whereby phones are set to automatically receive free Wireless Emergency Alerts issued by government agencies. "App" stores have numerous free and paid weather programs that provide emergency alerts;
- CTs for crowdsourcing. A general architecture of crowdsourcing applications is based on a campaign organizer that monitors the targeted spatial data collection effort. Participants contribute to the geo–crowdsourcing campaign by sharing geospatial data using their own mobile devices. End users then process the data collected by the participants according to their needs;
- crisis mapping, which uses locations from street maps and volunteered geographic information (VGI) sources for areas at risk of disaster and matches them to geo–parsed real–time tweet data streams.

The global WMO network of Regional Specialized Meteorological Center (RSMC) and World Data Centers (WDC) provide critical data, analysis and forecasts that allow the Meteorological and Hydrological Services (NMHS) to provide early warning systems and guidelines for various natural hazards, such as cold waves, winter storms, tropical cyclones, heatwaves, floods and droughts. For example, the WMO network turned out to be very effective in 2004, during one of the most intense hurricane seasons in the Atlantic and Caribbean regions. Atmospheric data gathered in situ and by space–based instruments were transmitted to the United States National Hurricane Center, one of the WMO RSMCs (RSMC–Miami), where forecasts and hurricane advisories were issued round the clock. These notices were transmitted via the Global Telecommunication System (GTS), telefax and the Internet at 3–6 hour intervals to the NMHS of the countries at risk. NMHS meteorologists used these warnings on hurricanes to produce their own warnings on national hurricanes, which were immediately sent to newspapers, radio and television stations, emergency services and other media outlets. As a result of this information, many lives were spared through timely evacuations.

It is crucial to ensure that all countries, in particular less developed ones, have the systems, infrastructures, human capabilities and organizational structures to create and use early warning systems to reduce natural hazards disasters (WMO, 2012).

Another example of an operational monitoring and early warning system for agricultural purposes is the Agricultural Risk Information System (ARIS), which employs a high resolution grid data (1x1 km) and issues daily updates, including 10–day forecasts. ARIS identifies potential weather–based cropping risks in Austrian agricultural regions by using adequate agro–climatic indicators, such as for drought and heat impact and the drought status of main crops (Fig. 9.10). Users of the monitoring system include farmers, agricultural research and advisory services and insurance companies.

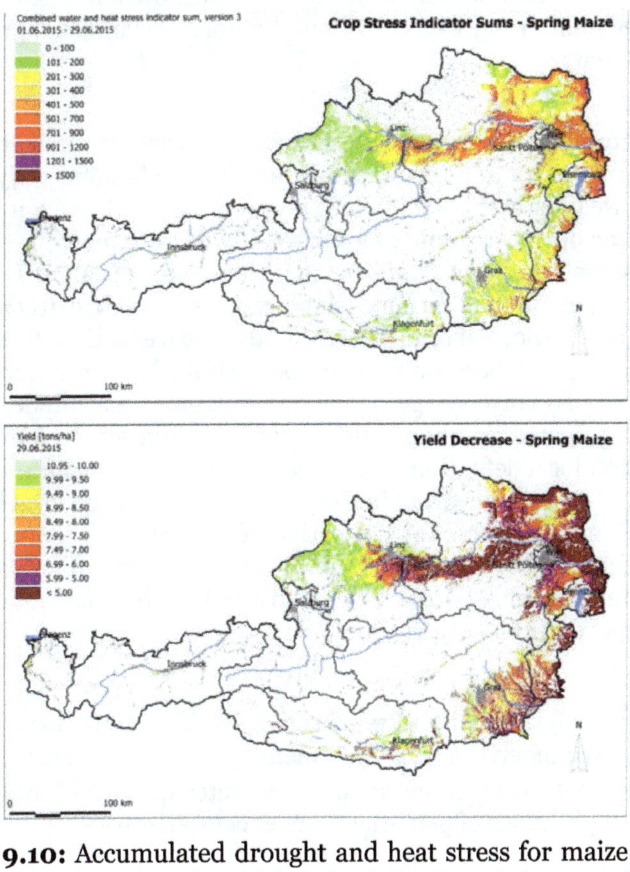

Figure 9.10: Accumulated drought and heat stress for maize (above) and related expected grain yield (below) for Austria in 2015 (https://warndienst.lko.at/mais+2500+++6577?typ=YIELDCLASS).

10

Assessment and Management of Risk

In many parts of the world, weather and climate extremes are among the most important risks affecting production and impacting the performance and management of agricultural systems. In particular, extreme meteorological events such as droughts, floods, cyclones, extreme temperatures and strong winds strongly hamper sustainable agricultural development.

Meteorological extremes are classified as rare events with respect to their frequency of occurrence or duration. Definitions vary, but an extreme weather event would normally be considered rare if it falls into the 10th or 90th percentile of a probability density function estimated from observations (www.climatechangeinaustralia.gov.au/en/). There is no generic definition of extreme climate events, because extremes in one location may be normal in another. Moreover, extreme climate events are multidimensional, because they can be described by several attributes, such as rate of occurrence (probability per unit time), magnitude, duration and timing, spatial scale and multivariate dependencies. Because of these differences, local definitions are necessary. A heat wave, for example, is considered an extreme event in a certain region if it persists at least two consecutive days during the hot period of the year with thermal conditions recorded above given thresholds (maximum, minimum and daily average temperatures). In particular, an event can be characterized by its magnitude, duration, severity and extent (WMO, 2016).

Over the last few decades, extreme weather events have been increasing in number, duration and intensity as a consequence of climate change (WMO, 2007). Many adaptation strategies specifically aim to assess the implications of potential changes caused by weather and climate

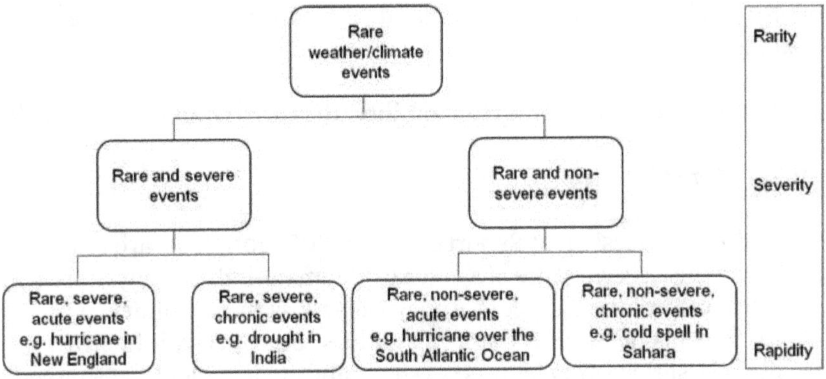

Figure 10.1: Scheme of rare climate events. Stephenson *et al.* (2008).

events and to measure their impact on human society. In the context of such events, not only the hazards caused by them but also exposure and vulnerability to these hazards contribute to determining risk. Therefore, adaptation strategies and risk management practices also depend on a rigorous understanding of these dimensions of exposure and vulnerability, as well as a proper assessment of changes in those dimensions (Cardona *et al.*, 2012).

In order to better understand the existing interrelations among the concepts of hazard, vulnerability, exposure and risks, some definitions are needed.

10.1 Risk

Although a number of definitions can be found in the literature (Brooks N., 2003), risk is often represented as the probability or likelihood of occurrence of hazardous events or trends, multiplied by their impact if these events or trends occur (Cardona *et al.*, 2012).

For example, according to the ISO Guide, "risk is the combination of the probability of an event and its consequences" (ISO, 2002). Risk defined in this way refers to actual risk and therefore can be objectively assessed based on scientific data. Nevertheless, things are more complicated in that the hazard event is not the only driver of risk: the levels of adverse effects are also partly determined by the vulnerability and exposure of societies and social–ecological systems (Cardona *et al.*, 2012).

Risk, then, is a complex concept. In the end it is determined by the interaction of social and environmental processes, and by the combination of physical hazards and the vulnerabilities of exposed elements (Cardona *et al.*, 2012).

10.2 Hazard, exposure and vulnerability: the three components of risk

10.2.1 Hazard

In the 4th Assessment Report of IPCC (IPCC, 2007), hazard is defined as "the potential occurrence of a natural or human–induced physical event or trend or physical impact that may cause loss of life, injury, or other health impacts, as well as damage and loss to property, infrastructure, livelihoods, service provision, ecosystems and environmental resources". In the context of climate change, the term "hazard" usually refers to climate–related physical events or trends or their physical impact. In these terms, climate or weather extreme events such as drought, heavy rainfall and floods, frost and heatwaves tornadoes or strong winds are climatic hazards: they are potentially destructive random factors. Such events mainly depend on natural factors and are characterized by a relatively well–known probability law (Guilard, 2016).

Although in everyday language risk and hazard are often used synonymously, today it is widely accepted that the hazard is a component of risk and not risk itself (Cardona *et al.*, 2012).

10.2.2 Exposure and vulnerability

Likewise, exposure and vulnerability are often used as synonyms, but this is incorrect. Exposure can be defined as "the presence of people; livelihoods; environmental services and resources; infrastructure; or economic, social, or cultural assets in places that could be adversely affected", while vulnerability is "the propensity or predisposition to be adversely affected" (Cardona *et al.*, 2012). Exposure is more closely related to physical characteristics, such as location, while vulnerability can be (bio)physical and socio–economic in nature and relates not only to fragilities and weaknesses of the system but also to the lack of capacities to mitigate the adverse effects on the exposed elements.

Based on these definitions, it is possible to be exposed but not vulnerable; however, to be vulnerable to a hazardous event, it is also necessary to be exposed. As an example, living in a floodplain means being exposed to rainfall–related extreme events, but not necessary vulnerable if good practices for reducing potential losses are implemented. On the other hand, if one is not exposed the risk does not exist, so that to be vulnerable implies to be exposed.

Finally, vulnerability is also a function of the sensitivity of a system, intended as "the degree to which a system is affected, either adversely or

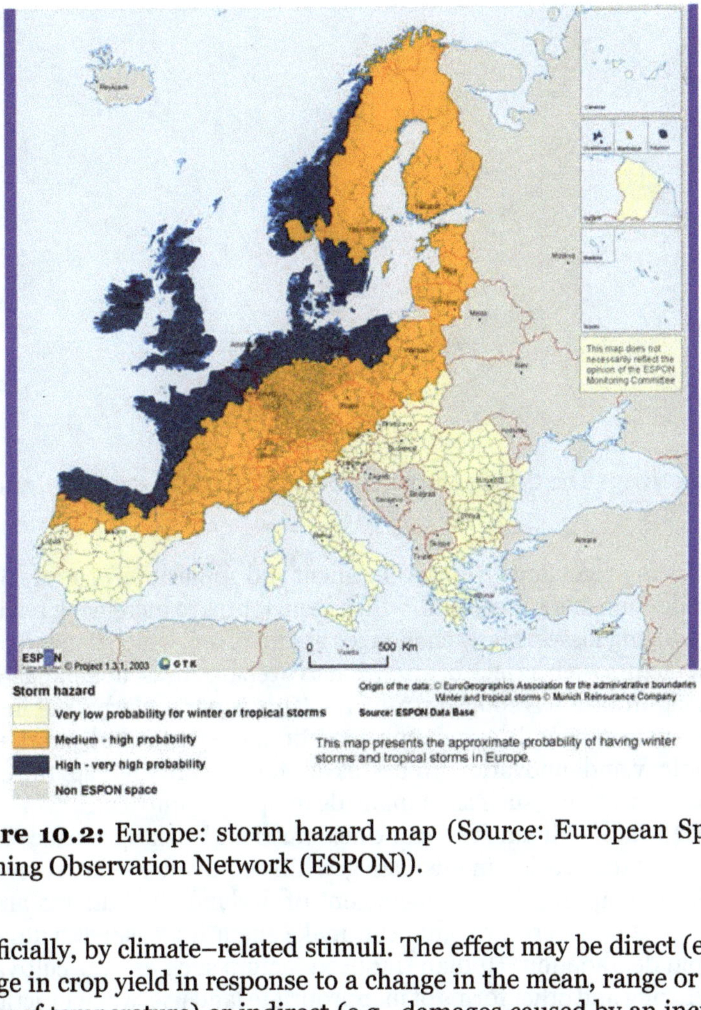

Figure 10.2: Europe: storm hazard map (Source: European Spatial Planning Observation Network (ESPON)).

beneficially, by climate–related stimuli. The effect may be direct (e.g., a change in crop yield in response to a change in the mean, range or variability of temperature) or indirect (e.g., damages caused by an increase in the frequency of coastal flooding due to sea level rise)".

10.3 Risk assessment and management

Risk assessment refers to a qualitative or quantitative approach to determine the nature and extent of risk. This is achieved by analyzing potential hazards and evaluating existing conditions of exposure and vulnerability that together could harm people, property, services, livelihoods and their environment. In order to assess risk, the first step is the identification of hazards and their main characteristics (probability of occurrence, intensity, etc.). Then exposure and vulnerability should also be analyzed. Finally, the effectiveness of capacities to cope with the identified risk scenarios should be evaluated.

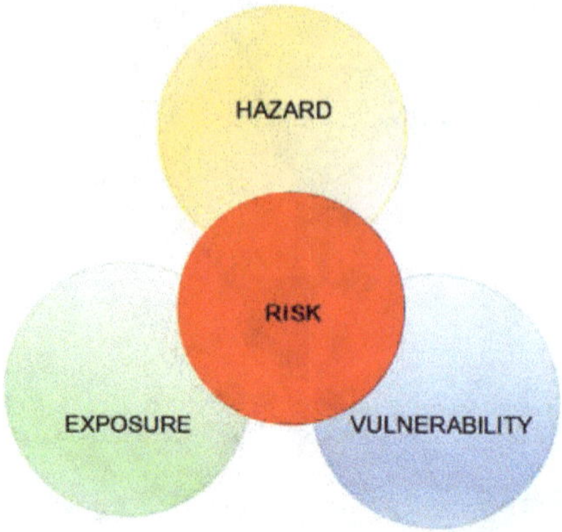

Figure 10.3: The three components of risk: hazard, exposure, and vulnerability.

Risk management is the development and application of strategies for risk reduction and prevention, which contributes to increasing resilience and reducing losses. Many challenges are involved, which depend on different factors such as lack of attention to preparedness, mismanagement and inappropriate response strategies. This is particularly true in areas where environmental degradation conditions are particularly marked and technology and innovation processes are too slow to mitigate the effects of such conditions (such as in many developing countries).

High rainfall intensities can cause nutrient leaching, surface runoff and finally soil erosion in mismanaged hilly cultivations (i.e. vineyards). Similarly, inappropriate management of drylands by humans and the occurrence of severe droughts can make them increasingly vulnerable to rapid degradation. Human–induced factors such as over cultivation, overgrazing and other forms of inappropriate land use, when practiced in dryland conditions, may result in significant degradation of vegetation, soil leaching and, finally, desertification.

As reported by the WMO (WMO, 2007), many provisions can be taken to assist in coping effectively with weather risks. One of the most important strategies is improved use of climate knowledge and technology, which includes the development of monitoring and response mechanisms to current weather. By providing information about the environment in which farmers operate or about the likely outcome of alternative or relief management options, uncertainties in crop productivity can be reduced. Quantification is essential and computer simulations can assist in delivering such information; in this regard, it may be particularly useful to quantitatively compare alternative management and relief options in areas where seasonal climatic variability is high and/or are prone to extremes.

Both structural and non–structural measures can be adopted to reduce the impact of the variability (including extremes) of climate on crop production. Structural measures include irrigation, water harvesting, windbreaks, frost protection, artificial and controlled climates (greenhouses), microclimate management and manipulation and other structural preparedness programs. Non–structural interventions include the use of tailored agrometeorological short–term weather forecasts, seasonal to inter–annual climate forecasts, improved application of medium–range weather forecasts and crop insurance. Crop insurance can be resorted to only when there is sufficient spatial variability of environmental stress (e.g. with hail) with limited recurrence. But it remains extremely difficult to implement for some of the major risks, such as drought, which typically affect large areas. One measure that has been adopted with regard to credit and insurance is to make them conditional on the implementation by farmers of improved risk–reducing practices, like early planting. Along these lines, a project regarding agricultural insurance was presented in Serbia by "KfW Development Bank" in collaboration with "Generali Insurance Serbia" in 2017. This project involved training in the adjustment of losses for those who had insurance. Training activities aimed at developing the capacity to underwrite and adjust agricultural losses for six selected crops: wheat, barley, corn, sunflower, soybeans and sugar beet. The training provided was based on local experience and practice in Serbia and regulated according to the best international practices (in Canada, United States, Germany, Switzerland, Austria and Turkey). Contingency planning is an important part of such strategies, as ways must be found to avoid, reduce, or cope with risks: such planning is indeed commonly made by governments in some drought–prone areas around the world as an effective strategy in the face of risk (WMO, 2007).

The impact of climate change on agricultural systems is of particular concern in relation to the risks associated with the decline in production expected for almost all crops and to the consequent economic and social impact. Within this context, the scientific community has developed two different approaches with which to face risks related to climate change. These approaches are differentiated both at the level of time scale and in terms of objectives; they take the names of adaptation and mitigation.

10.4 Adaptation and mitigation

The purpose of *adaptation measures* in agriculture is to manage potential risks related to climate change over the next decades (i.e., to counter negative impacts or take advantage of positive ones). Adaptation measures are actions designed to reduce the impact of climate change on crops and animals and, consequently, to preserve their productive status.

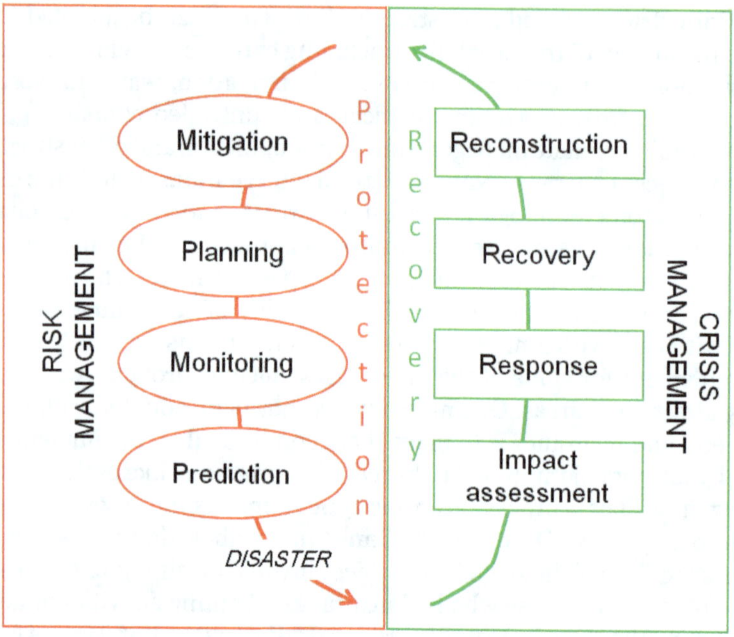

Figure 10.4: Example of risk management cycle. Adapted from National Drought Mitigation Center (Source: http://drought.unl.edu).

In the course of the last decades, numerous scientific studies have focused on the development of agricultural strategies that can be efficient in countering yield losses caused by climate change. These adaptations can be applied to different temporal and spatial scales and therefore cover short–term adjustments and long–term adaptations at the farm, regional or national policy level. Some adaptations for crop production are outlined below.

Short–term adaptation strategies can involve various aspects. Varying the sowing date and cultivars, for example, allows spring crops to increase the length of the growth cycle or to anticipate the most sensitive phases of development during periods in which there is less risk of water or thermal stress. Also, a different depth of sowing may be important, helping to ensure better germination of seeds. Another example is the implementation of good practices to conserve soil moisture. These practices can be included in the most common agronomic practices, such as soil tillage and irrigation. While tillage consists in particular of maintaining the crop residues of the previous year in order to increase the organic matter content in the soil and, consequently, to protect the soil from erosion and evaporation, irrigation allows correct management and use of water on the basis of the real needs of the considered crop.

By contrast, *long–term adaptations* that provide for structural changes to the whole productive system concern changes in land use, the introduction of resistant varieties, the use of more efficient cultivars and the development and use of more advantageous crop management.

Regarding changes in the use of the soil, efficiency can be obtained by sowing varieties whose production results are rather variable at the inter–annual level (e.g. wheat) within rotations that also provide for the growing of crops with more stable yields.

Genetic improvement, on the other hand, allows the choice of resistance factors to plant diseases and stress, while replacing crops helps maintain high yields through the use of varieties better capable of preserving soil moisture (e.g. the substitution of corn with sorghum, which is more resistant to high temperatures).

Finally, new management techniques such as precision farming, minimum tillage and mulching, inter–cropping, crop rotation and other innovative strategies, together with irrigation management and monitoring of soil moisture conditions can be considered long–term adaptation strategies capable of improving water use efficiency.

Figure 10.5: Multi–cropping system as a form of adaptation.

Mitigation strategies act on the causes of climate change and are aimed at the reduction or stabilisation of greenhouse gas (GHG) emissions into the atmosphere from agricultural activities. Globally, agriculture accounts for about 10% of global GHG emissions, of which, specifically, 50% of carbon dioxide (CO_2), 27,5% of methane (CH_4) and 22,5% of nitrous oxide (N_2O) (FAOSTAT, data from 2016).

The N_2O emissions come largely from nitrification and de-nitrification processes as a result of the application of fertilizers, grazing animal dung and other biological pro-cesses from the exposed soils. Instead CH_4 emissions come largely from livestock digestion and rice cultivation.

There is a lot of variation in emissions between Member States; in fact, Ireland holds the highest share of GHG emissions in agriculture (32.2%, 18.7 $MtCO_2e$) and Malta the lowest (3%, 0.088 $MtCO_2e$).

In absolute terms, the largest contribution of greenhouse gas emissions from the agricultural sector comes from France (18%, 79.0 $MtCO_2e$), Germany (15%, 66.1 $MtCO_2e$) and the United Kingdom (10%, 44.6 $MtCO_2e$). Together these three Member States account for just under 44% of total agricultural emissions in the EU–28.

According to EDGAR (Emissions Database for Global Atmospheric Research – http://edgar.jrc.ec.europa.eu) and FAOSTAT (www.fao.org/faostat), the largest source of emissions are emissions from enteric fermentation, while the US EPA (United States Environmental Protection Agency – https://www.epa.gov) says that emissions from agriculture soils are the dominant source.

All three databases agree that enteric fer-mentation and agricultural soils together account for about 70% of total emissions, followed by the cultivation of paddy rice (9–11%), biomass combustion (6–12%) and manure management (7–8%). Global manure emissions, such as manure on cultivated land grew between 1961 and 2010 from 0.57 to 0.99 $GtCO_2eq$ / year. Emissions have grown by 1.1% a year on average. Emissions from synthetic fertilizers had an average increase of 3.9% per year from 1961 to 2010, with absolute values increasing more than 9 times, from 0.07 to 0.68 $GtCO_2eq$ / year.

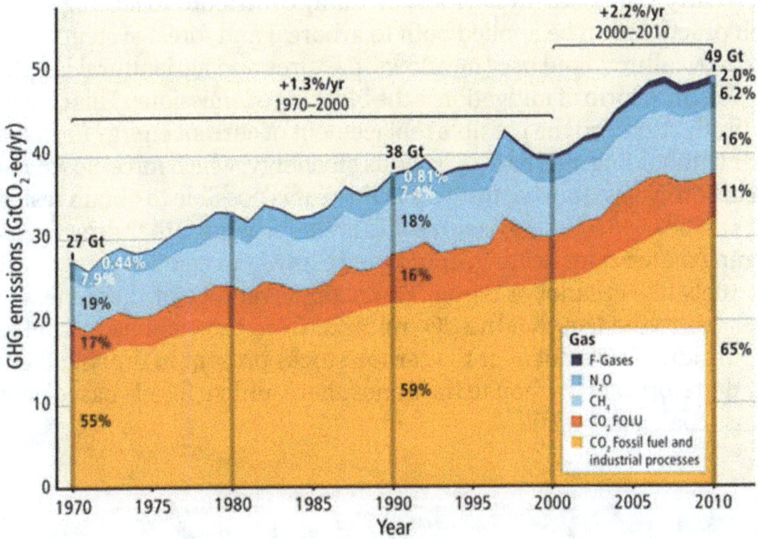

Figure 10.6: Total annual anthropogenic GHG emissions ($GtCO_2eq$/yr) shown for different groups of gases, 1970–2010. These include CO_2 from fossil fuel combustion and industrial processes, CO_2 from Forestry and Other Land Use (FOLU), methane (CH_4) and nitrous oxide (N_2O) (Cardona *et al.*, 2012).

Mitigation strategies differ from adaptation measures in that they act over the long term and their effectiveness depends on coordination at the national or global level. Currently, the international scientific community proposes three main ways for reducing and stabilizing GHG emissions and concentrations: the reduction of emissions, the increase in removals and the blocking of emissions.

Emissions reduction can be achieved mainly by increasing efficient management of the main GHG flows (in the form of CO_2, CH_4 and N_2O) of the carbon and nitrogen cycles from agricultural systems. This can be guaranteed by acting directly on the components that most influence gaseous exchanges between the crop system and the atmosphere through, for example, management of manure, the main source of methane flows, and the reduction of nitrogen fertilization, the principal cause of N_2O emissions. The latter can be further improved by considering the type of fertilizer and its method of distribution, in addition to its amount.

By contrast, the increase in removals concerns the improved efficiency of soil management. In fact, soil must be considered as the most important atmospheric carbon storage immediately after the oceans. Therefore, all management practices that aim to increase photosynthetic inputs or slow down the process of carbon returning from organic matter CO_2 (e.g. through respiratory processes, fires or soil erosion) contribute to storing carbon. Such practices can be applied both to arboreal and forest systems as well as to agricultural land use (meadows, pastures and agricultural land).

The third form of mitigation is the blocking of emissions. This concerns the production and the possible replacement of current energy forms with other withholding alternatives, such as bioenergy, which force closed natural carbon cycles. Indeed, from agriculture it is possible to obtain residues that can be used as energy sources. This can be done either directly, as in burning material for heating, or indirectly, as when residues are converted into fuels like ethanol or diesel. The main advantage of these measures is related to the fact that the CO_2 released from the combustion of these crop residues does not alter the carbon stocks present in the soil; they do not, therefore, add carbon to the atmosphere, unlike fossil–based energy forms (coal, gas, fuels).

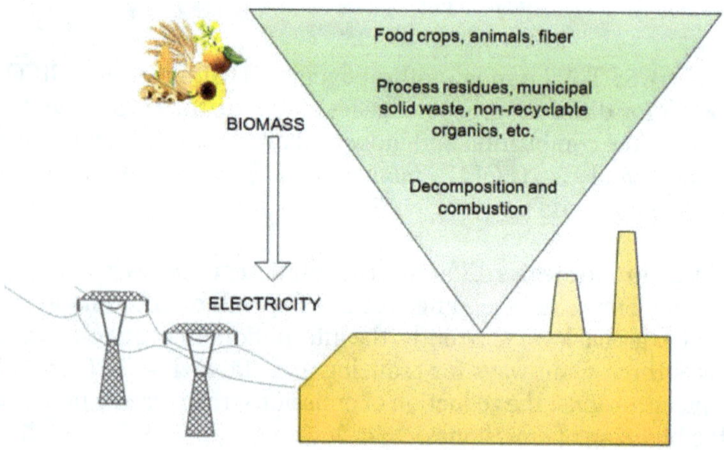

Figure 10.7: Residues and biomass can be used for energy production (Source: www.followgreenliving.com).

To reduce the production of greenhouse gases in the field agriculture it is also necessary to stop, or at least limit, practices such as indiscriminate deforestation of large vegetated surfaces (especially rainforests) and the removal of other non–agricultural vegetation. The reduction of these practices can indeed notably increase the sequestration of carbon and its storage in wood compartments for prolonged periods, thus avoiding further emissions into the atmosphere of carbon in the form of CO_2 and other GHGs.

Although these three methods constitute the fulcrum of mitigation strategies, they also involve a whole series of specific actions that have a significant influence on the seizure capacity of the analyzed system. Among these we should mention:

- the recovery of organic and degraded soils, if reactivated, would ensure an increase in their fertility with positive consequences in terms of carbon storage;
- better management of agricultural crops, meadows and pastures (elements, nutrients, sowing techniques, residue management and irrigation);
- greater attention to management of activities such as rice–growing and breeding, as these are major contributors to inputs of methane and nitrous oxide in the atmosphere;
- energy saving through construction of buildings with greater energy efficiency, and company logistics more attentive to environmental problems.

10.5 Climate smart agriculture: a way to manage climate change

Climate Smart Agriculture (CSA) is a project developed for helping countries to maintain a sustainable and adequate level of production and income for farmers. This approach, is based on a sustainable point of view, where the agro–ecosystem is primary and where evaluations on the use of resources and energy are important in agricultural and food systems. In the context of change climate is important to know how to plan investments, based on how climatic conditions will develop. For this reason, it is important to make reliable forecasting models to help in making choices. The final piece of the evidence base is focused on identifying institutional and financing needs to actual implement the priority actions. The approach is mainly based on dialogues between experts and non–experts to make decisions, in the form of public or private meetings or workshops. It is necessary to link climate finance to agricultural investments and have a certain ability to measure, signal and verify that the actions are actually generating adaptation and mitigation benefits (Asfaw and Lipper, 2016).

The three focal points of the CSA are productivity, adaptation and mitigation. Indeed, CSA maintains ecosystems services, including clean air, water, food and materials. CSA has multiple entry points at different levels: connects the food system, landscape, value chain or political levels. CSA is context specific, in fact the local management are different due to the different climate. Finally, CSA engages women and marginalised groups. It is committed to involving all local, regional and national actors in the decision–making process, so as to allow a broader and more sustainable operation (FAO, 2013).

Figure 10.8: Climate smart agriculture (CSA) (Source: www.fao.org/climatechange/climatesmart).

11

Models of Agrometeorology

11.1 Modelling approaches

During the past decades, several software tools using meteorological data have been developed for agricultural research and decision making purposes. Models can describe system processes using simple to very complex approaches. They may take into consideration only parts of a multifarious system, such as soil temperature or soil water balance, or more complex, interacting systems, such as crop growth, accounting for the most relevant interactions of the soil–plant–atmosphere system. They can further be limited to physical processes and bio–physical interactions, or extend their "system" to human interactions, by linking, for example, biophysical with socio–economic models (such as farm models in agriculture or land–use change models on a global scale).

In agriculture, crop and whole–farm system modeling, pest and disease warning models or algorithms, models for irrigation scheduling and agro–climatic indices or algorithms can significantly help farmers in making decisions for crop management options and the application of related farm technologies.

For research purposes, models can be used to simulate and analyze the complex interactions in the soil–plant–atmosphere system. For example, they can be applied to analyze climate change impact on crop water balance and crop yields. Nevertheless, these modeled systems include many uncertainties and limitations resulting from unknown trends in future technology and human activities, oversimplified representations of reality, lack of knowledge of system responses, or lack of calibration data.

11.2 Type of models

Two main types of models can essentially be distinguished: descriptive (empirical, statistical) and process oriented (dynamic, mechanistic) (Fig. 11.1).

Descriptive crop models (empirical or statistical) rely on site specific yield and phenology information. The model forecast is therefore based on broad trends coming from observed data and correlations between, for example, yield and phenology from one side and most important factors affecting it (weather, management). Main weakness of descriptive models is that their inherent statistical relationships are location and crop specific and should not be applied on other crops on same location or same crop on other locations with different environmental conditions.

Process oriented dynamic models consist on system of time dependent differential equations, which are used to calculate, in each time step, variables important for description of plant growth and development (e.g. intensity of photosynthesis, biomass production, etc.). However, these models always contain empirical elements as well, such as for phenomena which are not well known or are too complex to simulate at the process level (such as crop phenology).

Process–oriented crop models have been applied in research for more than 50 years. The three most important "schools of development" come from Australia, the Netherlands and the United States; these are, respectively:

- APSIM models (Keating *et al.*, 3003);
- SUCROS based models (such as WOFOST) from the "School of De Wit" (van Ittersum *et al.*, 2003);
- the DSSAT family (such as CERES) of crop models (Jones *et al.*, 2003).

It should be noted that there are links between these modelling groups.

However, a number of other well–known models (often connected to one of the three "modelling schools") are also used in climate change impact studies for different crops and environments and on different scales. In Germany, for example, AGROSIM (Kersebaum *et al.*, 2007) has been applied for effects on field–based yield and water balance, while HERMES (Kersebaum, 2007) has been used to analyze effects on yield, nitrogen leaching, adaptation and fertilization. Furthermore, MONICA, an extended model based on HERMES and on the algorithms of the Danish model DAISY (Nendel *et al.*, 2011), has been adopted for climate change impact studies. Other crop and ecosystem models developed in the USA include CROPSYST (Stockle *et al.*, 2003), EPIC (Cabelguenne *et al.*, 1999) and DAYCENT (Parton *et al.*, 1998). DAISY is a crop model developed in Denmark (Abrahamsen and Hansen, 2000). In Europe, the SIRIUS model

(Jamieson *et al.*, 1998) has been used to assess drought–related reduction of potential yield as well as to predict wheat yields, while SIRIUS–QUALITY (Martre *et al.*, 2006) additionally considers grain quality of cereals.

An example from France is the STICS model, which is widely used at INRA and other organizations for several crops (Brisson *et al.*, 2003). New developments that have been reported mainly regard the application of crop models within whole–farm system models and spatial applications in combination with GIS and remote sensing methods. At the Macaulay Institute in Scotland, for example, the LADSS (Land Allocation Decision Support System) farm–scale integrated modelling framework and the PALM model consist of a core of biophysical simulation models overlaid by financial, social and environmental accounting modules (Matthews *et al.*, 1999, 2006). The crop module is a version of the CROPSYST model developed at Washington State University. A model, that involves the hydrological processes, is SWAT (Arnold *et al.*, 1998) using the EPIC approach to consider vegetation.

The most frequently used process–oriented crop models for research or operational applications in Europe are CERES, WOFOST and STICS, though each is used to differing degrees in different countries (for further details see Nejedlik and Orlandini, 2008). WOFOST is the only model which is operationally integrated at the European level for the European crop yield prediction system (MARS), covering all countries.

11.2.1 The model scaling problem

The first basic condition of any ecosystem modelling approach is the scale of application with regard to space and time.

Optimum scaling is determined by:

- the aim of the model application;
- the required details of the simulated processes;
- knowledge of the simulated processes;
- the required spatial representativeness of the model;
- the availability of input data on temporal and spatial scales.

Figure 11.2 shows different application examples of ecosystem models regarding optimum scales. Depending on the application, models may need high spatial and temporal resolution of simulated processes (such as with plant physiological models) or–especially when applied on large, global scales–low spatial and temporal resolution (such as global climate or land surface SVAT [Soil–Vegetation–Atmosphere–Transfer] models). A general problem, especially for large area model applications, is the availability of input data, which need to be available on the applied modelling scale. There-

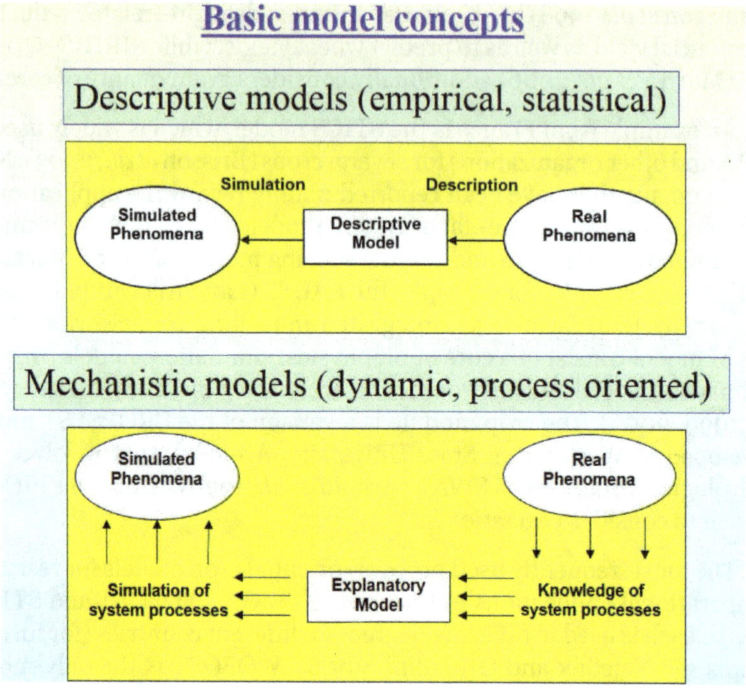

Figure 11.1: Basic model concepts in agrometeorological modelling (Source: BOKU-Met).

fore any specific model is always a compromise between available input data and scaling needs to address specific processes in a suitable manner.

11.3 Model characteristics

11.3.1 Model structure

Process–oriented agrometeorological models, such as crop growth models, involve all significant dynamic processes for crop growth in a specific context, such as crop/plant type, environmental conditions and agronomic (crop management) conditions. An important aspect of crop models developed in recent years is the simulation of the effects of crop management practices and climate on crop growth and yield. However, their use for predicting yields over large areas is limited by the difficulty in obtaining representative information about local field conditions, management or crop characteristics at any given point. Some crop or soil features may be considered as constant (e.g., for a group of genotypes or soil types in a given region), but others depend on temporal changes in local conditions (e.g. Guerif and Duke, 1998).

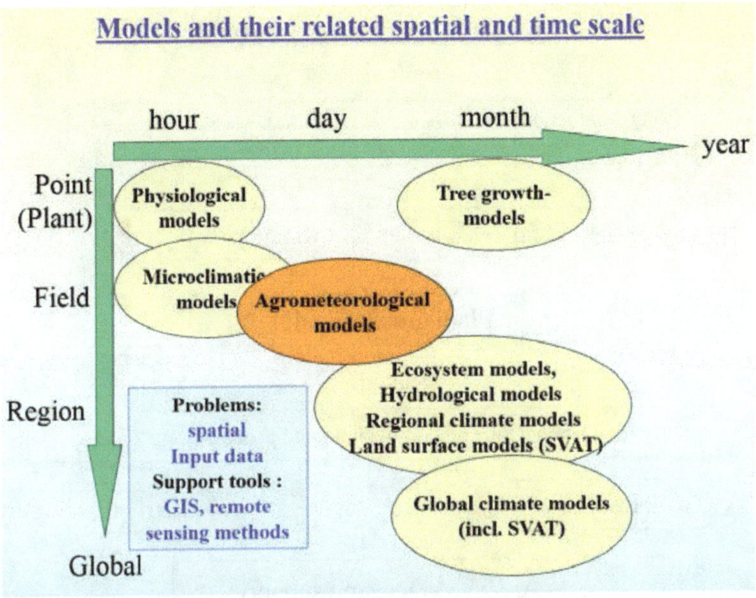

Figure 11.2: Spatial and temporal scales of ecosystem model applications (Source BOKU-Met).

Testing crop models over a range of environmental conditions (called calibration and validation) is required to establish confidence in applying them (e.g., Goudriaan and Van Laar, 1994). Crop models are available for almost all economically important crops, and on many occasions they have been successfully used in research. In the future, models may be useful for improving the efficiency of agricultural systems and may be used by farmers as tools in their efforts to improve the profitability of their farms (e.g. Jacobson *et al.*, 1995). Nevertheless, a relevant pre–requisite is, that models must be calibrated and evaluated for each climatic region where they are intended for use in decision making.

Crop simulation models enable the summary of scientific knowledge on the biological processes that regulate plant growth; they therefore require permanent improvement and testing. Integrating the work of experts in different fields and places, models represent very powerful tools because they allow the simulation of many scenarios by any change of settings or inputs (Fig. 11.3). With respect to field experimentation, they lower costs and save time; yet field experiments are still necessary to gather new findings to further improve parameterization and calibration. Models simulate final variables of the crop cycle, such as grain yield, but also forecast the evolution of many intermediate variables. As they are generally built with an analytical purpose, models are often used as a predictive tools, applied to, for example, climatic or crop management scenarios.

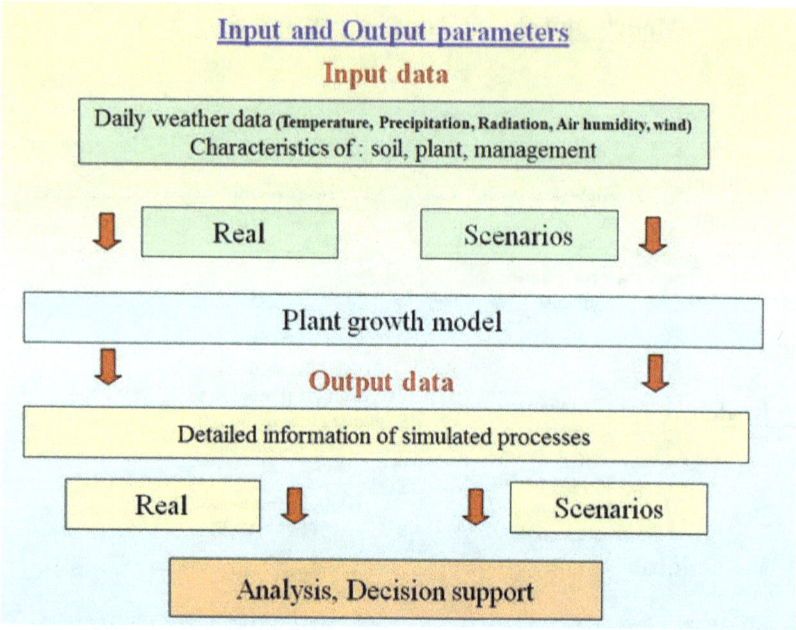

Figure 11.3: Crop model inputs and simulation outputs, which allow the generation of many kind of scenarios (Source: BOKU–Met).

A simplified, basic scheme of a crop growth model is shown in Figure 11.4. This basic scheme of soil–crop–atmosphere interaction is present in all crop models, differing only in details of simulated processes in relation to crop type, application purpose or the available scientific background of model developers. Many of the simulated processes indicated in Fig.11.4, which describe impacts of climatic parameters on processes, have already been presented in previous chapters (such as the impact of temperature and solar radiation on photosynthesis rate and calculation methods for soil water balance).

11.3.2 Model inputs and parameters

All crop growth models are based on weather input variables (mostly at a daily time step), which describe changes in weather conditions during the simulation (growing) period of crops. The most important weather variables are temperature, solar radiation and precipitation (called also the minimum weather input data set). Depending on the method applied for calculating evapotranspiration, air humidity and wind speed may also be needed. Observed weather needs to be representative for the scale of application. Weather variables have different spatio–temporal variability, which has to be considered when evaluating their suitability for site–spe-

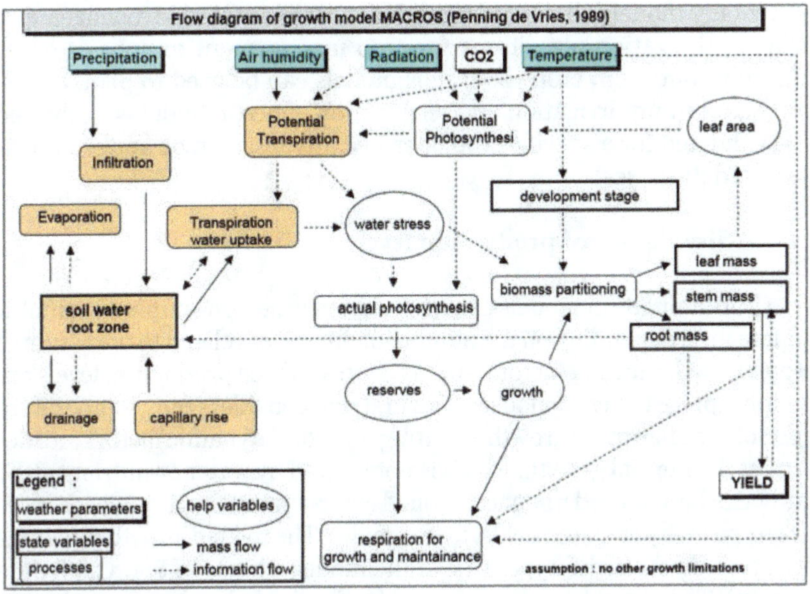

Figure 11.4: Generic scheme of crop growth simulation in dynamic crop growth models according to the MACROS approach (schematic, after Penning de Fries, 1989).

cific modelling (Kersebaum *et al.* 2015). Applying a crop model to reflect observations at field scale requires weather data, which should be observed as close as possible to the experimental site. Another important factor is the actual carbon dioxide (CO_2) concentration of the atmosphere, in the case that the calculation of the direct photosynthesis rate is required in the model. (It is usually set as a constant over selected decades or in stepwise increments over future periods).

Other inputs of crop models–which are considered to be constant and not variable during the simulation period–are layer–specific physical and chemical soil characteristics (at least over the maximum rooting depth), especially crop–available water storage capacity as a minimum requirement (although some models allow this calculation as well to be made from soil texture values). We further need inputs of crop characteristics (either crop type or cultivar specific) which describe crop responses (of specific physiological processes) to their environment. Finally, we require weather input variables that are specific to the model.

For agricultural crop production as a managed ecosystem, we additionally need crop management inputs, especially the sowing date, the fertilization and irrigation regimes (timing and application amounts) and the type of soil cultivation. Other data may additionally be required, depending on the capabilities of the model and the aims of the application. Many models allow also the simulation of automatic management, using optimization algorithms for crop management options in order to receive optimum crop yield: thus models can be used to plan, for example, optimum irrigation and fertilization. Once a model is calibrated and validated for a specific crop or crop cultivar, it can be applied for its designed purposes.

11.3.3 Simulation of production levels

Crop simulation models can be designed based on simple or complex approaches, depending on the number of processes related to the assumed crop–growth limiting factors. In this context, three production levels are mainly applied in crop models. The first one considers only temperature and solar radiation as growth–limiting temporal dynamic factors and describes a potential growth, which is not limited by water or nutrients (the latter can be achieved under controlled environments such as glasshouses or on optimally irrigated fields, for example). The related growth processes are given on the right upper side of the model scheme of Fig. 11.4 (white section). The second production level further includes the effect of water availability on crop growth, which requires an additional calculation of the soil water balance (coloured part in Fig. 11.4). This condition could already be fulfilled under rain–fed field conditions, without any other growth limitations. The third production level includes calculation of nutrient balance as a crop–growth limiting factor, especially for nitrogen, phosphorus and potassium (not shown in Fig. 11.4). This level is especially necessary under limiting nutrient conditions and for calculating optimum fertilization regimes or N–leaching.

Further levels can be added as well, such as pest and disease effects on growth and final yield. However, pest and disease models are usually not integrated into crop growth models but are rather used separately or in combination with them (Donatelli *et al.* 2017). In general, we must remember that the more complex the model, the greater the number of input parameter requirements and the potential for higher uncertainties in modelling results. Moreover, more data for calibration and validation are needed.

11.4 Crop model calibration and validation

In general, crop model calibration is related to the fitting of mathematically described processes within the soil–crop–atmosphere system by qualitative data sets, such as from controlled field experiments conducted over several years and, ideally, from different sites as well (Kersebaum *et al.*, 2015). Crop model parameters to be calibrated concern especially crop type or cultivar–specific response functions of simulated processes to environmental conditions, such as development rate vs. temperature (and many more, depending on model complexity). These are usually defined as calibration coefficients in model inputs or settings (such as crop genetic coefficients).

By contrast, model validation is the test of the calibrated model (and its simulated outputs) on data sets, which are not dependent on the calibration data set, for example, from different years of the calibration site or from different sites. Model validation should therefore prove whether a model still performs suitably under conditions that are different from the calibration site conditions. This allows application of the model under a wider range of similar conditions, with some degree of certainty that model results will be reliable. However, such certainty is never absolute, especially if a model was validated just in a limited range of environments.

To test or evaluate model performance by validation – for example, by comparing observed/measured versus simulated output parameters – numerous statistical tests have been developed and are used in studies with crop model applications. Each crop model output can theoretically be calibrated/validated on measured ones, if these are available, but the most important factors are crop yield or biomass (at harvest), crop phenology entry dates, soil water and nitrogen content, and leaf area index. These parameters are also often available from past field experiments and for different years and sites. To cover the impact of climate variability on specific sites, at least 3–10 years of observed data, especially for main phenological phases and final crop yield, are recommended for a sound model validation (Grassini *et al.*, 2015).

The most frequently applied statistics for crop model validations are:

- mean absolute error (MAE), is the mean absolute difference between model estimates and observations (also calculated as the mean absolute percent error);
- mean square error (MSE), or equivalently its square root, the root mean square error (RMSE). Both MAE and RMSE express average model prediction error in units of the variable of interest. Both metrics are indifferent to the direction of errors. They are negatively–oriented scores, which means that lower values are better. Since the errors are squared before they are averaged, the RMSE gives a relatively high weight to large errors;
- standard deviation (σ) of observed (σ_o) and simulated values (σ_s). The simulation is considered as good if i) RMSE < σ_o, and ii) $\sigma_s < \sigma_o$;
- relative root mean square error (RRMSE). Model accuracy is considered excellent when RRMSE <10%, good if 10% < RRMSE < 20%, fair if 20% < RRMSE <30%, and poor if RRMSE >30%;

- index of agreement (d) is a dimensionless measure of model performance (Willmott *et al.*, 1985). The simulation is considered as good if the index of agreement is close to 1;
- mean bias (MB) or mean bias error (MBE) is the mean difference between model estimates and observations. It can convey useful information, but should be interpreted cautiously because positive and negative errors will cancel out (which is not the case when using MAE);
- a linear regression between estimated and observed values (or vice versa) has some limitations and needs more careful interpretation. The hypothesis is that the regression passes through the origin and has a slope of unity. The r coefficient (coefficient of correlation) and its counterpart r^2 (least–squares coefficient of determination) in isolation, however, do not indicate model accuracy and do not imply that the estimated regression line is a good fit for the model estimation (see Bellocchi *et al.*, 2006);
- comparison of the distribution of the observed and estimated values has also been utilized to identify model adequacy for stochastic and deterministic models. For example, a common Kolmogorov–Smirnov's D test has been used to assess the probability that two data sets (observations and simulations) have the same distribution.

In summary, many authors believe that there is no robust statistic or graphical representation which can be used to draw conclusions in model evaluation. Therefore, several methods need to be used together to obtain a reliable assessment (e.g., Yang *et al.*, 2000). Many of these statistics can be calculated using standard statistical software, such as "Excel" or "R". The use of multiple metrics (such as those listed above) is recommended for measuring crop model performance, because the different statistics describe "performance" from different aspects (see recommended Literature).

11.5 Crop model sensitivities to weather extremes and related uncertainties

As different crop models use different approaches for simulating specific processes, the results of simulated results (such as final crop yield) may also differ. Sensitivity analysis of models and model comparison can point to model weaknesses or application limitations as well as aspects in which models may need further improvements.

Adverse agroclimatic extremes considered most detrimental to crop production include drought, heat, heavy rains/hail and storm, flooding and frost, and, in particular, combinations of them. However, there is still a mismatch between advances made regarding deeper understanding of abiotic stress physiology and its incorporation into ecophysiological models in order to more accurately quantifying the impacts of extreme events at crop system level (Rötter *et al.* 2018).

Fig. 11.5 shows the simulated yield deviations of seven different crop growth models for maize with a significant drought and heatwave from May to August at a site in a semi–arid region in Austria. Significant differences between the considered crop models can be detected in the final simulated yields in the case of drought stress during summer (for maize here), whereas the differences in the model sensitivity are much smaller in the case of extreme temperatures. This supports the assumption that the simulation of soil water content and crop drought stress are still major factors of uncertainty in crop simulation results. The reasons for this disparity are manifold. It could be due to the weak representativeness of simulated growth processes or soil water content, or to differences in model parameterization (e.g., critical limits to drought or heat stress for crop growth processes).

The effect of soil characteristics (such as bulk density and crop–available soil water storage capacity) on simulation results can be significant as well. For example, soil bulk density and soil water storage capacity are influenced by soil cultivation. The model sensitivity to these applied

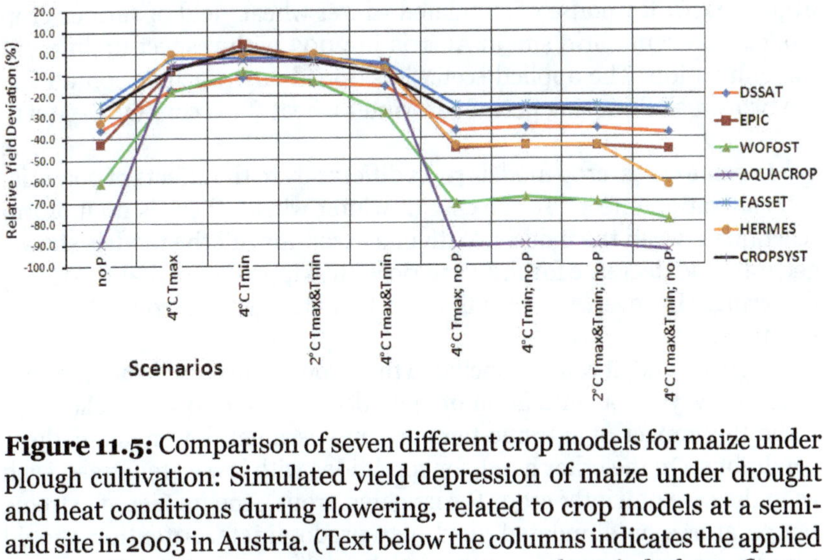

Figure 11.5: Comparison of seven different crop models for maize under plough cultivation: Simulated yield depression of maize under drought and heat conditions during flowering, related to crop models at a semi-arid site in 2003 in Austria. (Text below the columns indicates the applied changes of weather parameters during a two-week period of crop flowering, compared to the real weather: noP=no Precipitation, T=Increases in daily air temperature), (Source: BOKU-Met).

changes for certain scenarios with drought effects are shown in Fig. 11.6 for seven selected crop models. Moreover, differences in the model approaches regarding rooting depth and root distribution results in distinct differences in the response of models to dry conditions (Wu & Kersebaum, 2008).

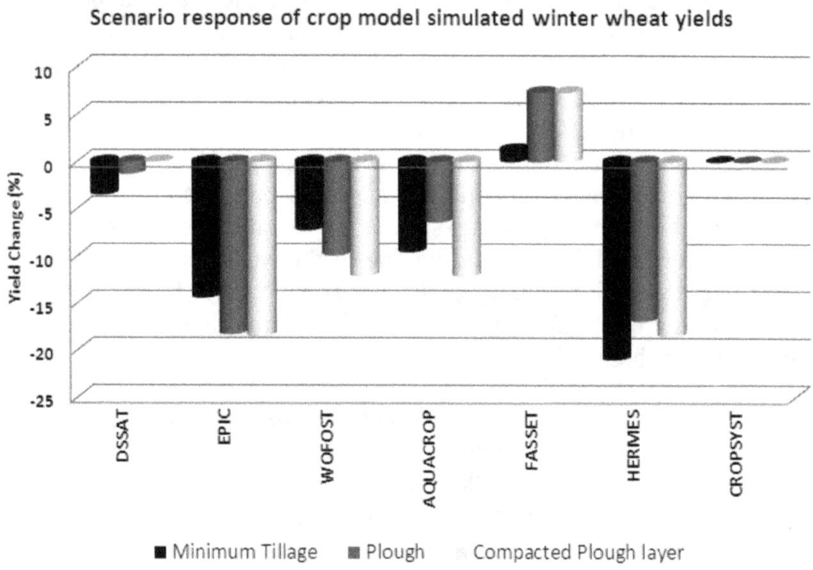

Figure 11.6: Response of simulated winter wheat yield of various crop models at a semi–arid site in Austria in 2003 with respect to different soil cultivation. The applied scenario skipped any precipitation during the two–week flowering period for winter wheat (Source: BOKU–Met).

In some cases, crop models react differently to the effects of potential soil cultivation effects. In the case of winter wheat, there is no uniform reaction between the models. With one exception, all the models show a general yield decrease for the scenario with skipped precipitation during flowering. However, the results indicate a change with soil cultivation practices.

On one hand, it can be concluded that crop models can behave in very different ways in a simulation of yield decrease due to water shortage, while the impact of extreme temperatures is much lower under these conditions. On the other hand, for conditions with low water stress, such as in the case of winter crops (or irrigated crops), the impact of extreme temperatures on simulated yield may predominate. Attention should thus be paid when applying crop models in climate change impact studies, especially for local or regional studies where sound calibration and validation is a precondition. For example, a comparison of climate change impact researches based on different crop models has only limited reliability if the limitations are not considered in the interpretation of results.

11.6 Crop model applications

While crop models can be applied in agronomic practice, their main application area is in the field of research, especially in assessing climate change impact on future crop yields and yield potentials, environmental impact, various cropping risks or the use of water resources for crop production. Practical applications mainly regard the short term, such as irrigation and fertilization planning or short– to mid–term yield forecasting. On the other hand, they are rarely used for early warnings or the mitigation of damage from extreme meteorological phenomena and processes, where agrometeorological indices are more common (see Chapter 11.8). Examples of crop model application results can be found in chapter 12 on the impact of climate change on agriculture.

Most crop simulation models in Europe are applied to annual crops, especially cereals and maize, the continent's economically most important crops (Fig. 11.7). However, permanent grassland, potatoes, sugar beet, oilseeds and others also play an important role regionally, which has led to the creation of specific model applications.

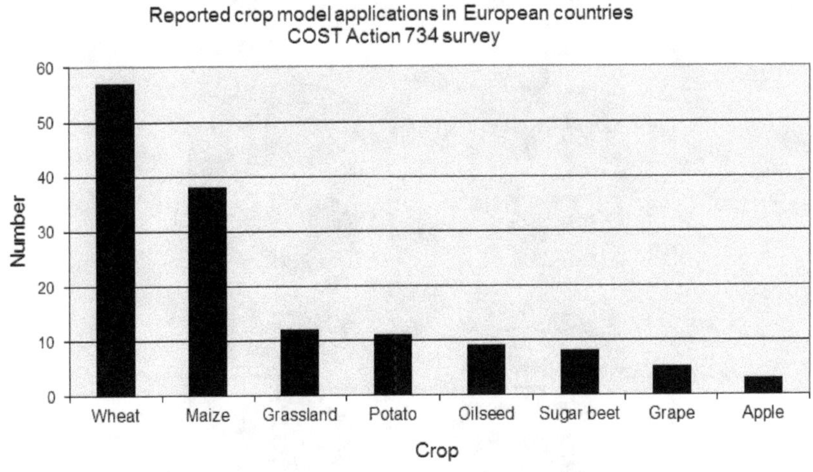

Figure 11.7: Reported crop model applications (operational and research, one count per model and country) according to COST734 survey (Nejedlik and Orlandini, 2008).

Crop model applications are characterised by several uncertainties which determine limitations of their use in research and practice (e.g., Eitzinger *et al.*, 2008). The main reported limitation for the application of crop models in Europe is related to input data. The most frequently reported problems are the unavailability or the low quality of the soil input data (especially for spatial model applications), the lack of long–term biophysical crop data for model validation and calibration and, in some cases, the unavailability or high costs of meteorological data. This is related to differences in socio–economic conditions and the local administration of data among the different regions of Europe. The reliability of data for climate scenarios or seasonal forecasts is another crucial point for the use of such models for operational purposes or for making long–term strategic decisions.

11.7 Pest and disease models or algorithms

Pest and disease models are specific algorithms describing the conditions of pest and disease development in order to detect critical limits for farm measures. Pest models often use the temperature sum approach for calculating pest phenology (see chapter 8). Disease models are often more complex and multivariate in that they need to consider several conditions for calculating disease development cycles. Often micro–meteorological data with high temporal resolution is required (Donatelli *et al.* 2017). An example of such an algorithm scheme for the detection of risk of secondary infection

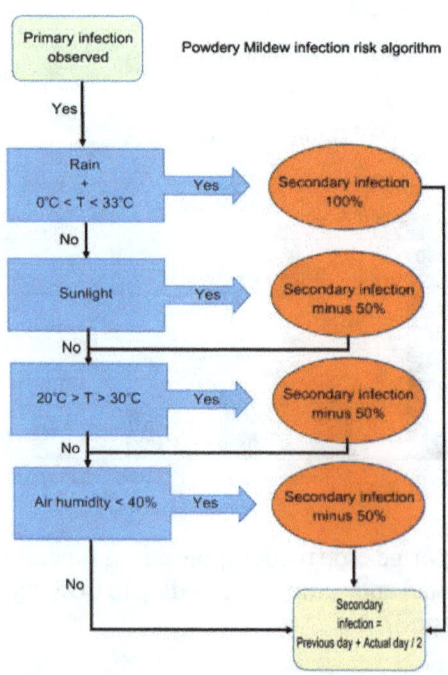

Figure 11.8: Generalized scheme of an algorithm for secondary infection risk of a specific type of powdery mildew, based on daily critical micro-meteorological conditions (Source: BOKU-Met).

of fungi is given below (Fig. 11.8). It shows the stepwise consideration of critical micrometeorological conditions for a (not further calibrated) specific powdery mildew type. Application example is presented in Chapter 8.

Tab. 11.1 presents examples of the most important pests and diseases for which algorithms are available for forecasting and warning. Several software packages are available, such as PESTCLIM, which utilizes the already proven algorithms of ECAMON (Trnka *et al.*, 2007), and CLIMEX (Svobodová *et al.*, 2014), a new efficient data handling and calculation procedure enabling sophisticated optimization and improved performance. BAHUS the biometeorological system for messages on the occurrence of diseases in fruits and vines is calibrated for apple scab (*Venturia inaequalis*), fire blight (*Erwinia amylovora*) and grapevine downy mildew (*Plasmopara viticola*) (Mihailovic et *al.*, 2001, Lalic *et al.*, 2016).

Algorithms for	Applied for	Disease name
Disease risks	Potatoes, apples, grapes	– Potato blight (*Phytophtora investans + Alternaria solani*) – Apple scab (*Venturia inaequalis*) – Fire blight (*Erwinia amylovora*) – Grapevine downy mildew (*Plasmopara viticola*)
Pest risks	Field crops, Orchards	– European Corn Borer – Western Corn Root worm – Wireworms (*Coleoptera, Fam. Elateridae*) – Codling moth (*Cydia pomonella*) – Plum Fruit Moth (*Grapholita funebrana*) – Colorado Potato Beetle (*Leptinotarsa decemlineata*)

Table 11.1: Examples of algorithms applied for pest and disease management in agricultural practice.

11.8 Agroclimatic indices and algorithms

Agroclimatic indices, either crop specific or not, are widely used in operational forecasting to project agroclimatic risks related to specific weather phenomena. Well known examples regard cold ("frost") risk, drought forecast, and sowing and harvest conditions. These indices have already been used in agroclimatic research to express conditions under climate change, such as drought and extreme heat.

Weather–related (abiotic) cropping risks can be quite different in their nature and seasonal frequency. They may include all weather parameters and phenomena directly and indirectly affecting crop growing conditions as well as yield and damage potentials. Agroclimatic indices describe drought and heat effects on assimilation or yield forming processes, overwintering conditions of winter crops (especially great temperature variations and snow cover conditions), cold risks at different phenological stages, risks for sowing and germination (erosion, soil hardening and inadequate soil temperatures), bad harvest conditions leading to yield loss, hail, wet periods, strong winds and thunderstorms (leading to lodging, N–leaching and erosion), high humidity and leaf wetness (causing diseases), high air temperature (triggering pests) and many other factors (Tab. 11.2). A crop's vulnerability to the severity and duration of these phenomena differs by species and variety, phenological status or the occurrence of other crop stresses at the same time. Thus crop specific effects need to be calibrated for the relevant index or algorithm.

For example, during a crop growing season several weather–related cropping risks can occur at different times, at the same time or can overlap. Until now the crop modelling community has made great efforts in the MACSUR and AgMIP projects (www.macsur.eu, www.agmip.org) to address combined drought and heat stress effects, which are considered the main weather–related crop production risks under climate change in Europe. However, many other weather–related risks are still insufficiently accounted for in the crop models due to the complexity of related processes (Rötter *et al.* 2018); in these cases, indices can serve as a complementary information source to better assess the impact of adverse weather conditions on agriculture in general.

Indicator(s) for	Applied for	Hot spots
Drought	All selected crops	All regions with less than 800mm annual precipitation and high air temperature and wind; relevant for crop stress and pest development.
Heat	Annual crops	Mainly lowland regions in central, eastern and southern Europe; heat stress and fertility impact on several crops; related increased ozone levels leading to yield losses.
Soil conditions for crop (root) growth	Annual crops	All regions with loamy–clay soils (critical soil temperatures and wetness) and sandy soils (drought).
Heavy precipitation	Field crops	Especially humid regions (mountainous regions); relevant for lodging and diseases as well as soil surface hardening.
Snow cover conditions	Annual winter crops	In all regions with extreme snow cover variation (too much, too long or too little).
Overwintering conditions	Selected crops	Frequency and duration of mild/cold fluctuations during winter; weakening frost hardiness, chilling conditions.

Humidity and leaf wetness	Selected crops	Humid crop growing regions; important conditions for many diseases.
Harvest conditions	Selected crops	Crop specific, include all weather parameters.
Soil workability	Arable crops	For soil cultivation and crop management; danger of soil compaction.
Suitable conditions for crop management measures	Selected crops	e.g., number and frequency of dry/calm days within certain periods, crop specific.
Drying conditions	Annual crops	Field drying conditions, relevant for yield quality and diseases (e.g. fusarium).
Frequency of rain days	Selected crops	Relevant for diseases and pests; biomass accumulation (low radiation and cloudiness).
Cold ("frost") damage	Many crops, but especially orchards	Combination of frost occurrence with crop vulnerability assessment (phenology model). Especially spring crops and orchards; huge damage potential for sensitive crops; great impact of orography.
Soil erosion	All cropping systems with periods without surface cover	Direct damage and long–term effects on soil fertility (and nutrient and water storage capacity); strong effect of soil cover and orography.
N–leaching	Annual crops	Cereals and high yielding crops such as maize; great impact of soil conditions and crop management, high spatial variability.

Table 11.2: Examples of agroclimatic indicators and pest/disease algorithms.

An example of available software for calculating agroclimatic indicators is AGRICLIM. It is a tool that allows one to calculate more than 200 agrometeorological parameters and indices; it was especially designed to allow for a large number of calculations (Trnka et al, 2011). It is particularly suited to analyse the occurrence of agrometeorologically adverse events which are capable of significantly affecting yields but are which often not considered by dynamic crop models and are therefore neglected in related climate change impact and adaptation studies.

An example of a spatial GIS–based application of an agroclimatic indicator for operational monitoring and forecasting is presented in Fig. 11.9 – 11.10 for field working days for the harvest period of spring barley. This indicator is calculated as follows:

Field Working Days Indicator

Accumulation of days with meteorological and soil conditions suitable for field work.

if ($p_{n-3} < p_{thresh3}$ & $p_{n-2} < p_{thresh2}$ & $p_{n-1} < p_{thresh1}$ & $p_n < p_{thresh}$ & $rss_n < rss_{thresh}$): $fwd_n = 1$
else: $fwd_n = 0$

Version1 : accumulation of fwd from June 1st to June 30th
Version2 : accumulation of fwd from July 1st to July 31st

fwd_n	day n with meteorological and soil conditions suitable for field work [0,1]
p_n	sum of precipitation at day n [mm]
p_{thresh}	threshold value of the precipitation sum at day n [mm]
rss_n	relative soil saturation at day n [%]
rss_{thresh}	threshold value of the relative soil saturation [%]

Variable	Threshold value
rss_{thresh}	70 % of maximum soil water holding capacity
$p_{thresh3}$	20 mm
$p_{thresh2}$	10 mm
$p_{thresh1}$	5 mm
p_{thresh}	0,5 mm

Figure 11.9: Calculation procedure of Field Working Days indicator for harvest periods (Source: BOKU–Met).

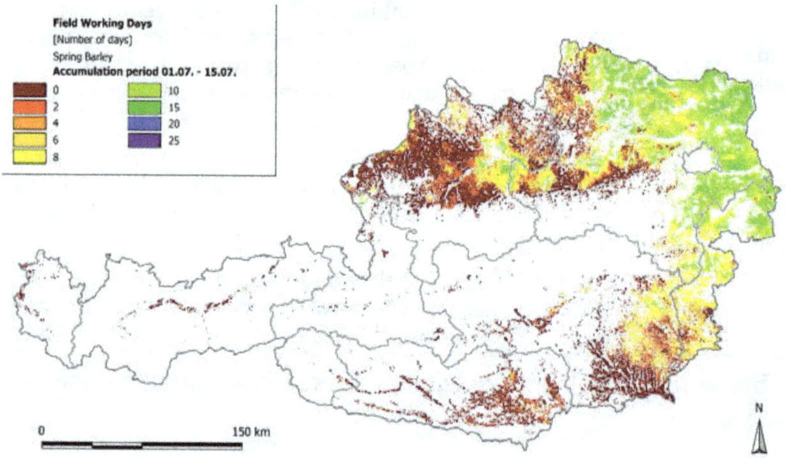

Figure 11.10: Field Working Days for soils/cells of land use type "arable land" (Commission of the European Communities, 1995) in Austria in July 2015. Simulation for phenological phases of harvest of spring barley at a spatial resolution of 500 m. The image shows the accumulated days from July 1st to July 15th, 2015 (within the accumulation period of July 1st to July 30th) with soil conditions suitable for harvesting (Source: BOKU–Met).

12

Relationship Between Climate, Agriculture and Climate Change

12.1 Climate and climate change – an introduction

Climate can be defined by different expressions, but in general it represents the mean conditions of weather variables over an area of different extent in the certain period. Climate differs from "*weather*", which only describes the short–term conditions of weather variables in a given place (e.g. changing from day to day).

In other words, climate represents the composite or generally prevailing weather conditions of a region, as temperature, air pressure, humidity, precipitation, sunshine, cloudiness, and winds, throughout the year, averaged over a series of years. The related spatial scale can range from continental or regional to local or site. For example, we may describe regional climates (e.g. Pannonian lowland climate), local climates (e.g. climate of a valley or a city) or even microclimates (e.g. the climate within a plant stand such as a vineyard at a specific site).

Climate features are related to atmospheric or weather variables expressed by different statistical measures, such as averages (e.g. mean annual air temperature), sums (e.g. mean monthly precipitation), extremes frequency (e.g. mean number of heat waves per year), probabilities (e.g. probability of heavy precipitation), percentiles (probability of deviations from a normal) etc. In that sense, climate is the statistic analysis of weather over long periods from many different potential viewpoints.

An important prerequisite of the definition is, to consider a suitable length of period, in order to catch the statistical normal or statistical distribution of weather variables. Indeed, this length of period depends on the climate itself and its statistical features.

However, for making it possible meaningful to compare climates, e.g. of different regions or sites, a climate Normal period of 30 years have been defined by WMO (World Meteorological Organization) for comparable ranges of years (e.g. 1961–1990 or recently 1981–2010), (http://www.wmo.int/pages/prog/wcp/wcdmp/GCDS_1.php).

Climatological Normals have long filled two major purposes. Firstly, they form a benchmark or reference against which conditions (especially current or recent conditions) can be assessed, and secondly, they are widely used (implicitly or explicitly) as an indicator of the conditions likely to be experienced in a given location.

This is especially important for the field of climate change impact research, in order to compare different climate normal periods in one hand, and to make comparable different research studies using the same reference climate normal periods.

While "climate" refers to the mean state of the climate system or the statistics of weather, "climate change" refers to the change of statistics of weather over time.

Accordingly, climate change is the change in climate patterns on global or regional scales. In order to conclude that climate is changing one should compare statistics of weather variables, for example averages of weather variables over two 30–year periods (Climate Normal period). Recently, in climate change studies, as the most important measure of anthropogenic caused climate change, increase of air temperatures at various scales has been used (e.g. increase of global mean air temperatures by 3 °C, 4.5 °C, etc.).

12.2 Climate classification and regionalization

Historically, a great number of climate classifications have been developed since Hellenistic times; interest in the subject has especially grown since meteorological measurements have been available globally.

The purpose of climate classification is to organize and synthetize climate data in order to allow its more effective communication and to identify regions with the same or similar climate characteristics. Even there are three basic types of classification used in climatology in order to identify climate zones, however many classifications are of mixed nature:

- empirical climate classifications are based on observable features of climate, such as weather variables or type of vegetation or river types (as a response to climate). Köppen classification, presented below, is an example of this type of classification based on temperature and precipitation;
- genetic climate classifications are based on climate elements which are considered as a cause of climate and they include geographic determinants, surface energy budget elements, wind conditions, pressure systems and other elements derived from air mass

analysis. These classifications are constantly improving since our knowledge about climate and its causes is ongoing;
- applied climate classifications are focused on particular climate related problems and can be of empirical and/or genetic nature. Thornthwaite classification, for example, is based on potential evapotranspiration and can be seen as applicable for assessing regional water balance or potential water demand for human activities.

Climate zones associated with specific regions are commonly presented on geographical colored maps or hatching. Sometimes, climate maps show the climatological means of elements and parameters on which classification is based.

Whether individually or in combinations, parameters used for determining climate classes are manifold. Some examples of commonly used are listed here:

- incoming solar radiation (polar, temperate and tropical zones), (genetic/empirical type);
- geographical regions with "similar" climates (empirical type);
- hydrological characteristics (e.g. the type of seasonal river water flow), (empirical/applied type);
- potential evapotranspiration (e.g. from Thornthwaite in 1931 and Budyko in 1948), (empirical/applied/genetic type);
- the main atmospheric circulation systems (e.g. by Flohn in 1950 and Alissow in 1954); (genetic type);
- a combination of meteorological variables and vegetation types (as climate indicators); (e.g. by Köppen in 1900), (empirical type);
- combinations of meteorological factors only, such as seasonal (monthly) temperature and precipitation (e.g. by Walter and Lieth in 1958), (empirical type);
- tailored climate mapping schemes for use in agrometeorology and agriculture (such as vegetation or crop–specific growing periods, monthly mean and extreme temperatures and the mean duration of snow cover), (applied type).

Climate classifications often employ indicator values based on the variables used for calculation as a basis for defining climate zones and mapping climate types (tropic, sub–humid, etc.); alternatively, directly calculated mean meteorological variables are used, such as mean annual or monthly temperatures and precipitation, from weather station data.

Different classification schemes are often drawn as maps at different regional resolutions and printed for climate atlases or associated to weather stations as climate diagrams. Since the age of digitalization, the possibilities and options provided by computers for creating climatic schemes or maps are endless. Still, a few "old" (but updated) climatic classification schemes continue to be used as a common standard, such as the Köppen–Geiger climate classification (Fig. 12.1) and the site–based climate diagrams of Walter and Lieth (Fig. 12.2).

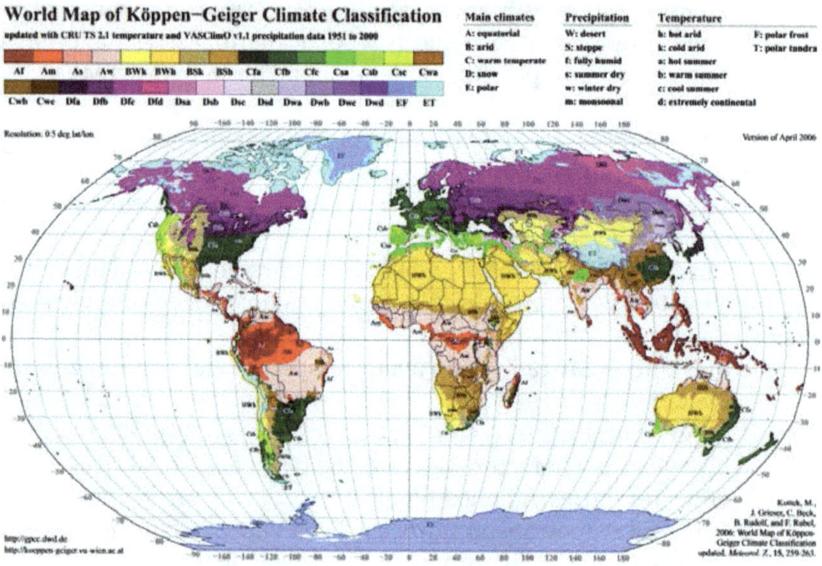

Figure 12.1: Global Köppen–Geiger climate map for 1953–2000 (Source: http://koeppen–geiger.vu–wien.ac.at/ which includes an animated map showing developments under future climate scenarios).

An alternative source for the FAO Köppen–Geiger climate map and related data base can be found at: http://www.fao.org/nr/climpag/globgrids/KC_classification_en.asp

Walter (Walter and Lieth, 1967) diagrams are based on monthly mean temperature and precipitation sums, in which the scaling of 1°C is equal to 2 mm precipitation; they indicate a negative water balance (dry season) when the precipitation line lies below the temperature line.

Diagrams as originally designed can contain a number of additional information, such as annual precipitation or extreme values in various months (for details see http://www.zoolex.org/walter.html). Simplified Walter diagrams from weather stations all over the world can be found online at https://climatecharts.net/. As an example, the dry summer Mediterranean climate of Florence, Italy, is shown in Fig.12.2.

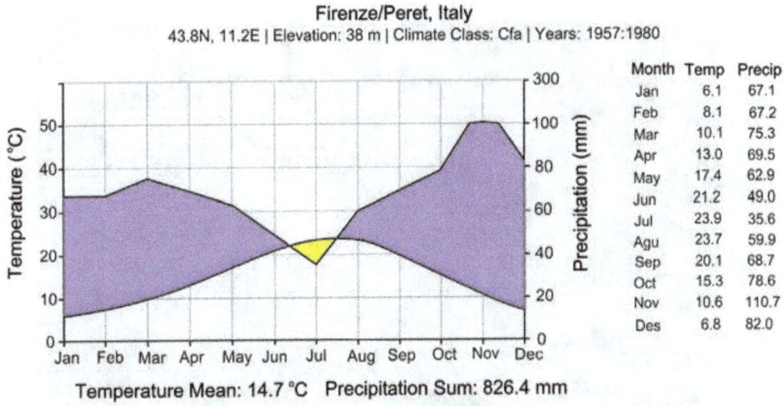

Figure 12.2: Climate diagram (based on Walter and Lieth) of Florence, Italy (modified, based on data from: https://climatecharts.net/).

Even simple climate diagrams such as those of Walter and Lieth can provide valuable information on seasonal climatic characteristics for agriculture, especially crop production conditions.

Other types of climate diagrams may combine other climatic factors, which maybe of interest for agriculture. As an example, Fig. 12.3 shows the combination of mean daily and annual temperatures, including day lengths of two different sites, reflecting distinct growing conditions.

Farmers, however, often need site–specific local climate conditions with high spatial resolution, which may differ considerably from those of distant weather stations or large–scale mapping values because of, for example, orography and land use impact. Wine producers, for example, require small–scale "climatic terroirs". In such cases, additional data must be generated to assess topographical impact on local climatology (Fig. 12.4). Different approaches can be used, such as transect measurements of microclimatic parameters over a limited region: these include remote sensing data and algorithm derivation for such relationships as small–scale air temperature modification due to sea level, slope and the slope orientation of hills.

For example, a related study in the wine–growing region of Carnuntum, Austria, has been carried out to generate maps of terroir characteristics (climate, soil and rock–bed) as major viticulture functions. The study produced various thematic layers and geo–data analyses which describe the geo–environmental properties and variability of the wine growing region and which delimit homogenous multilayer mapping units, using a Geographic Information System. These results have been converted to multilayer web services, which are presented with an interactive web map application (http://www.geologie.ac.at/en/research–development/mapping/substrate–floor/natural–region–carnuntum/).

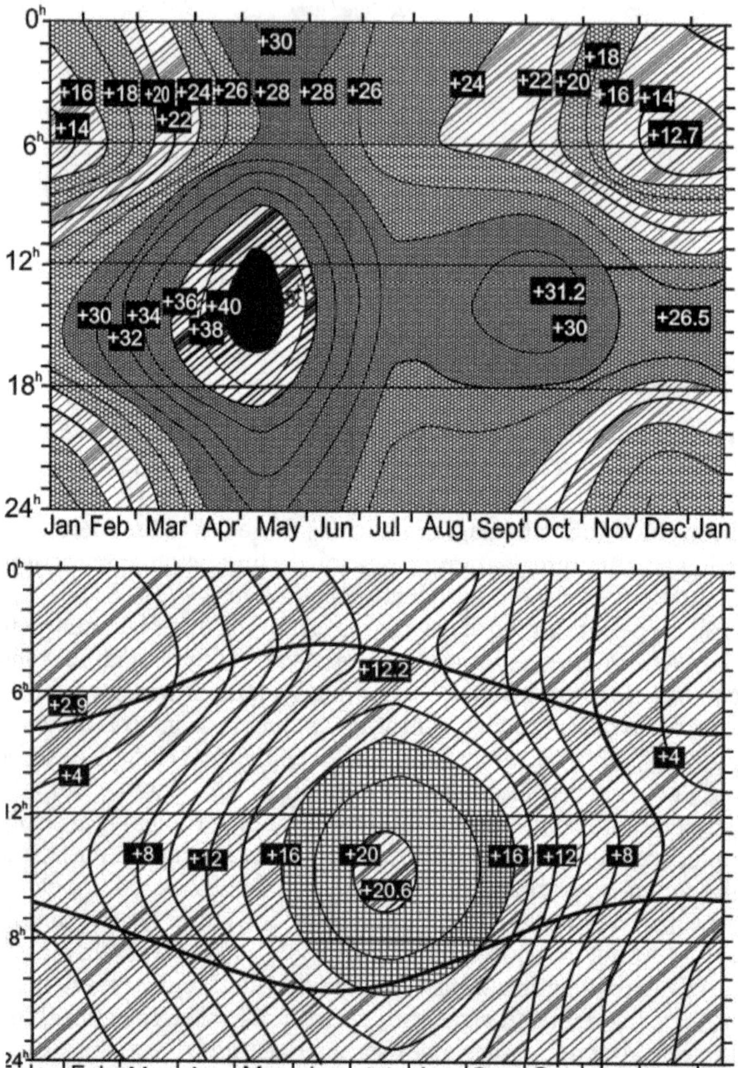

Figure 12.3: Combined daily and annual temperature (in °C) course and day lengths of Nagpur (left, tropical climate) and Oxford (right, Atlantic climate) as important crop growing factors (Troll, 1955).

Sophisticated environmental maps which take an extended number of parameters into account have been published recently; these might be used as a reference for climate change impact studies. For example, a map of environmental stratification over the European area (Metzger *et al.*, 2005) considers, in addition to climate factors (monthly temperatures, precipitation and sunshine duration), geomorphology, geology, soils, vegetation

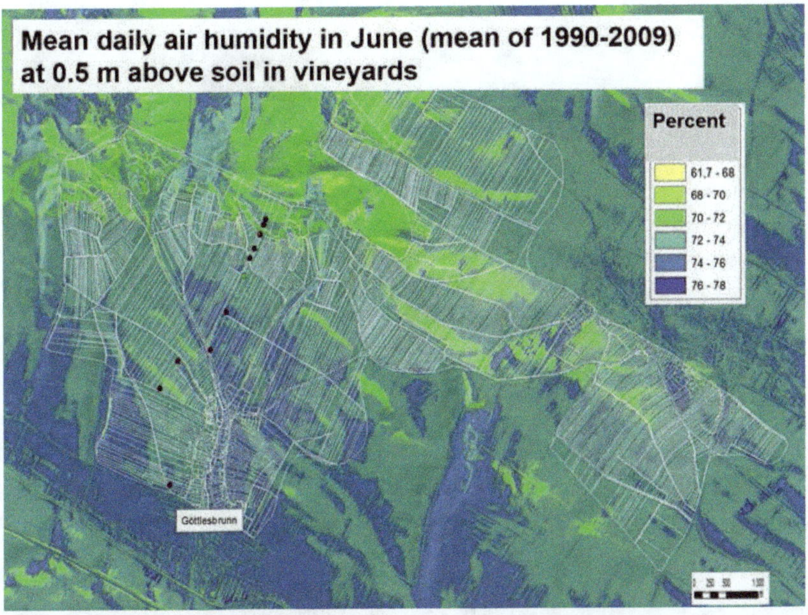

Figure 12.4: Small-scale mean (1980-2005) monthly air humidity (parameterized to microclimatic conditions in the vineyards) over a vineyard region in Austria in June (white lines represent vineyards), (Source: BOKU-Met).

and fauna, assuming these as the main ecosystem components. This map (Fig. 12.5) has been used as a reference in Tab. 12.1 in this chapter as well.

12.3 Climate change

The Earth's climate system is a dynamic system comprising the atmosphere as well as the hydro–, cryo–, bio–, and lithosphere. It is characterized by variations which are not predictable in detail. It is necessary to emphasize that global climate change is always an ongoing natural process caused by natural and, increasingly, human factors. The Earth's climate has exhibited marked variations and changes, which are evident on time scales ranging from millions of years down to one or two years (IPCC, 2013).

In the very long term (e.g. over many thousands of years, the time frame that determines the major ice ages, among other phenomena), the Earth´s bio–geochemical cycles and three orbit factors (e.g., the Milankovic´s cycle) are the main causes of climate change. These natural factors influence in particular the distribution and amount of energy

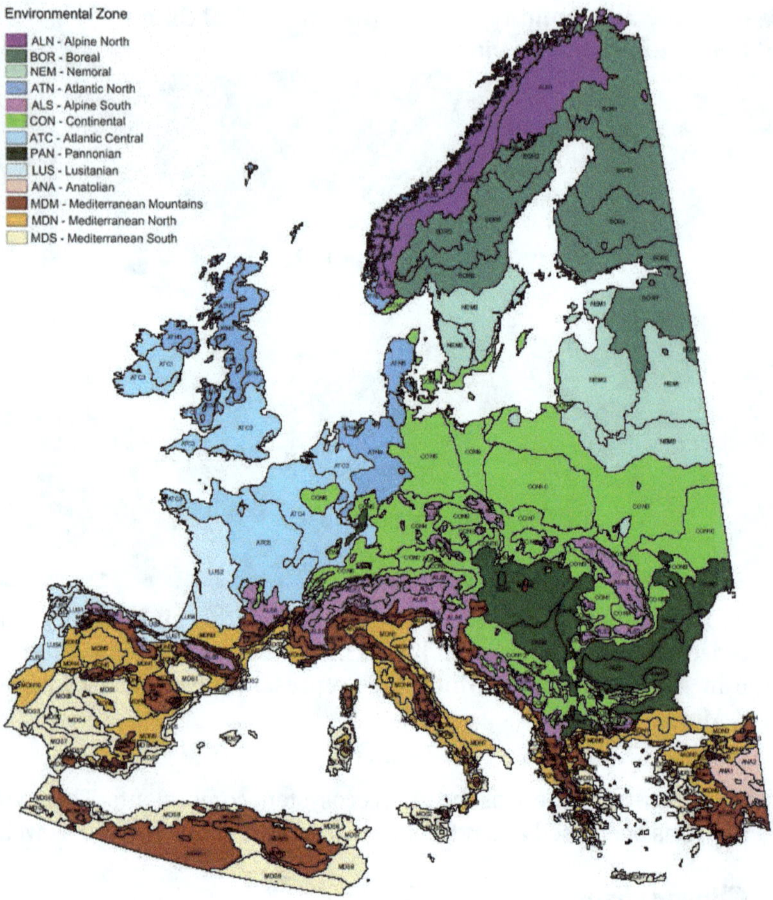

Figure 12.5: Environmental zone stratification over Europe (Metzger *et al.*, 2005).

absorbed by the Earth´s surface, which in turn drive (and change) global atmospheric circulations, oceanic stream flows, the cryosphere, etc., thereby affecting climate variability and change. Over periods of several years, fluctuations in global surface temperatures of a few tenths of a Celsius degree are common.

Some of these fluctuations are related to the Sun's activity, major volcano eruptions, El Niño (Southern Oscillation–ENSO), North Atlantic Oscillation (NAO) and other natural phenomena. Besides anthropogenic factors, solar and volcanic forcings have been the two predominant natural contributors to global climate change during the Industrial Era (APCC, 2014).

An example of natural climate change over the past 500 years is shown in Fig. 12.6, which covers the "small ice age" during the early modern period in Europe (three periods with an increased number of very cold years, called the Spörer, Maunder and Dalton Minimums). Since the beginning of industrialization in the 19th century, a marked additional signal of warming by human activities, mainly the burning of fossil energy, is evident. There is a large consensus, that the annual mean surface warming since the 20th century has reversed the long–term cooling trends of the past 5000 years in mid–to–high latitudes in the Northern Hemisphere, as was also reported by IPCC (2013). This does not mean that humans did not influence climate in earlier times as well, by, for example, deforesting or causing other land use changes, burning and air pollution. However, these changes remained mainly local to regional (e.g., cooling effect on climates in the middle ages due to wide spread deforestations) rather than global.

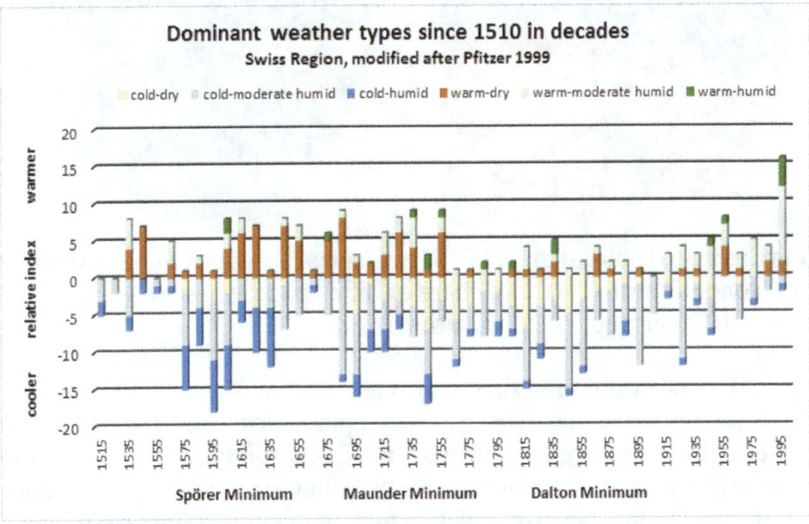

Figure 12.6: Scheme of annual weather conditions by decade from 1505 – 2000, classified as: warm–dry (red), warm–wet (dark green), warm (light green), cold–dry (yellow), cold–wet (dark blue) and cold (light blue); (Pfister 1999, modified from Eitzinger *et al.*, 2009).

The basic problem we face today is that warming, which is not a unique phenomenon in climate history, is occurring at a relatively fast rate, affecting many ecological processes (the so–called ecosystem services). In combination with other human activities, warming could have negative direct or indirect effects on human living conditions on the Earth in the future. Most scientists consider the emissions of carbon dioxide (CO_2) in particular, but also of methane (CH_4), nitrous oxide (N_2O) and other so called greenhouse gases (GHGs), as the main drivers of the current global warming trend. These GHGs (Fig. 12.7b) are emitted into the atmosphere by human activities in different economic sectors (Fig. 12.7a) and may cause irreversible climate change. It is extremely likely that human activities caused more than half of the observed increase in average global surface temperatures from 1951 to 2010. It is an assertion which is supported by sound evidence from multiple studies using different methods (IPCC, 2013).

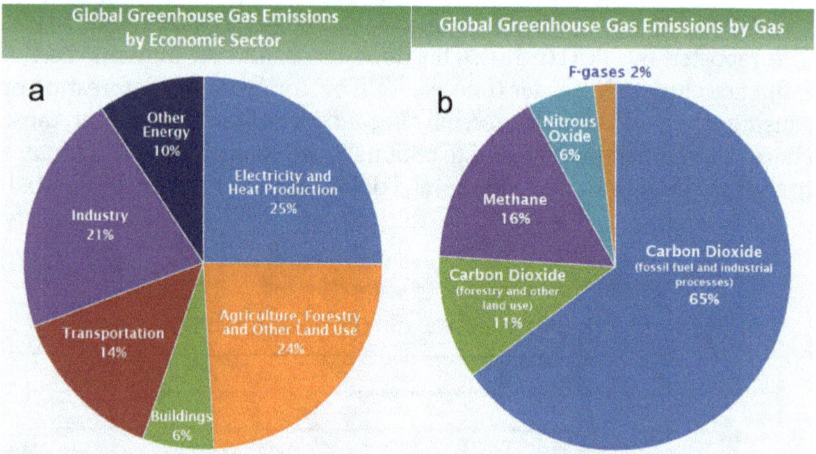

Figure 12.7: Global shares of GHG emissions (based on CO_2–equivalents) by economic sector (a), and share of different greenhouse gases in the total GHG "budget" (b) (IPCC, 2013).

CO_2 concentration in the atmosphere has increased by around 33% since preindustrial times (to about 407 ppm in 2017, and at a rate that is continually increasing), chiefly from fossil fuel combustion and land use change (e.g. aforestation). Increases in greenhouse gas concentrations since 1750 have led to a positive radiative forcing of climate (increased by 2.2 W m^{-2}), which tends to warm the surface and produce changes in climate.

Some European agroecosystems and economic sectors have shown particular sensitivity to recent trends in temperature and (to a lesser extent) precipitation (Alexandrov *et al.*, 2011). Climate change–related phenomena that have been studied include (IPCC, 2013; APCC, 2014; Eitzinger *et al.*, 2009):

- upward altitudinal shift of the tree line;
- phenological changes (earlier onset of spring events and lengthening of the growing season);
- increasing productivity and carbon sink of forests;
- changes in high mountain vegetation types and new occurrence of alpine vegetation on high summits;

- Northern Europe: increased crop stress during hotter, drier summers; increased risk to crops from hail;
- Germany: advance in the beginning of growing season for fruit trees;
- Britain, southern Scandinavia: increased area of silage maize; more favorable conditions due to warmer summer temperatures;
- France: lengthening of growing season for grapevines; changes in wine quality.

Global climate change will impact all economic sectors, but agricultural production is perhaps the most sensitive and vulnerable, as climate is the primary determinant of agricultural productivity. World agriculture, whether in developing or developed countries, remains very dependent on climate resources. Although agriculture in Europe only accounts for a small portion of GDP (Gross Domestic Product), such that the vulnerability of the overall economy to changes that affect agriculture is relatively low, local effects on society might be great. There is no doubt that the question of global and regional climate variability and change as well as the related impact on ecosystems is a major and important environmental issue facing the world at the beginning of the 21st century (e.g. IPCC, 2013).

The future development of climate can be calculated for a given evolution of radiative forcing or of its main causes (emissions and land–use changes) with Earth system models. These are extensions of classic climate models simulating the atmosphere and oceans with predictions of biogeochemical processes. Results of such computations, carried out with different models by various research groups, are presented in the 5th IPCC report (APCC, 2014; IPCC, 2013). Besides using different models, researchers modified initial conditions especially for the ocean in order to provide an ensemble of projections allowing for an estimate of uncertainties.

Four different "emission scenarios", related to socio–economic development scenarios, were outlined in the 5th assessment report (*Representative Concentration Pathways, RCPs*), which indicate trends similar to those presented in the 4th assessment report (called SRES – Special Report on Emission Scenarios), but with increasing differences beyond 2040 (Fig. 12.8).

Only for the lowest of these pathways global warming would most likely remain within the politically defined objective of 2 °C maximum warming. This pathway supposes a rapid and massive reduction of greenhouse gas emissions. The highest pathway leads to a temperature rise of about 4.5 °C (global mean) from the second half of the 19th century to 2100 (with further increases thereafter). Until about 2040, the differences between the various paths are smaller than the bandwidth of the

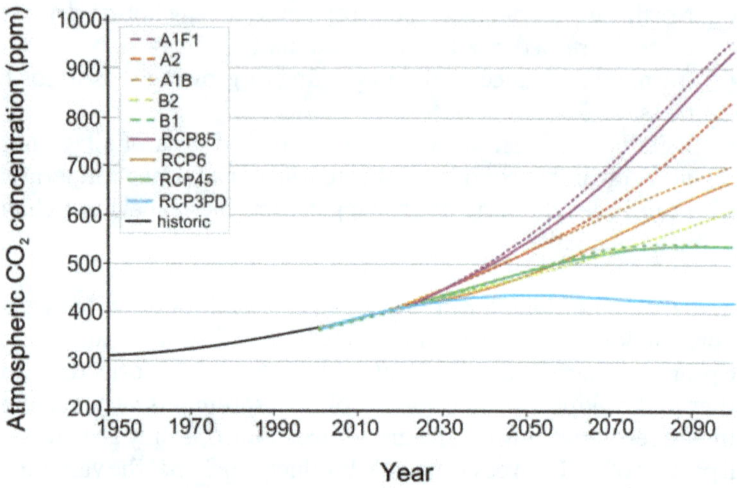

Figure 12.8: Atmospheric CO_2 concentration of the emission pathways of the past SRES (4th IPCC report) and the current RCPs of the 5th assessment report (modified, Kotlarski and Truhetz, 2016).

different scenarios. Warming differs regionally and globally, especially between land and ocean surfaces (Fig. 12.9).

Precipitation changes are inhomogeneous as well: moist regions and seasons will probably receive more precipitation and dry ones less (Fig. 12.10). Dry regions may spread pole wards. Ocean warming and melting of inland ice will lead to sea–level rises expected to range between 0.25 m and 1.0 m in 2100. Moreover, sea levels are predicted to rise over the next centuries in any case because of the inertia of oceans and ice shields.

Based on these emission scenarios, climate predictions (available as numerical data) were generated from the outputs of a number (an ensemble) of Earth system models. However, only for temperature, and to a lesser extent for precipitation, changes are considered reliable enough for further use in any kind of impact study and impact model (for example, for crop models to predict future crop growing conditions).

Uncertainties of simulated results from global climate models are significant on a local scale. Global climate models represent the Earth´s surface on large grids (e.g.,>100 km) and are often not able to catch important feedback from weakly represented topography or land use. As a result, statistical downscaling models and regional climate models of higher spatial resolution have been developed to reduce uncertainties of future predictions for smaller regions. This is especially important for better detecting the development of extreme weather events. In the future, however, the spatial resolution of Earth system models will also be improved with increasing computation capacities, down to, for example, a 10 km grid.

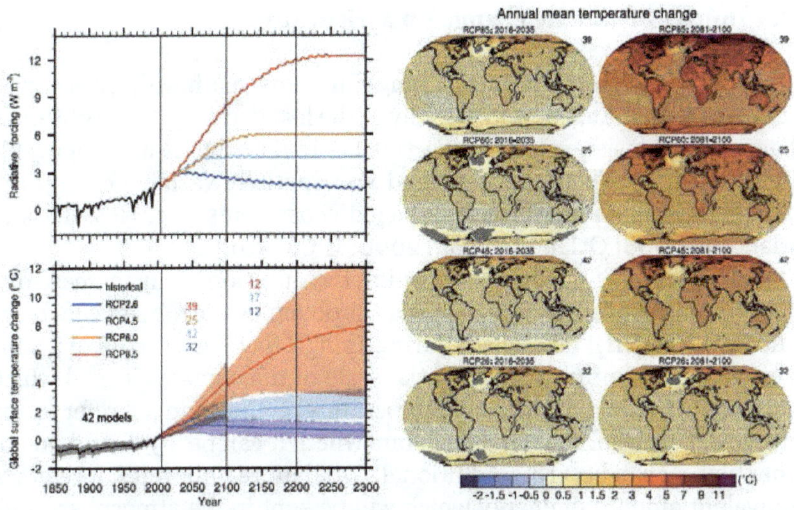

Figure 12.9: Radiative forcings of emission scenarios (top left) and related simulated mean global surface temperature trends towards 2100 from Earth system model ensembles (bottom left: shaded area with range of model responses). Distribution of simulated ensemble mean surface temperature changes over the Earth for two periods and four different RCPs (right). (IPCC, 2013).

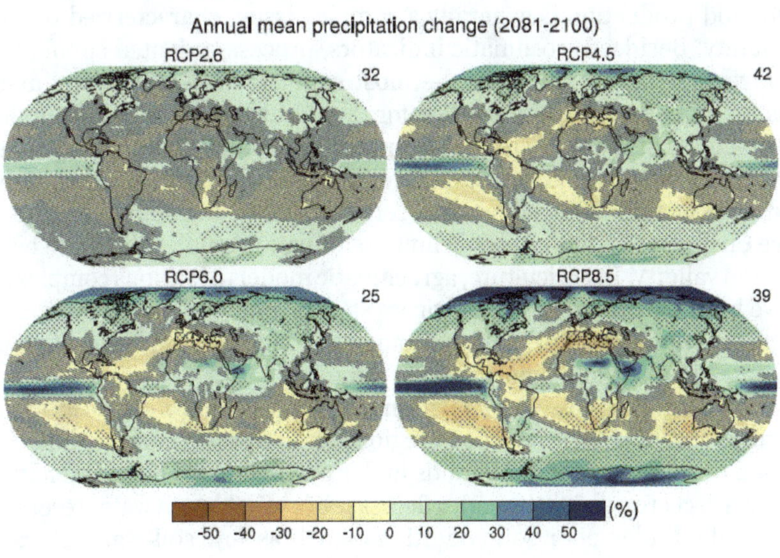

Figure 12.10: Distribution of simulated ensemble mean surface precipitation changes over the Earth for 2081-2100 and four different RCPs (right). (IPCC, 2013).

12.4 Impact of climate change on agriculture

Many types of impact from ongoing climate change have been observed and reported for Europe and globally in the last decades. These observed changes are consistent with projections of impact due to anthropogenic climate change. The warming trend and spatially variable changes in rainfall have already affected managed ecosystems (e.g. Nejedlik and Orlandini, 2008; Orlandini *et al.*, 2008; IPCC, 2013).

Often from the public view of point, the question "Why increase of 3 °C in air temperature over 30 years is a problem, while the same increase is not a problem if it takes place during a single day?" is raised. This, however, is related to the applied spatial scale of the climate Normal which indicates that global increase of temperature, actually do not refer to any particular location. It means that somewhere it can be +5 °C and on another place it can be +1 °C. Additionally, and more important, it refers to equivalent amount of energy which will be kept in the atmosphere and, potentially, spent on different processes (increased ice melting and water bodies evaporation, increase temperature of ocean etc.) with high spatial variability, affecting whole climate system. Finally, also small changes of temperature over a certain period of time can affect living organisms strongly due to cumulative effects, such as accumulated stress days or temperature sums driving phenology.

Indeed, the potential impact of future climate change on agricultural and food production is generally a complex issue, characterized by uncertainty. Besides agroclimatic indicators, process–oriented simulation models (see Chapter 12) are the most sophisticated tools for climate change impact studies and are therefore mainly used to assess the complex interactions of the soil–crop–atmosphere system at different scales. Yet even process–oriented models still represent only a simplification of the different factors involved, as they rely on defined boundary conditions (see Chapter 11). The results of climate simulation studies therefore have limited validity. In agriculture, agroclimatic models of various complexity have been used mainly to investigate specific phenomena, such as crop yield or yield risks, under defined boundary conditions, including various climate, land use and management scenarios.

In general, results from such impact studies indicate that in many Central and Southern European regions an increasing number and duration of summer drought periods under most climate change scenarios could affect crop yields through a decrease of available soil water reserves, especially under poor soil conditions (such as low soil water storage capacity). On the other hand, negative yield impact through higher temperatures and shortening of the growing period in many locations may be partly offset by the effect of increasing CO_2 levels in the atmosphere via an increase of the photosynthesis rate and higher water use efficiency.

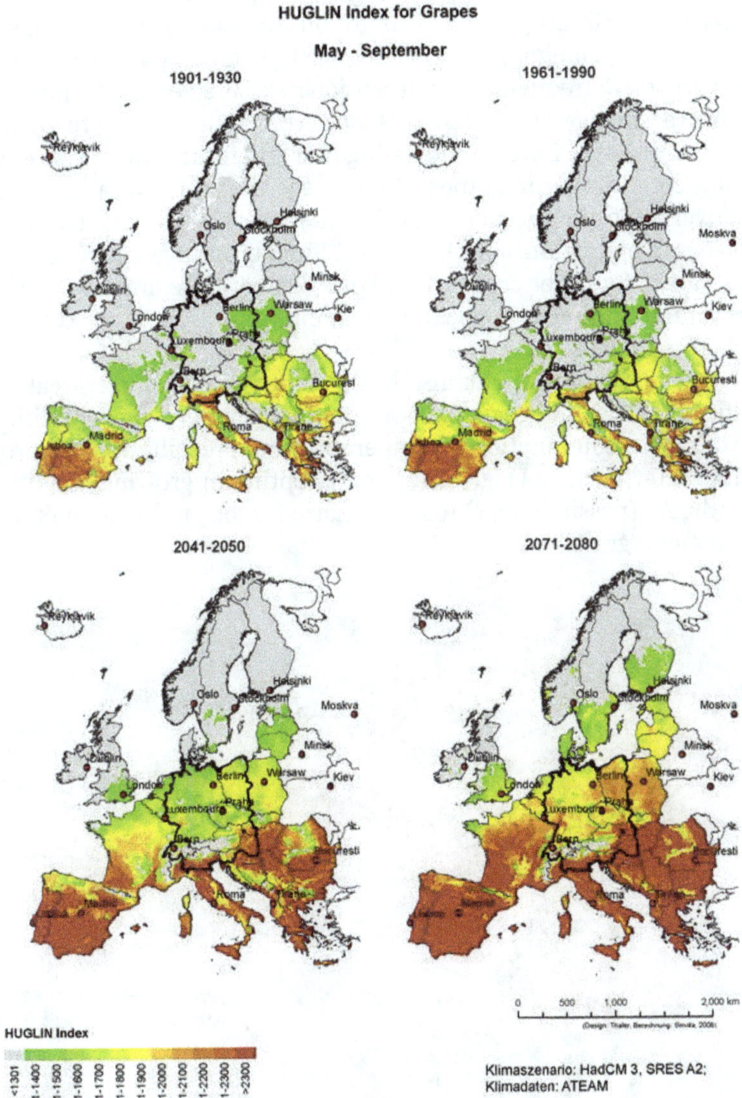

Figure 12.11: Huglin Index (for grapes) mapped over Europe for different past and future climatic periods (green: low temperature sum; red: high temperature sum); (Eitzinger *et al.*, 2009).

In regions with increased precipitation or no water limitation for crops, higher temperatures may boost the production potential for many crops (e.g. by longer vegetation periods for permanent crops or by removing temperature–limiting growing conditions).

Agroclimatic indicators (see Chapter 11) provide insights into the changes in specific crop growing conditions or abiotic/biotic risks, such as the potential change in growing degree days, the impact on crop phenology, frost risk frequency or soil erosion risk. As an example, Fig. 12.11 presents the change of the HUGLIN index (based on temperature sums; see Chapter 8) over Europe, indicating past and future climatic summer growing conditions for grapes. We can see that the overall warming trend over the future decades will shift the potential climatic suitability for grapes to higher latitudes in Europe. Similar findings are reported in the literature for other crops with higher temperature demand, such as silage and grain maize, soybean and sunflower, based on process–oriented crop models.

Agroclimatic indices can also be combined to describe critical crop specific climatic limitations for optimum seasonal growing conditions. For example, a combination of temperature and precipitation in a crop–specific water balance (Fig. 12.12) shows optimum growing conditions under the A2 (pessimistic) climatic scenario for the production potential of permanent grasslands.

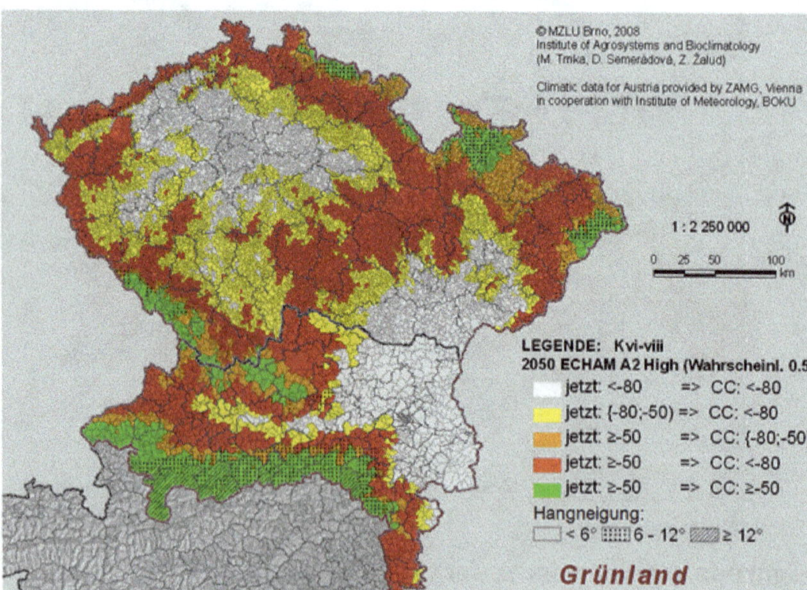

Figure 12.12: Indicator–based change of water balance in permanent grassland growing period due to climate change (A2 scenario 2050s) in Austria and the Czech Republic. Green indicates no change or improving conditions, while other colors express increasing negative water balance during summer, leading to decreasing yield potential (Eitzinger *et al.*, 2009).

Similarly, Figure 12.13 presents a drought indicator (actual vs. potential grass evapotranspiration) of a hypothetical grass surface under changing climate on a larger, European–wide scale. Across the continent, the drying/wetting trends recorded in various regions over the past century are clearly shown, together with predictions of their future climates. Southern European (Mediterranean) regions in particular will become significantly drier.

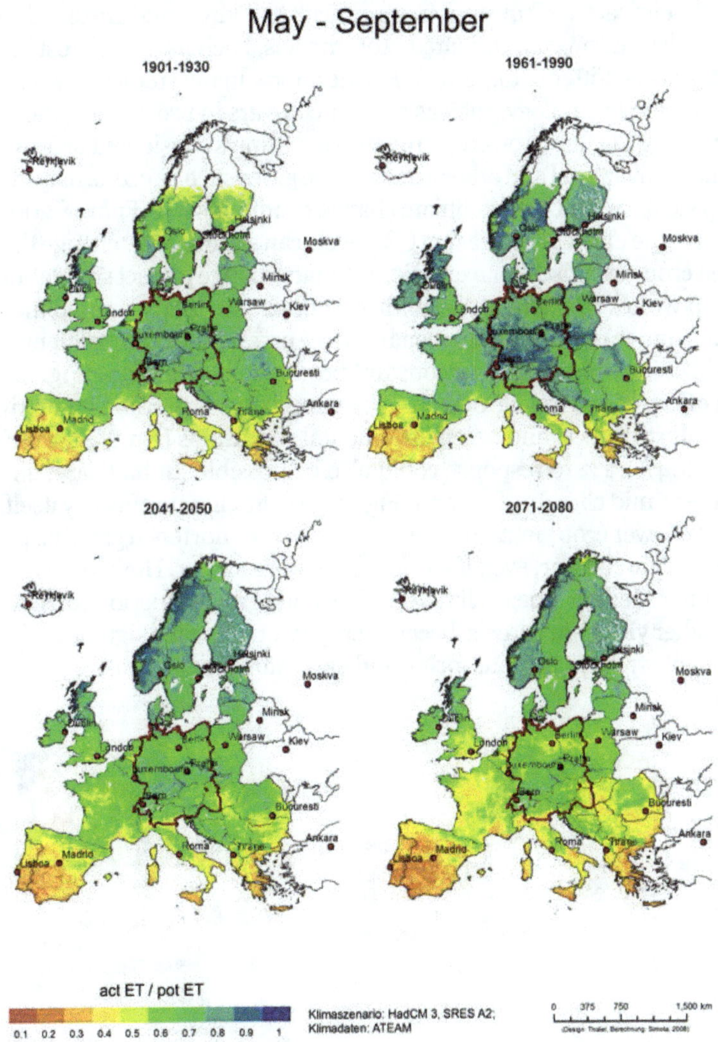

Figure 12.13: Relationship of actual to potential evapotranspiration of grass from May–September as a drought indicator (0.1 – very dry, 1.0 – wet) over different past and future (scenario–based) climate periods across Europe (Eitzinger *et al.*, 2009).

The impact of climate change on crop production in Europe has been investigated in many studies. Using process–oriented crop models, these studies have been conducted on different spatial scales and for different climate and management scenarios (White et al, 2011; Olesen *et al.*, 2011). Most of these crop simulation studies mainly considered climate effects, including extreme weather impact, such as the consequences of drought and heat, which can be predicted by process–oriented crop models. Several potential forms of extreme weather impact are, however, more difficult to address and are therefore often not or less in process–oriented crop models, such as cold ("frost"), hail or flood risks (Rötter *et al.*, 2018). Furthermore, many studies rely on current production technology, such as current cultivars in use. Finally, the effect of increasing CO_2 levels on crop growth is not always considered or is uncertain at the cultivar level and other atmospheric gases such as ozone may influence crop yields as well under climate change conditions (e.g. Fuhrer, 2003).

Yet the effect of increasing CO_2 levels can significantly change the simulated crop response. For example, a climate change impact simulation studies on winter wheat (e.g. Alexandrov *et al.*, 2002, 2011) showed increasing yields under future climate scenarios in Austria due to the influence of CO_2 fertilization only, assuming constant production technology (Fig. 12.14). As the effect of increasing CO_2 levels is uncertain–depending on cultivar, on long–term effects and on environmental conditions (e.g. Manderscheid *et al.*, 2014) – a zero response could also be possible. In that case, under the semi–humid climate of Austria (Fig. 12.14), the climate effect by itself would lead to lower crop yields, mainly because of the shortened growing period of the specific cultivar, even if sowing dates were earlier. However, in the drier summer Mediterranean climate, a shortened growing period could also lead to higher yields for cereals, because crops would escape significant heat and drought stress in the late spring and early summer months.

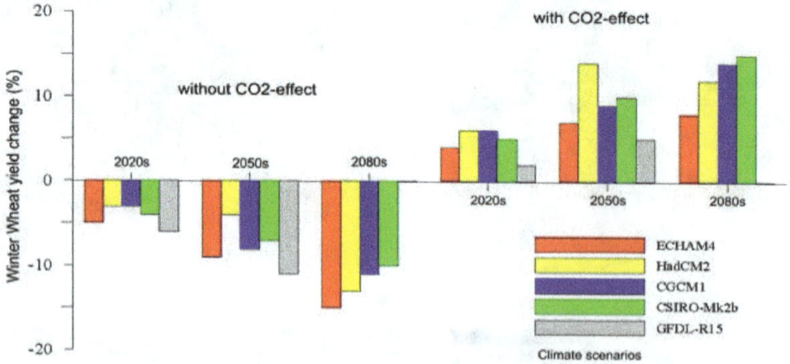

Figure 12.14: Simulated impact of climate change scenarios on winter wheat yield in semi–humid eastern Austria with and without effect of increasing CO2 fertilization, with constant production technology and adapted sowing dates (Source: BOKU–Met).

In Europe, the effect of climate change on water balance parameters and water stress of crops has been investigated in several studies. For example, winter wheat production in northeastern Austria (e.g. Thaler *et al.*, 2012) has been studied under Central European conditions using the CERES–Wheat model, including the fertilization effect of increasing CO_2 levels. Results show that despite higher yield levels, crop transpiration dropped compared to current conditions through the simulated increase in water–use efficiency and reduced total potential evapotranspiration (related to the shortened growing period) under the applied $2xCO_2$ climate scenarios. In the end, this pointed to an outcome of less water stress for the crop.

Sensitivity analyses show that soil water storage capacity can have a strong impact on the yield potential under climate change, especially at sites where water is a limiting factor during the growing period, such as in northeastern Austria. Figure 12.15 shows simulation results for winter wheat for the same region, indicating the range of potential yield reactions by considering different climate scenarios, soil types and the direct CO_2 effect. It shows that negative yield trends are simulated mainly in soils with low soil water storage capacity, but also that such trends would increase considerably especially with a lower CO_2 effect.

Similar findings of mainly positive yield effects (except in sandy soils and dry regions) have been reported by other studies on cereals in Europe, mainly due to the simulated effect of enhanced atmospheric carbon dioxide levels (e.g. Kersebaum and Nendel, 2014). Similar results have also been obtained for non–cereal crops, such as potato.

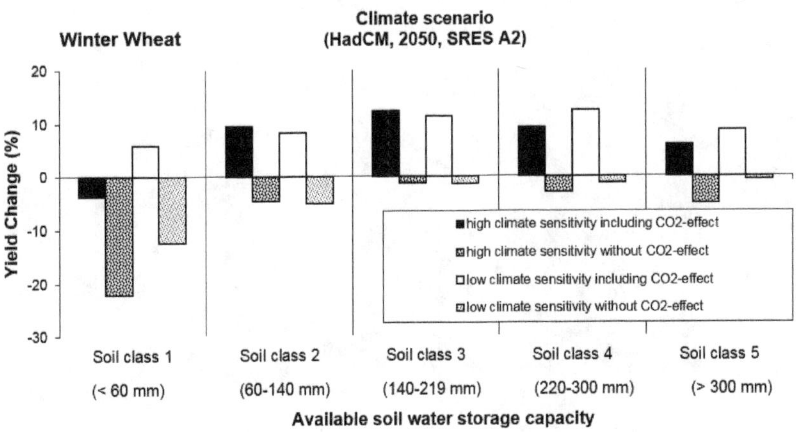

Figure 12.15: Simulated winter wheat yield response under various climate scenarios, CO_2 fertilization effects and soil types of different available soil water storage capacity in north–eastern Austrian region Marchfeld (Source: BOKU–Met).

Compared to that of C3 crops, the response of C4 crops, such as maize, to further increasing CO_2 levels (fertilization effect) will be weak. Most studies on maize therefore indicate increased yield potential in cool and/or humid regions due to higher temperatures (e.g. Northern Europe, the Alpine region) and decreased production potential in most Southern European regions, mainly because of increasing summer drought and heat stress conditions. However, experimental results for maize indicate, that the beneficial effect of CO_2 through an improved water use efficiency is still not sufficiently considered by most crop models (Durand *et al.* 2017). Significant negative yield effects for several other crops and additional water demand for irrigation might be expected in Southern Europe in general or in regions with low soil water availability (Lobell and Field, 2007).

As many simulation studies show, rain–fed summer crops are in general at a higher yield risk because of increasing summer droughts and heat stress under most climate scenarios. They will be more vulnerable and dependent on soil water reserves, as the soil water or higher ground water tables during the winter period cannot be utilized as much as by winter crops. Evapotranspiration during summer due to higher temperatures will increase significantly. Many studies further conclude that if the frequency and duration of droughts increase, as recent research indicates (Lobell *et al.*, 2009), or if soil– and groundwater reserves decrease (e.g. by decreased summer river flow from the Alpine region), water shortages during summer will become more common, which could also limit water availability for crop irrigation regionally.

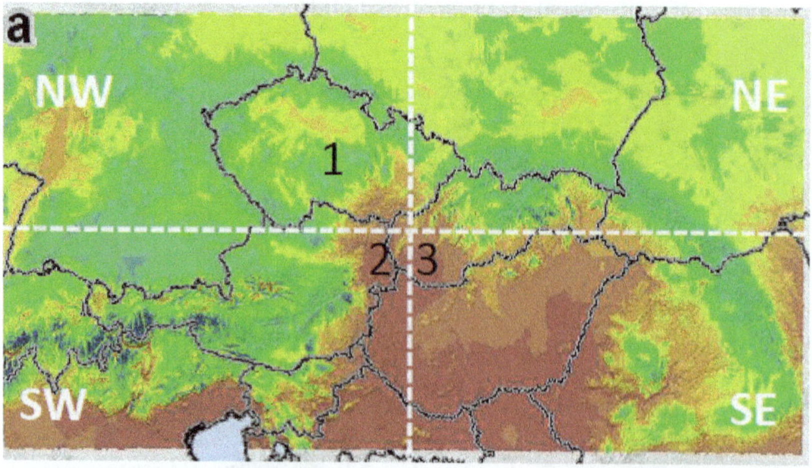

Figure 12.16 Map of mean annual effective global radiation for1961-1990 (a - sectors of the considered simulation domain; NW - northwest; NE - northeast; SW - southwest; SE - southeast; 1 - Czech Republic; 2 - Austria(Marchfeld), 3 - Slovakia (Danubian and Záhorie lowlands) (Eitzinger *et al.*, 2013).

A summary of climate change impact based on simulation studies of different modeling approaches was produced for Central Europe in a past study (Eitzinger *et al.*, 2013). Figure 12.16 presents the reference for effective global radiation as a basic primary production factor, while Tab. 12.1 presents simulated changes under climate change scenarios of this and several other agrometeorological indicators, as well as crop yields.

Crop response factor	CC scenario and time horizon	Sector of the domain[a]	Crops affected	Trend (+/o/-)	Comments
Effective global radiation	All 2050	Northwest	All	++	Especially Pannonian lowlands affected negatively.
		Northeast		+	
		Southwest		-	
		Southeast		-	
Drought/ water stress		Northwest	All	+	Water deficit during summer increases over the domain.
		Northeast		-	
		Southwest		+	
		Southeast		-	
Huglin index		All	Grapes	+	Improved growing conditions.
Winter conditions		All	Winter crops and perennials	+	Improved winter conditons. Little change for vernalization conditions and late frost risk. Increased soil erosion risk.
Spring conditions		Southeast	All crops	+	Spring conditons improve or decrease depending on region. Fall and harvest conditions will mostly improve over the domain.
		Northwest		-	
		Northeast		++	
		Southwest		-	
Nitrate leaching (crop model)	All 2035	Austria – Marchfeld	Winter wheat	++	Especially on sandy soils.
			Spring barley	+	Especially on sandy soils.
Pest pressure –Corn borer	All 2050	Northwest	Maize	--	More infestation of maize due to recent presence of pest in still nonestablished areas. Additionally, increase in number of generations in regions with long–term presence of pest.
		Northeast		--	Similar to northwest region.
		Southwest		-	Modest growth of number of generations.
		Southeast		-	Similar to southwest region.
		Northwest	Potato, tomato	--	Increase in number of generations.

		Northeast		--	Similar to northwest.
Pest pressure –Colorado beetle	All 2050	Southwest		+	Croatia and northern Italy – recession of pest as reaction to high temperature stress.
		Southeast		+	Serbia, Hungary similar to southwest area.
Crop yield (indirect effect)	SRES–A2 2050 high ECHAM	Czech Republic – arable land	Winter barley, winter rye, winter rape	-	Lowlands mostly affected negatively.
	SRES–A2 2050 high HadCM			--	
	SRES–A2 2050 high NCAR			-	
	All 2050		Spring barley, spring wheat, oat	--	Especially within drought–prone regions (e.g. southern Moravia).
	All 2050	Czech Republic – arable land (1)[b]	Winter barley, winter rye, winter rape	++	Especially regions within higher altitudes with quality soils will be affected positively.
			Spring barley, spring wheat, oat	+	Steady positive effect through all included altitudes.
Crop yield (combined effects assessed by crop models)	All 2035	Austria–Marchfeld (2)[b]	Winter wheat	+	Soil–type dependent (most enhanced yields on medium soils); additional water demand 30–40 mm.
			Spring barley	o	Most limited in sandy soils; additional water demand.
	SRES A1B ARPEGE 2021–2051 2071–2100	Slovakia – Danubian and Záhorie lowlands (3)[b]	Winter wheat	+	Soil–type dependent (most stable yields in calcaric chernozems).
			Spring barley	+	Soil–type dependent (most stable yields in calcaric chernozems);higher yield variability in 2010–2051.
			Grain maize	-	Drought periods during growing season reduced yield on all evaluated soils; higher yield variability in 2010–2051.

[a] = sectors of the considered simulation domainin this Table refer to those of Fig. 12.16.
[b] = refers to the crop modelling case study regions as indicated in Fig. 12.16.

Table 12.1: Summary of climate change impact on crop production in Central Europe, derived from agroclimatic indicator, pest model and crop simulation studies (Eitzinger *et al.*, 2013).

12.5 Addressing uncertainties in climate change impact studies on agriculture

Often it is not feasible to consider all potential important environmental factors, types of crop management or forms of socio–economic feedback in impact simulation approaches. A lack of data, methods and information can indeed represent sources of uncertainty for assessments of future climate change impact. Such sources can be detected at all scales of model application, including uncertainties based on: a) the model representation of involved processes and b) model inputs. Examples of the first type at the crop production level is the potential effect, mentioned above, of increasing atmospheric carbon dioxide concentration on crop growth processes and yield, or the ambiguous representation of root growth in crop models. Uncertainties in crop model inputs, meanwhile, maybe due to measurement errors or inexact inputs (especially those based on other model outputs, such as climate change scenarios from global climate models).

Model application, then, is always a compromise between model simplification and input data demand: on one hand, the uncertainty of simulation results increases with the growing simplification of simulated processes; on the other, in more complex models inaccuracy of simulated results increases because of a greater number of input parameters, which are not always available or contain a high degree of uncertainty in the data itself.

Many uncertainties are in general related to the scaling problem. Significant differences may result between model inputs according to their regionalization at the field level of the individual farmer. For example, soil input data from a low resolution soil map or weather input data from distant stations or from climate models may not be accurate when used for a field–level simulation. Another problem is that impact simulation models only consider and use changes in mean weather elements (e.g. of temperature and precipitation) from climate scenarios, often neglecting potential changes in climate variability. In addition, many extreme weather events–such as hail–that can have additional negative impact on crop yields are not directly represented by modeled climate scenarios. Regional climate scenarios can differ considerably from Global Climate Models (GCM´s) on a regional basis and can represent local conditions much better (e.g., seasonal variations of temperature and precipitation); at the same time, they often contain a higher degree of uncertainty. Downscaled methods should therefore be used for regional crop yield simulations when available, but only if they are reliable. Bias correction of variables might be necessary to reflect historical weather series. However, bias correction has to consider the inter–relations between weather variables.

12.6 Expert assessments as an additional information source

In order to reduce the above mentioned potential uncertainties and shortcomings from simulation study results, complementary information sources should be used. The first option is to combine results from different modeling approaches (such as the agroclimatic models/indices, pest and disease models and process–oriented crop models presented in Tab. 12.1) for impact assessment. A second, additional option is to gather expert assessments for a specific target area, in order to assure or improve the representation or limitations of simulated results. For example, a European–level expert assessment was carried out in the COST 734 action, addressing selected cropping risks by assessments from local experts (Fig. 12.17).

a	Growth Duration	Overwintering	Frost	Suitable harv.	Seasonal variability	Drought	Heat stress	Hail	Pest& Diseases	Weeds	Soil erosion	Nitrogen losses
ALN												
BOR	1.0											
NEM	0.0		1.0	0.5	1.0	1.0	0.0	0.5	1.0	1.0	1.0	1.5
ATN	1.3		1.0	0.0	0.5	0.3	0.0	1.0	1.5	0.0	1.0	1.0
ALS	0.0		0.0	1.0	1.0	1.7	1.3	1.0	1.8	0.5	1.0	0.5
CON	-0.9		-0.8	0.4	0.7	1.2	0.6	0.3	1.3	0.6	1.0	0.6
ATC	0.3		-1.0	1.0	0.7	0.7	0.3	0.7	0.7	0.7	1.3	0.3
PAN	0.0		0.0	0.4	1.7	2.5	1.5	1.0	1.0	1.5	0.8	0.5
LUS	2.0		-1.0	1.0	1.0	2.0	2.0	0.0	0.0	1.0	1.0	0.0
MDM	-1.0			0.0	0.0	-1.0	-1.0	-2.0	0.0	0.0	0.0	1.0
MDN	-0.8		-0.7	-0.3	0.3	1.3	1.0	-0.1	0.5	0.8	0.8	0.8
MDS	-1.0		-0.3	0.3	0.0	0.3	0.3	0.3	0.5	0.5	0.3	0.3

b	Growth Duration	Overwintering	Frost	Suitable harv.	Seasonal variability	Drought	Heat stress	Hail	Pest& Diseases	Weeds	Soil erosion	Nitrogen losses
ALN	2.0	2.0	0.0	-1.0	2.0	0.0	1.0	0.0	1.0	1.0	0.0	1.0
BOR	2.0	0.5	0.0	0.3	1.6	0.5	0.5	0.0	0.5	0.5	0.5	0.8
NEM	1.3	1.0	-0.3	0.5	1.0	1.0	0.5	0.3	0.8	0.0	0.0	0.3
ATN	1.6	-0.6	0.0	0.5	0.8	1.0	0.4	0.0	0.3	1.0	0.3	0.5
ALS	1.1	0.0	0.0	1.1	1.3	1.8	1.3	0.5	1.0	0.3	0.5	0.7
CON	0.7	-0.4	-0.1	0.6	0.6	1.2	1.0	0.0	0.5	0.6	0.3	0.4
ATC	0.8	0.0	-0.3	0.0	1.0	0.5	-0.3	0.0	-0.3	0.3	0.0	0.3
PAN	0.7	0.0	0.0	0.3	1.5	1.3	1.0	0.5	0.7	0.0	0.0	0.3
LUS	1.0		0.0	1.0	1.0	1.0	0.0	0.0	0.0	0.0	0.0	0.0
MDM	0.0	0.0	0.0	0.5	0.5	1.0	0.0	0.0	0.0	0.0	0.0	0.0
MDN	0.5	-0.3	0.0	0.3	0.8	0.8	0.0	0.3	0.0	0.5	0.5	0.5
MDS	0.5	0.0	0.0	0.3	0.3	0.8	0.0	0.0	-0.3	0.2	-0.3	0.0

c	Growth Duration	Overwintering	Frost	Suitable harv.	Seasonal variability	Drought	Heat stress	Hail	Pest& Diseases	Weeds	Soil erosion	Nitrogen losses
ALN												
BOR												
NEM												
ATN	1.5	-1.0	-1.0	1.0	1.0	0.5	0.0	1.0	1.5	0.0	1.0	1.0
ALS	0.0	-0.5	0.0	0.0	1.0	2.0	2.0	1.0	1.5	1.0	0.0	0.0
CON	-0.2	-0.8	-0.3	-0.2	1.3	0.6	0.4	0.5	1.0	0.3	1.2	1.0
ATC	0.3	0.0	-0.3	0.3	1.0	0.7	0.3	1.0	1.0	0.5	1.0	0.5
PAN	0.3	-0.7	0.0	0.4	1.7	0.4	0.5	1.0	0.8	1.0	0.8	0.5
LUS	1.0		1.0	1.0	1.0	0.0	0.0	1.0	0.0	1.0	1.0	0.0
MDM	-1.0	0.0	0.0	-1.0	0.0	-1.0	-1.0	-2.0	-2.0	0.0	0.0	1.0
MDN	-1.0	-0.3	0.3	-0.3	0.3	1.0	0.8	0.3	0.0	0.8	0.8	0.8
MDS	-0.3	0.3	0.8	0.5	0.3	0.6	0.4	0.8	-0.3	-0.1	0.5	0.3

Figure 12.17: Expected impact of climate change on a range of crop production limiting factors for three selected crops over main European environmental zones (see Fig. 12.5): a) grain maize; b) grassland; c) grapevine. The scale used for colour–coding reflects the degree of positive (green) or negative effects (red), while gray indicates areas without data (Nejedlik and Orlandini, 2008).

As can be seen in the expert assessment above, the overall main cropping risks under climate change in all environmental zones for these three crop types are seasonal changes (climate variability), drought, heat, pest and disease risks, soil erosion and hail; at the same time, different regions present distinct risks.

Predominantly positive trends are seen in overwintering conditions, frost risk, growth duration and harvest condi-tions. The agreement of such expert assessments, which are mainly based on observations of ongoing regional trends, with simulation study results will render the latter more reliable. These assessments can also be used to fill gaps in simulation studies for risks that are not considered or to help to find reasons for regional biases. Finally, impact assessments are the basis for developing adaptation options and measures in agriculture in order to reduce vulnerability to ongoing climate changes and to reduce production risks, as will be outlined in the next section.

12.7 Adaptation of agriculture to climate change

As the most important production factors in agriculture, weather and climatic conditions largely determine farm management options. Famers try to adapt to these conditions in order to maintain maximum productivity and resource use efficiency while reducing production risks.

Environmentally–friendly agricultural systems should aim to keep food production sustainable at both the farm and the regional level as well as to protect biodiversity and other ecosystem services. Farming methods and technologies–whether newly generated, already known or developed over many centuries–offer many opportunities for adapting to existing climatic and weather conditions in crop production. Farm technologies optimized for the relevant socio–economic framework influence the adaptation process in both the short and the long term.

Because of ongoing climate change, the optimization of farming methods and technologies becomes even more important for ensuring sustainable productivity of various agricultural production systems at different farm input levels. The use of available farm technologies should secure sustainable production within given climatic and weather conditions through the proper management of natural resources or conditions fitted to specific farming systems: water, soil (including nutrients), crops and microclimatic conditions (Fig. 12.18).

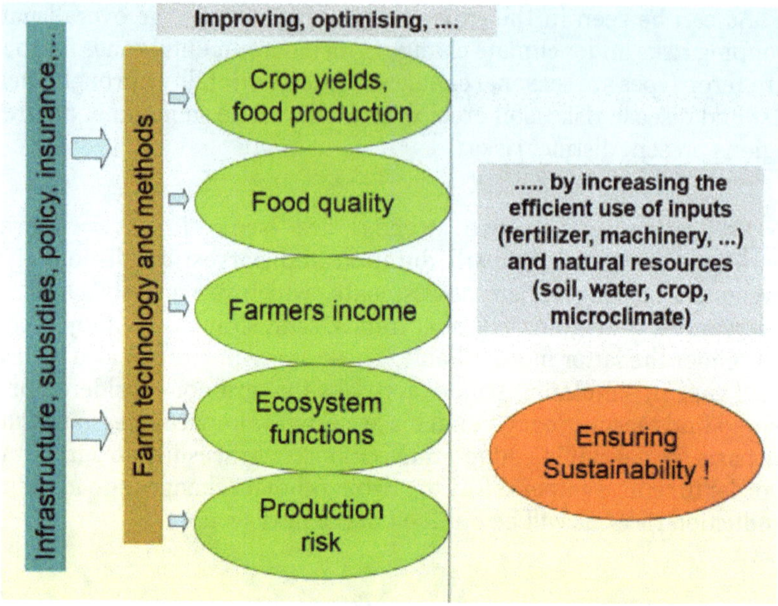

Figure 12.18: Farmers' interests and tasks in the context of farm optimization, adaptation and influencing framework conditions (Eitzinger *et al.*, 2009).

Assessing the effects of adaptation options in agriculture under future climate scenarios is even more complex because of the uncertainty of human and other factors. However, some simple and effective adaptation measures can be easily implemented. Production potential could be maintained under a warmer climate by changing to crops or cultivars with higher optimum growth temperatures or by sowing spring crops at an earlier time (Fig. 12.19a). The introduction of new cultivars, resistant to climate change related stresses (drought and heat stress, ozone, pests, diseases, etc.) is another option. For example, under warmer temperatures farmers could breed or use cereal cultivars with a longer corn filling period (Fig. 12.19b) or change to maize cultivars with a higher growing degree day demand.

Future options include not only measures regarding crop management, but also long–term adaptation of land use and farming systems. Additionally, the rural socio–economic conditions and other boundary conditions beyond the farm level, such as infrastructure, available production technologies, insurance options and agricultural policy, play a significant role for feasible adaptation options. Several studies indeed indicate that suitable management and socio–economic conditions can outrange the impact of pure climate change.

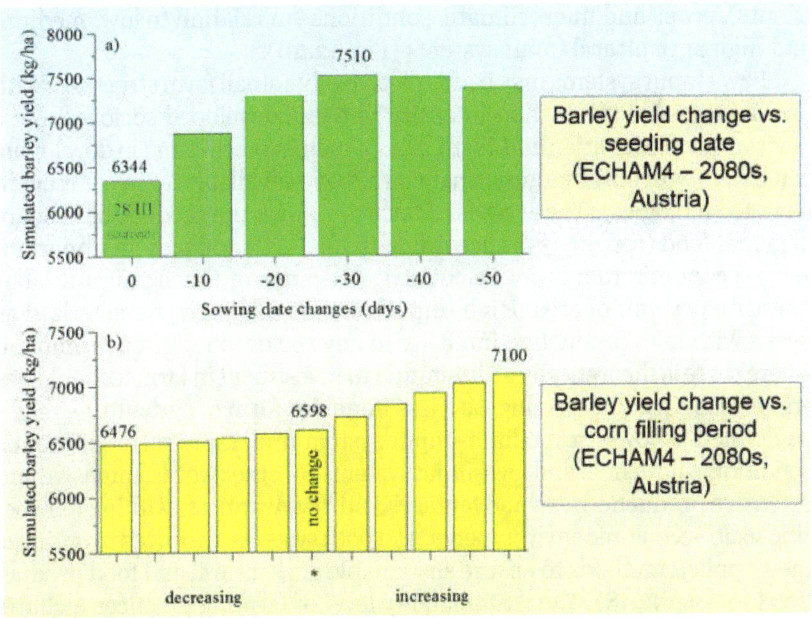

Figure 12.19: Impact of changed sowing dates (a) and genetic–determined corn filling period (b) of barley on barley grain yield, as simulated by DSSAT crop model (Alexandrov, *et al.*, 2002).

In all agro–ecosystems, farmers have been developing specific strategies for centuries, mainly with the application of different farm technologies and related management options to survive in a given environment. For various reasons, however, these have not always been carried out with sustainability in mind. The development and improvement in farming methods and technologies has been responsible for most of the increase in productivity and yields in agricultural production over the past decades, especially in developed countries. Nonetheless, a large yield gap between potentially achievable and actual farmers yields still exists, especially in less developed countries and agricultural regions (www.yieldgap.org). In that sense, adaptation to climate change in agriculture not only aims to ensure existing yields but also to close yield gaps in less developed farming systems by sustainable intensification (Mueller *et al.* 2012).

12.7.1 Optimizing farm technologies with respect to agricultural system

Analyzing optimization strategies in various agricultural systems logically requires considering the most important and most climate–sensitive natural agricultural resources to be managed–water, soil (including nutrients), crops and microclimatic conditions–in relation to low, medium and high agricultural input systems (Fig. 12.20).

Low–input systems may be characterized as small farm structures with low or extensive production intensity in a less developed socioeconomic environment (Maharjan and Joshi, 2013), such as those found in developing countries (with only marginal financial reserves available for investment in farm technologies). These types of farms produce the predominant share of global food (roughly 75%, according to the FAO) and play an important socio–economic role in many countries because of the significant labor intensity per unit of area. High–input systems might be characterized as farms with high production intensity in any socioeconomic environment, where there is theoretically no limitation to investment in farm technologies and inputs. Organic farming, as a less intensive form of agriculture, might be defined as low– or medium–input system under any socio–economic conditions. Both high and low–input systems (in other words, intensive and extensive production) have advantages and disadvantages within their specific socio–economic environments, and both systems are forced to optimize their applied methods to ensure sustainable agricultural and food production (see Fig. 12.18). The sustainability level of farming practices includes several factors, such as maintaining or improving local natural resources, ecosystem functions and socio–economic conditions, as well as becoming less dependent on external inputs (e.g. local closing of nutrient cycles).

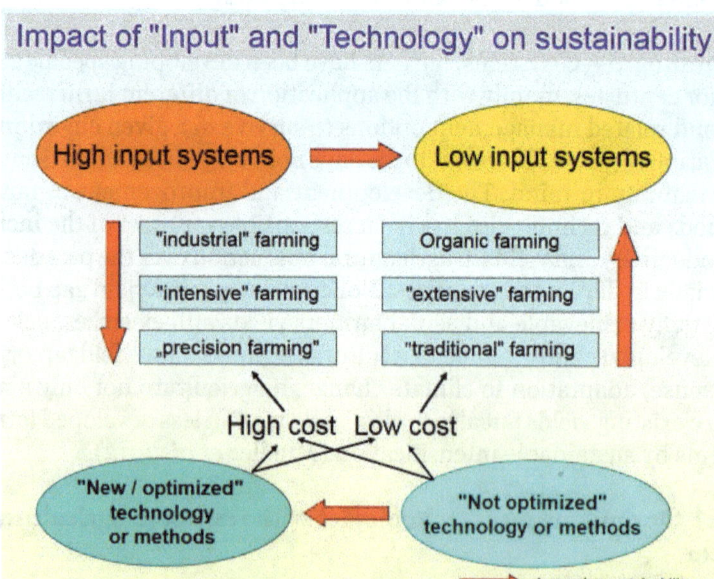

Figure 12.20: General scheme of high– and low–input farming systems and their relation to overall level of sustainability in their farming practices (Eitzinger *et al.*, 2009).

12.7.2 Adaptation options addressing water resources

Water availability could be the most important agricultural constraint in Mediterranean agriculture in the future and in many other agricultural regions worldwide. In this context, assessments of crop water use efficiency, such as crop water footprints, which are simulated with help of crop simulation models, are increasingly used for developing adaptation options (Höckstra *et al.* 2009, Kroes *et al.*, 2016). Studies on crop water footprints (expressed as crop water consumption per unit of produced and harvested crop yield) have shown, for example, that the blue water footprint (the crop water demand met by irrigation water) would increase in many semi–humid or drought–prone regions because of climate change (Fig. 12.21). This will lead to a decrease in available water from precipitation (green water footprint). Options for more efficient water use in crop production and irrigation agriculture will therefore be crucial for protecting natural water resources.

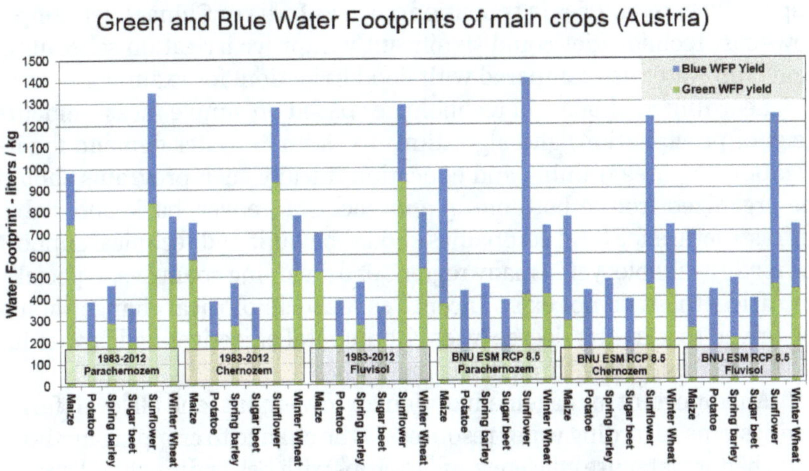

Figure 12.21: Green and blue water footprints of irrigated crops depending on soil conditions, and RCP 8.5–related climate scenario for the 2050s in a semi–humid region of Austria (Source: BOKU–Met).

In the past decades, many new technologies for improved agricultural water management have been successfully implemented and have increased agricultural productivity. For instance, modern sprinkler and drip–irrigation systems, which can significantly reduce crop water use, have been introduced–at great expense–in some Mediterranean regions of Europe. Adaptation studies have shown that only a few regions can expand irrigated land in a changed climate because of a shortage of irrigation water. Adapted crop cultivars, irrigation and drainage technology and irrigation methods will therefore need to be implemented. The introduction of deficit irrigation and similar methods to improve irrigation efficiency needs to be adapted to local conditions, taking into account climate variability.

Water management investments for irrigation include optimizing channel designs, water distribution systems and pumping devices. Furthermore, irrigation advice services call for government investment in trained personnel, as well as laboratory infrastructure and supporting technological facilities. As the total expenditure involved is quite high, however, only developed countries are able to afford these potential measures.

In developing countries, expensive new technologies are for the most part unaffordable without external support and are therefore not applicable to low–input agricultural systems with weak infrastructure and poor socio–economic conditions. The adaptation and use of traditional methods should be recommended in these cases. Beside the revival of original or modified traditional or indigenous methods, new low–cost technologies are an increasingly promising option for low–input farming systems, especially for countries in transition, such as India or China. Even simple low–cost technologies could significantly improve irrigation scheduling and crop water use compared with flood irrigation, for example.

Learning methods and technologies based on simple measurements (e.g. of precipitation) and algorithms to estimate water demand for irrigation requires training and education; ideally such programs should be organized from a bottom–up approach. Moreover, basic and stable infrastructure for local companies should be built and technical support be made available; yet in many regions of developing countries, especially in Africa, these do not exist. Incentives for technological change should be driven to a greater degree by environmental objectives and farmer innovations, operating through the market.

An important management option for low–input (and all level) farming systems regarding water resources is the change to cropping methods with better water use efficiency and to crops with better drought tolerance. This is especially important in regions where pressure on water reserves is increasing owing to human activities, climate change and variability, such as during the drought season in India. For example, changing from wetland rice to dry land rice or other crops can have enormous effects on agricultural water reserves, as has been demonstrated in northern China.

12.7.3 Adaptation options addressing soil resources

Soil types that have developed over many centuries are determined by climatic conditions, among other factors. Soil conditions and functions under prevailing crop management permanently interact with climate and climate variations. In this sense, farm technologies and soil and crop management have to be adapted to maintain important soil functions in order to secure sustainable agricultural production, which is basis for food security and the welfare of many countries. In many regions with extreme weather conditions, for example, soil functions can react very quickly to agricultural practices. Unfortunately, this can lead to rapid and irreversible degradation of soil functions and exacerbate desertification, which has become a significant problem in many agroecosystems in the world.

With regard to irrigation methods, improper irrigation schemes and the use of saline irrigation water can lead to the increasing salinity of soils, making them unusable for agricultural production. Likewise, some land use practices, such as overgrazing, endanger soils or soil functions, leading to wind erosion and desertification processes. Crop production in warm semi–arid zones with frequent strong winds is often subject to wind erosion triggered by soil degradation. In tropical regions, high soil temperatures combined with high precipitation leads to sustained decomposition and leaching rates; under these conditions, an inappropriate change in soil use for agricultural crop production can lead to rapid soil degradation. In climates with frequent extreme precipitation events, such as the Asian monsoon regions, soil water erosion, especially in hilly terrains, has already caused enormous soil degradation. This is the case where soil surface cover is lacking or where no terrace systems are in place. Under climate change and changing climate variability, these problems will become a more significant threat for soils in many agroecosystems, if the above mentioned measures are not implemented on farms (e.g. Klik and Eitzinger, 2010). Perennial crops in various climatic regions, such as vineyards, orchards, tea or coffee, which are often grown in hilly regions, are also subject to water erosion, especially during extreme precipitation events. Mulching technologies, such as the use of grass or straw mulch or other crop residues, are therefore often applied and are sometimes mandatory. In some cases, even more costly or manpower–intensive terrace systems have been re–established in order to stop long–term soil erosion.

The first important aim of soil cultivation is to control weeds and to optimize root growth conditions. This is still an important argument in favor of ploughing in many agricultural areas and in ecological farming. However, because soil cultivation is an important cost factor, less intensive methods have been developed, such as reduced soil cultivation or minimum to no soil cultivation and tillage systems. These methods also reduce the risk of soil water and wind erosion. They can also increase soil water–holding capacity and water infiltration. Finally, it has been demonstrated experimentally that increasing soil water–holding capacity by reducing soil cultivation in combination with mulch has significant positive yield effects. At the same time, this method also has disadvantages, as herbicides are often used for weed control under minimum soil cultivation systems, which can lead to increasing chemical emissions.

Although climate change and climate variability directly affect soil erosion, changes in the crop types grown or in crop rotation driven by climate change may indirectly increase soil erosion risk in vulnerable regions. For example, greater precipitation and decreasing soil cover due to the increased share of maize exacerbate soil erosion in many regions, such as Central Europe. In any case, soil erosion leads in the end to a decrease in soil fertility and hence to a reduction in crop productivity because of the loss of organic matter and nutrients and a lower water–holding capacity. This can be the start of self–forcing desertification processes and the permanent degradation of fertile land.

In high–input agricultural systems, heavy machinery is often used. In combination with excessive soil water content, this can contribute to soil compaction, which decreases water infiltration and increases runoff, thereby causing water erosion. In humid regions of Europe, these problems are apparent for summer row crops such as sugar beet and maize, as soils are not or only partly covered by canopy for weeks in the early growing season. In the case of late harvests, heavy machinery has a devastating and often irreversible effect on soil structure during wet harvest periods in autumn. This problem accelerates with the increasing slopes of fields, which are frequently found in Europe. Weather–dependent planning and forecasting of machinery use can be a helpful measure to avoid such damage to soils.

12.7.4 Adaptation options addressing climate resources

Changes in climate variability and climate affect microclimatic conditions in many ways. However, the design and management of crop stands can greatly mitigate this impact. In semi–arid low–input systems, for example, measures have been taken not only to improve water resources but also to optimize the temperature and radiation regimes of crop stands. A classic example is the oasis agroecosystem: complex crop mixing and patterns permit efficient use of radiation within a small area, increasing air humidity for shaded crops and avoiding extreme diurnal temperature variations. Likewise, windbreaks or hedgerows are often used in semi–humid areas of Europe to reduce wind and evapotranspiration.

Agroforestry systems (Fig. 12.22) are a widely used option for sustainable farm management (Louise *et al.*, 1998) to improve overall microclimatic conditions and not only to reduce wind and evapotranspiration. They are mostly applied in subtropical and tropical climates with extreme temperatures and/or weather variability. As crops respond especially to climatic extremes, any measure to reduce these extremes in most cases has an accumulating positive effect on crop yield. For example, heat stress on crops can be reduced by shading, which has been reported as a significant yield factor for many crops. Different types of agroforestry systems related to specific climates and agroecosystems and their effects have been analyzed. These systems were optimized for the specific characteristics of the relevant agroecosystem (climate, soils, crop production, farm input level and socio–economic framework).

Figure 12.22: Agro–forestry system with poplars in a semi–arid region of China (Eitzinger *et al.*, 2009).

Tree shading can also prevent frost damage to crops and reduce nocturnal radiation cooling on crop surfaces. Other frost protection methods, such as covering plants with sheets or foil, are used in small plots in low–input systems (see chapter 9). For orchards or large fields, methods such as frost irrigation, foil covering, or aerosol application are costly and are therefore found mainly in medium– and high–input farming and cash crop production. The selection of crop stand location in relation to orography (avoiding cold air lakes) is very important for protection against radiation frost damage. These measures are often ignored, especially in areas where frost seldom occurs, but the effect on perennial crops can be more devastating than hail damage, as entire plantations can be destroyed.

12.7.5 Adaptation options addressing plant/crop genetic resources

Crop yield and crop production within a certain territory can be seen as an interaction of many factors. However, crops adapted to certain conditions are an important local resource for crop productivity with a significant influence on yield risk. Crop physiological processes normally respond nonlinearly to changes in their growing conditions (especially temperature); they further exhibit threshold responses and are often affected by combinations of stress factors (see e.g. Chapter 10). Higher temperature and precipitation variability increases the risk of lower yield, as many experimental and simulation studies have shown.

Farmers over the centuries have selected the best cultivars for their use, creating locally well adapted crops. Some of these are still in use in agricultural systems and are an important genetic resource for modern crop breeding. Every year arable farmers have the option of selecting crop types and cultivars and also of modifying crop management (e.g., choosing the sowing date according to expected seasonal weather).

12.7.6 Adaptation options for better crop management

Weather forecasts represent the most important information for farmers for planning their farm work and timing their crop management. Short–term standard weather forecasts are already available worldwide; when it comes to tailored products for agriculture, however, there is still much room for improvement. Seasonal weather forecasts will especially have great potential in the future, considering progress in weather forecasting models and data bases.

In subtropical regions, seasonal weather forecasting has already been successfully applied. For example, seasonal precipitation patterns (the onset of rain, the duration of the rainy season, the distribution of rain during the crop growing period) are among the most important data for farmers in monsoon–type climates using rain–fed cropping. Such climate conditions often occur under low–input systems in developing countries. Efficient seasonal rain and weather forecasts and the related information transfer or warning to local farmers enable them to adapt their sowing dates and crop selection. Although already successfully applied in developed countries such as Australia, there are still problems when it comes to making such information useable for farmers in developing countries.

However, upcoming new technologies ("precision farming") and modern communication technologies will allow significant progress in developing new tailored forecasting products for agriculture, including effective and real–time transfer of information to farmers. For example, technologies such as remote sensing (ground based, satellite, drone) will permit monitoring of crop conditions on a much smaller scale, such as for a single field and even smaller areas. These technologies are under permanent development and incorporate methods such as GPS (Global Positioning System) and GIS (Geographical Information System). Because of the still relatively high costs, precision farming technologies are applicable at the farm level only for high–input farming. In any case, specific applications, such as the observation of spatial variabilities of crop conditions–the nitrogen content of leaves, for example–in crop fields for fertilization optimization, are becoming more and more common.

An increasingly used method is the combination of agrometeorological monitoring and near term to seasonal forecasting (e.g. Lalic *et al.*, 2015, 2018). These methods can be applied not only for specific sites, but as spatial GIS applications, with spatial resolutions up to only a few meters, which well covers the field scale required by farmers. This method allows farmers, for example, to observe drought status for irrigation scheduling and disease/pest occurrence for better pest management. Drought and other cropping risks can also be monitored by applying various risk algorithms (see Chapter 11). Using the observed information, farmers can take measures based on the conditions of specific sites. This can significantly help farmers in decision–making, decrease costs for irrigation, fertilizers and chemicals, and enhance crop yield and productivity. An example of operational crop–specific risk monitoring/forecasting as applied in Austria is shown in Fig. 12.23.

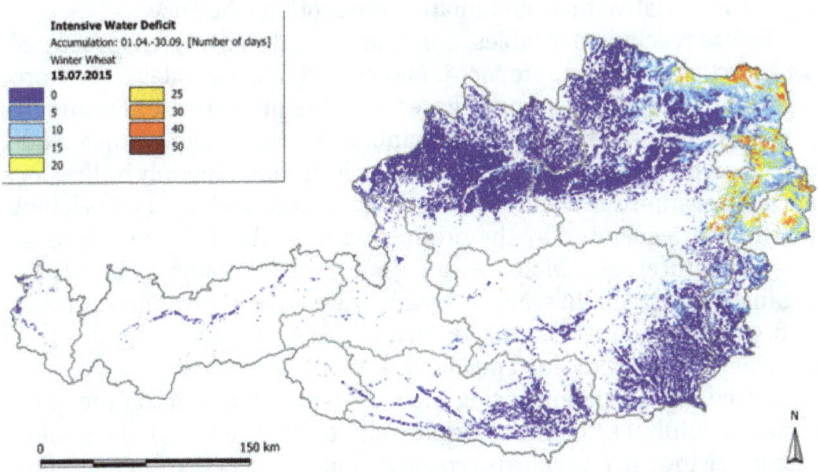

Figure 12.23: Example of daily updated operational Agricultural Risk Information System (ARIS) of Austria: accumulated intensive water deficit days for winter wheat are indicated (spatial resolution: 1x1 km) (Source: BOKU–Met).

In summary, adaptation in the form of implementing systems for early recognition of risks in the short (warning) and long term (advice) is crucial for all stakeholders in the entire food production chain. It has been proved that anticipatory, precautionary adaptation is more effective and less costly than forced, last–minute, emergency adjustments or retrofitting. Regional climate change impact and adaptation studies demonstrate the high spatial variability of potential effects. As impact depends on specific agroecosystems, soil conditions, climate regions and farming systems, regionally–fitted adaptation measures are of great importance.

Adaptation in agriculture at a global level is influenced by many factors, which may not readily observable at the farm level. As agriculture has to do its part in contributing to mitigation, all feasible adaptation options for the needs of single farms should also meet global mitigation aims (the reduction of greenhouse gas emissions).

12.8 Mitigation options in agriculture

Greenhouse gas emissions by human activities (see Fig. 12.7a–b; EEA, 2018) are the primary cause of anthropogenic climate change. Agriculture contributes significantly to global greenhouse gas (GHG) emissions, especially to emissions of carbon dioxide (CO_2), methane (CH_4) and nitrous oxide (N_2O). To simplify calculations (IEEP, 2017), all GHG emissions of different gases are converted into so–called CO_2–equivalent units, according to their relative thermal radiation absorption efficiency.

Before reaching our tables, our food is produced, stored, processed, packaged, transported, prepared, and served. At every stage, food provisioning releases greenhouse gases into the atmosphere. Farming in particular releases significant amounts of methane and nitrous oxide, two powerful greenhouse gases. Methane is produced mainly by livestock during digestion due to enteric fermentation and is released via belching. It is further generated by the production of wetland rice and by other wet agricultural soils. Methane can also escape from stored manure and organic waste in landfills. Nitrous oxide emissions are an indirect product of organic and mineral nitrogen fertilizers (EEA, 2016), but also from cultivated partially drained peatland.

Based on calculations using the CO_2–equivalent unit, on the global scale agricultural activities contribute 24%to total global GHG emissions (according to 2010 estimates). A considerable part of its share (about 10% of total GHGs) is caused by forest burning (which occurs in land use changes from natural forests into arable lands, especially in tropical rainforests), wetland rice cultivation and cattle production (mainly methane emissions). This estimate does not include CO_2 that ecosystems remove from the atmosphere by sequestering carbon in biomass, dead organic matter and soils, which offsets approximately 20% of emissions from this sector.

In European countries, agriculture contributes to a lower share of total GHG emissions, about 10%. About 4% is related to enteric fermentation (methane emissions by animals, especially ruminant cattle), 1.5 % to manure management and another ca. 4% to emissions from agricultural soils (mainly CO_2 and N_2O). Peatland drainage and conversion of grassland into arable land use are significant sources of CO_2 emissions from soils. Fossil fuel burning by agricultural machines or traffic play a marginal role. In agriculture, however, gases other than GHGs, such as ammonium (NH_4) emissions from manure management, can have a considerable negative impact on the environment.

A significant decline in livestock numbers, more efficient application of fertilizers, and better manure management reduced the EU's emissions from agriculture by 24% between 1990 and 2012.However, agriculture in the rest of the world is moving in the opposite direction. Between 2001 and 2011, global emissions from crop and livestock production grew by 14%. The increase occurred mainly in developing countries, due to a rise in total agricultural output (EEA, 2016).

Almost all measures to reduce greenhouse gas (GHG) emissions or to adapt to climate change in these sectors have also had socioeconomic or ecological consequences other than the intended ones. The negative feedback of these measures also affects the GHG reduction potentials of climate–change mitigation strategies. One example is the GHG emission reductions associated with the substitution of bioenergy for fossil fuels, which has had direct and indirect systemic effects with regard to land use, such as changes in forest areas that may have resulted from the modified use of cultivated areas.

Even so, a multitude of options exist in the agricultural sector to reduce GHG emissions, in particular with regard to ruminant feeding, manure management, reduction of nitrogen losses and increased nitrogen efficiency. Increased production of agricultural bioenergy can help to reduce GHG emissions, especially when implemented according to an integrated optimization of food and energy production as well as a cascadic use of biomass.

Due to the high carbon stocks in forests, the forestry sector is a key player in land–use related GHG mitigation strategies. Forestry can contribute to climate–change mitigation through carbon sequestration as well as through the provision of low carbon resources (materials, energy). The systemic interdependencies between a forest's production and sequestration functions, as well as its delivery of other ecosystem services, need to be considered. Socioeconomic as well as ecological and climate outcomes can be improved through an integrated optimization of forest production and cascades of biomass use.

A wide range of mitigation actions are already available to the agriculture sector, but they have yet to be adopted at the scale and intensity necessary to deliver lasting emission reductions.

The major examples of mitigation options in agriculture for developed countries are listed below (EPA, 2018):

- Land and Crop Management:
  - adjusting methods for managing land and growing crops (e.g., adapting soil cultivation methods and timing, crop rotations);
  - fertilizing crops with the precise amount of nitrogen required for crop production, since over–application of nitrogen can lead to higher N_2O emissions without enhancing production;
  - draining water from wetland rice soils during the growing season to reduce CH4 emissions;
  - avoiding drainage of wetland soils;
  - optimization of water table in cultivated wetlands to minimize CO_2, CH_4, and N_2O emissions;
  - avoid conversion from pasture to arable land.

- Livestock Management:
  - adjusting feeding practices and other management methods to reduce amount of CH_4 resulting from enteric fermentation;
  - improving pasture quality to increase animal productivity, which can reduce the amount of CH_4 emitted per unit of animal product. Also, increased productivity in livestock can be introduced through improved breeding practices.

- Manure Management:
  - controlling the way manure decomposes to reduce N_2O and CH_4 emissions;
  - capturing CH_4 from manure decomposition to produce renewable energy;
  - handling manure as a solid or depositing it on pasture rather than storing it in a liquid–based system such as a lagoon. This would likely reduce CH_4 emissions but may increase N_2O emissions;
  - storing manure in anaerobic containment areas to maximize CH_4 production and then capturing the CH_4 to use as an energy substitute for fossil fuels.

The main global task fields of adaptation and mitigation can be summarized as follows (based on FAO recommendations):

- seasonal changes and sowing dates of crops;
- different/new varieties or species, crop rotations;
- water supply and irrigation systems;
- input management (fertilizing, tillage methods, grain drying, other field operations);
- pest and disease management;
- microclimatic modification: promotion of agroforestry;

- reducing food security risk;
- identifying present vulnerabilities;
- adjusting/strengthening agricultural research priorities;
- strengthening agricultural extension and communication systems;
- improving training and education (in rural areas);
- adjustment of commodity and trade policies;
- improving rural infrastructure and market access for farmers;
- fighting (rural) poverty.

13

Agrometeorology: Measurement Methods

13.1 Basic sensor technologies – introduction

- *Mechanical sensors*

In the past, mechanical sensors–without or with electronic components–were used for taking certain measurements: for example, mechanical clocks measured time, while conventional thermometers indicated temperature. Such instruments are still used today for specific applications: they may come in the form of handheld instruments or of such everyday devices as thermometers and hygrometers used to perform simple, cost effective applications in households.

Values measured by mechanical sensors (based on mechanic sensor principles) can be shown on a scale (as with a thermometer) or also by electronic driven pointers on a scale. In the case of permanent observations in weather stations, for example, they may also be written down by someone making observations or be logged on paper automatically (e.g. by a thermo–hygrograph). These types of recordings are also called analog measurements.

Before the 1980s, most weather stations worldwide, including national weather station networks, worked with mechanical instruments. Still today, these types of measurements are used in weather stations in many developing countries. In this case, an observer has to read data from sensors three times a day at specific times in the morning, afternoon and evening (7, 14 and 21 at local time) and write down the values in tables, in accordance with World Meteorological Organization (WMO) requirements. Many historical meteorological and hydrological records are available in this form. Today, though, records are for the most part already digitalized and therefore available as electronic values.

Examples of conventional mechanical sensors for atmospheric parameters:

- thermometers are based on volume extension of various fluids (mercury, alcohol etc.) in a glass capillary;
- psychrometers are a combination of wet and dry bulb thermometers to estimate wet and dry bulb temperature which are used to accurately calculate air humidity. The evaporative cooling effect is expressed by the temperature difference between the wet and dry bulb thermometers. Because of their accuracy, these sensors are still in use in modern weather stations for measuring water vapor pressure in the air or to calibrate electronic air humidity sensors;
- thermohygrographs measure air temperature by bimetal extension and air humidity by related hair length changes. The measurements are recorded mechanically as graphs on paper over a weekly period. They are used in analog weather stations;
- mechanical anemometers are wind instruments with propellers driven by wind that generate electric signals or show signals on a scale;
- mechanical air pressure (barometer) instruments measure air pressure based on the rise of the mercury column to a glass tube till it stabilizes at an equilibrium height when hydrostatic pressure of mercury on the pool inside the tube is equal to atmospheric pressure on the pool outside the tube;
- rain gauges collect rain water and other forms of precipitation such as snow (if heated): the amounts can be read on a scale;
- heliographs measure sunshine duration (e.g. Campbell Stocks): a glass ball focuses a radiation beam on a paper scale, which burns a strip, indicating the sun's daily course.

For more in–depth information on mechanical sensors, see the recommended literature.

- *Electronic sensors*

Electronic sensors have been increasingly used in weather stations since the 1980s, in the context of the development of computers and electronic data storage and transfer systems. In this case, an analog electric signal is digitalized and so can be stored electronically (as digital values) and transferred by modern communication technologies. This means weather stations can work automatically without the need of observers reading data from conventional instruments. However, stations need power (electricity), which is provided by a grid or by a battery or solar panel if no grid is available. Nowadays nearly all modern weather stations work in this manner, which will be explained in more detail in the following sections.

13.2 Measurements methods

Agrometeorological measurements normally address specific microclimatic environments and therefore do not need to adhere to the WMO standards for climatic weather stations that are followed by national weather station networks. These standards include pre–defined requirements for measuring heights (above ground) of temperature and wind, etc., surface characteristics of weather station sites, and minimum technical requirements of sensors, to name just a few (WMO, 2008, 2011). However, it is recommended that organizations and researchers who continuously collect data and conduct scientific research use sensors in WMO standards for agricultural meteorological measurements.

Before taking any environmental measurements, two basic requirements must be borne in mind:

- measurements must be taken in such a way that their success is ensured and their aim is achieved;
- the measuring process must be as efficient as possible in regard to costs, time and labor efforts.

The characteristics of the sensors (response time, drift, resolution etc.) should be chosen according to the method to be used and the expected data quality to be obtained.

Furthermore, two main purposes of agrometeorological measurements can be distinguished:

- for scientific applications;
- for practical or operational applications (applied agrometeorology).

13.2.1 Scientific applications

In addressing scientific questions through measurements, measurement systems generally require more technical flexibility and higher technical quality (e.g. accuracy) than "ready–to–use", designed agrometeorological weather stations for practical applications, which have already been fitted to specific applications (such as irrigation scheduling, microclimate conditions within canopies for disease management: these will be treated in the next section).

In line with the scientific question, the following aspects need to be considered to "design" a suitable measurement system:

- the number and type of variables to be measured;

- the minimum technical accuracy of sensors with respect to the measured parameters;
- the time resolution of measurements (e.g. seconds, hours, days);
- the number of stations required when information over spatial scales is needed;
- the time period of measurements (short–term or long–term experiments);
- data storage and transfer requirements (e.g. data logger capacity, costs of online data transfer);
- the energy demand and source for the measurement system;
- maintenance requirements of the measurement system;
- needs for data management and analysis (expertise, labor costs);
- selection of the place to measure.

There are manifold examples of scientific applications of agrometeorological measurements; a few examples are presented below:

• *Measuring microclimatic conditions of crop stands (see also "Further Reading" for more in–depth information at the end* examples:

- measuring evapotranspiration (of soil and atmosphere), heat and water fluxes and radiation balance (Fig. 13.1 a–d), using approaches such as the Bowen–Ratio Energy Balance (BREB) method (Fig.18b), the Eddy–correlation method (Fig. 13.1c) and the Scintillometer method;
- measuring gas fluxes (e.g. greenhouse gas emissions from soils).

Figure 13.1 a–d: Measuring of soil water and temperature profile at various soil depths (a, top–left); radiation balance by net radiometer (short and long wave) (b, top–right); Eddy correlation flux measurements consisting of high–frequency ultra–sonic (left) and water vapor and temperature sensors (c, bottom–left); and potential evaporation by an evaporimeter (d, bottom–right); (Source: BOKU–Met).

- *Measuring local climate phenomena*
examples:
  - transect measurements, which detect small–scale climatic variations, such as the effects of hedgerows or windbreaks and agro–forestry systems, or the effects of topography on canopy microclimatic conditions (Fig. 13.2 – 13.3);
  - air flows, such as cold air flows and inversions in cold air lakes.

13.2.2 Sensors and measurement methods and techniques for applied agrometeorology (practical applications)

Introduction

The first question involved in the practical application of measurement techniques in agrometeorlogy concerns the cost return (or cost–benefit) analysis: i.e., is a weather station which farmers buy for a specific purpose worth the expense? Besides the question of the "value" of agrometeorological information, such as weather forecasts for agrometeorological purposes, this issue has been investigated in several case studies. It is frequently reported that especially in the area of pest and disease warning and forecast and in crop irrigation scheduling, agrometeorological field

Figure 13.2: Transect measurement station of vineyard climate (air and soil temperature, relative air humidity, potential evaporation) in hilly Austrian terrain (compare derived results of Fig. 12.4 and Fig.8.11 in the other chapters); (source: BOKU-Met).

Figure 13.3: Transect measurements on wind and evaporation leeward of a hedgerow in Austria; (Source: BOKU-Met).

measurement–based methods can pay off within a few years, if applied correctly.

For example, since 1992, a network of agrometeorological weather stations which feed an Adcon addVANTAGE decision–support system has been running various models for downy and powdery mildew. The stations cover an area of well over 1,500ha. Data hosting, collection and processing are done by the local office of Austria's Chamber of Agriculture, whose staff sends out warnings by fax and email. Members pay a very low annual fee of only 10 Euros for this service. The experience of the last 16 years has shown that farmers using the service can achieve an average reduction of sprays of 35%. The number of treatments averages between five and six, and can even be as low as three in some low–pressure years, while standard treatment methods based on spray plans recommend 8 – 12 treatments per season. Even without the ongoing support of the Chamber of Agriculture, the purchase of this equipment would be quickly recovered, as even a small 5 ha farm could achieve a savings of approximately 600 Euros per year and 200 Euros per spray, thus paying the station off in less than four years.

While from a technical point of view the observation of micro–climatic data by means of an automatic weather station can be regarded as state–of–the–art, it is far from being an accepted, widespread practice. In the field of irrigation scheduling, even in drought–riddled countries such as Australia less than 10% of all farmers use weather stations and soil sensors to improve their daily irrigation practice. While most farmers try to get updated weather forecasts for their future spraying and irrigation decisions, only a very small percentage of farmers actually work with on–farm weather stations and soil sensors.

There are two basic agrometeorological methods to enhance water usage in crop production, which we present here in order of their popularity sorted by popularity respectively distribution:

a. weather stations to calculate evapotranspiration;
b. soil sensors to directly monitor soil moisture.

Both of these techniques has its distinct advantages and disadvantages, and each has its community of advocates and opponents. Upon closer look, though, it is not difficult to understand that they should be regarded as complementary technologies rather than as alternatives. For a systematic approach to the subject, we will need to categorize the equipment from various perspectives, including the usage of the equipment for purposes other than water management (e.g. pest and disease management).

Agrometeorological weather stations and sensors

The agricultural environment in its own right has proven to be a very challenging one for the instrumentation used with a weather station. The instrumentation is not only exposed to weather, which in itself already places quite a burden on electronics and sensors. It is also exposed to agrochemicals such as pesticides and fungicides and subjected to occasional showers by overhead irrigation systems. Furthermore, it is affected by dust and sand, which exerts not only an abrasive effect but also builds obstructing layers on sensor surfaces, thus further accelerating deterioration by adding more chemicals, such as fertilizers, to the mix. The effect of all this will be discussed further below, which treats problems encountered in daily operation.

Experience shows that in addition to these factors equipment also suffers from a lack of maintenance. Calibration of sensors is seldom performed, plausibility checks of data are rare, and even equipment cleaning is far from being a matter of course. Furthermore, both owners and operators of agro–met weather stations frequently lack the technical skills required to run a sophisticated weather station, to program new calibration factors into a data logger or to rewire a sensor.

These factors need to be taken into account when designing an agricultural weather station (Fig. 13.4). On the basis of these considerations, the prerequisites for an efficient station can be summarized as follows:

- equipment must be very robust, yet sufficiently accurate;
- equipment must be able to withstand extreme climatic changes through the seasons;
- equipment must be able to withstand agro–chemicals;
- equipment must provide long term stability, with little sensor drift over time;
- sensors must be easily and quickly replaceable, even by non–trained staff.

In light of these constraints, when building a station it is advisable not to focus only on the price of equipment, but rather at the price–performance ratio, and in particular at the "TOC", the total cost of ownership over an initial period of 10 years. This takes into account such questions as how many site visits will be necessary and how many recalibrations, sensors swaps, etc., will be required. More expensive high–performing equipment may in the long run prove more cost effective than cheaper devices.

The right sensors must therefore be chosen for the right application. As a guideline to sensor quality, buyers should look at the following characteristics of sensors:

- resolution: the smallest quantity in which information can be displayed (not to be confused with accuracy!);
- accuracy;
- overall accuracy, taking all potential errors into account, like hysteresis and temperature dependence;

Figure 13.4: Example of a commercial automatic and solar-energy driven agrometeorological weather station for measurement of standard meteorological elements and online data transfer; (Source: BOKU-Met).

- hysteresis: the capability of a sensor to yield the same result when multiple readings are taken under identical conditions;
- response time and power consumption, which are directly linked to each other: the faster a sensor's response, the lower its power consumption;
- maximum drift per year and over an initial 5–year period.

Sensors in use for agrometeorological applications:

Air temperature

While NTC (Negative Temperature Coefficient – higher heat conductivity with increasing temperature) thermistors (THERMally–sensitive resISTOR) are available at very low cost, the platinum–based pt100, pt500 or pt1000 sensors are definitely the better choice. These sensors offer linear output across their full temperature range and have proven to be fairly robust even under harsh conditions, avoiding the frequent need for recalibration due to sensor drift. The temperature range of these sensors should be wide enough to cover other applications as well, such as for frost warning and for running chilling hour models for pest control and other purposes. A common temperature range available from many manufacturers is -40°C to +60°C. Accuracy should be equal to or better than 0.5°C, ideally 0.1°C.

The temperature sensor needs to be protected from direct solar radiation by placing it into a radiation shield, which is usually of a non–ventilated type, such as the one shown in Fig. 13.5.

Figure 13.5: Passive radiation shields for temperature and air humidity sensors (Source: Pacher (left); BOKU–Met (right)).

Most manufacturers package their temperature probes together with a relative humidity sensor to reduce the cost for signal conditioners and amplifiers. For daily application this has not only a positive cost effect, it also facilitates installation and eliminates the need for a second radiation shield.

Relative Air humidity

Most of the formulas presented in this volume require the value of water pressure for their computation: this meteorological elements can usually be measured with sufficient accuracy by a standard relative humidity sensor. Most electronic sensors use capacitive sensor elements for this purpose. All such elements tend to lose accuracy at the higher end, towards saturation. Since most weather stations are not only used for irrigation purposes, but also for disease control, particular attention should be paid to the relative sensor selected.

Capacitive sensor elements are very sensitive to changing environmental conditions and are adversely affected by layers of dust and agro–chemicals in particular. They are in general subject to drift, usually tending to indicate higher humidity levels than actually present. Care should be taken that within an initial installation period of five years drift does not exceed a maximum of 5%. As it is of great importance not only for the calculation of evapotranspiration but also for disease models, it should be recalibrated or changed frequently, ideally on a biannual basis. Figure 13.6 shows a typical capacitance chip for relative measurement.

Figure 13.6: Capacitive humidity chip (Source: Pacher).

Figure 13.7: Sensor caps: mesh / paper (Source: Pacher).

As we have seen, temperature and humidity sensors are usually packed together in a common housing, thus requiring only one amplifier, cable and radiation shield for installation. To protect the delicate sensor elements from direct exposure to the environment, a protective cap shields the sensors, either by means of dense wire mesh (left) or special paper filters (right), as can be seen in Figure 13.7.

Sensor caps should be replaced frequently, as they tend to collect dust, agro–chemicals and moisture, thus quite frequently creating a microclimate of their own, which can distort measurements significantly. This is of particular importance when it comes to relative humidity, for which a sensor encapsulated in moist or even moldy filter paper will deliver inaccurate (usually too high) readings, often leading to an underestimation of evapotranspiration and an overestimation of disease pressure.

Leaf wetness

This type of sensor is relatively simple and should mimic the duration and extent of leaf wetness via various types of surfaces (Fig. 13.8). Leaf wetness duration information is crucial for inputs in many fungi–based disease models.

Solar radiation

Professional meteorology requires installation of highly accurate pyranometers to measure the full range of solar radiation, that is, wave lengths from 300 to 3000 nm, as defined in international standard ISO 9060. The standard formulas used to calculate evapotranspiration are also based on this type of sensor. The standard principle of measurement, the absorption of thermal energy and its conversion into an electrical signal, is usually performed by either a thermopile sensor (a stack of several thermocouples) or by a "black–and–white sensor". Both such sensors are very accurate, but also rather costly, which has advanced the so–called silicone pyranometer as the prime choice for agriculture. Silicone pyranometers

Figure 13.8: Leaf wetness sensor; (Source: BOKU-Met).

are not capable of catching the full wavelength spectrum, but only from some 300 to 1100nm; still, they deliver an approximation close enough for the targeted purpose. They can be manufactured at much lower cost and can provide a good alternative for agriculture.

A common mistake in sensor selection needs to be mentioned here. Quite frequently PAR sensors, measuring only photosyntetically active radiation, are installed rather than silicium pyranometers. Since the PAR spectrum ranges only from 400 to 700 nm, it largely reduces the amount of energy measured. Such sensors are thus not suitable for the application in question.

Pyranometers should ideally possess a hardened, abrasion–resistant lens (through whose opening radiation passes to the sensor element) and possess sufficient accuracy so as not to require recalibration more often than every 5-7 years. Figure 13.9 a-b shows a typical example of a silicon pyranometer and a Class 2 thermopile pyranometer.

Pyranometers need to be installed absolutely level to avoid monitoring errors by sensor surfaces tilted too far away from or too close toward the Sun. Ideally a pyranometer will face due south in the Northern Hemisphere and due north in the Southern. It needs to be installed clear of all obstacles that could block the Sun during its daily circle.

Wind speed and direction

Since a weather station in agriculture usually serves a variety of purposes, monitoring wind speed is of great importance. When winds are too high, overhead irrigation systems need to be shut off, the application of fungicides and pesticides is no longer possible (both for efficiency reasons as well as for increased health hazards caused by wind drift of chemicals), and vents on polytunnels and greenhouses need to be closed.

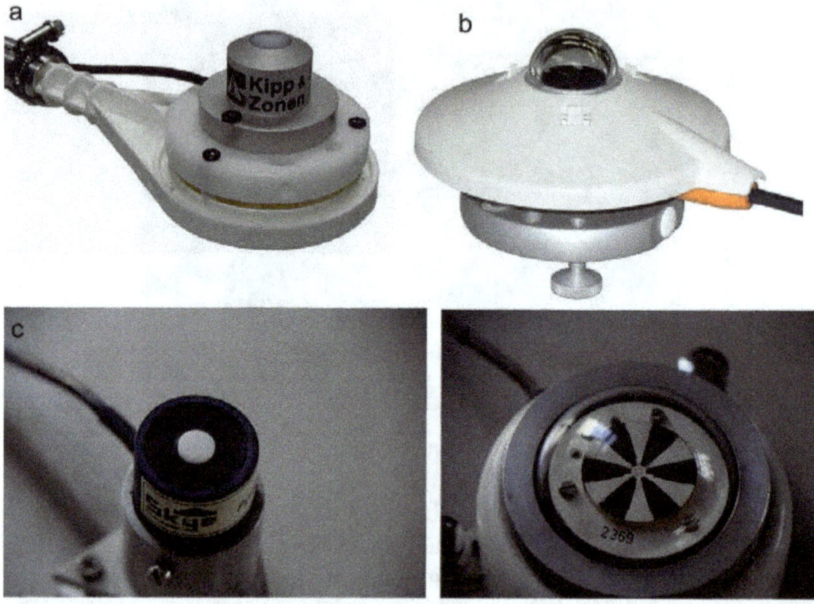

Figure 13.9: Types of pyranometers: Silicon pyranometer (a, top-left; c, bottom-left); Class 2 thermopile pyranometer (b, top-right) and star-pyranometer (thermophile, d, bottom-right); (Source: Pacher (a,b); BOKU-Met (c,d)).

For the calculation of evapotranspiration, wind speed is of even greater significance, as it replaces the moist air over the soil and the canopy by drier air, thus enhancing evapotranspiration.

Several constructive methods are used to build wind speed sensors, the most common being the cup anemometer, the propeller type and the ultra–sonic sensor. While ultra–sonic sensors excel with their extremely low starting thresholds and their non–existent inertia due to the lack of moving mechanical parts, they are still rather pricey and require a significant amount of energy for operation. For these reasons, agriculture cup anemometers and propeller sensors are predominant.

For agricultural purposes, the measuring range of a wind speed sensor can remain well below the maximum speed of 75 m s^{-1}, as indicated in standard WMO guidelines. For all agricultural applications, a top speed in the range of 35 to 55 m s^{-1} is largely sufficient. What is of greater significance is the starting threshold of the sensor, which should be well below 1 m s^{-1} (in general 0.2 m s^{-1}), as even such low wind speeds can already increase plant evapotranspiration.

As wind speed is of such great importance (the very popular FAO56 method, discussed below, relies also on wind speed), it is equally crucial to choose an accurate instrument that does not quickly wear out over time. Specifically, micro particles, such as those found in sand, fertilizers and agrochemicals, are the natural enemies of the bearings of a wind speed sensor. While solid particles that accumulate inside the instrument slow down its rotation, agro–chemicals degrade the lubricants, increasing wear and tear of the mechanical parts. Thus wind speed sensors should be controlled frequently with a reference instrument to make sure that they are performing properly.

Wind direction, by contrast, is of little significance for the intended purpose; indeed agricultural weather stations are quite frequently installed without such a sensor. As with solar radiation sensors, it is of utmost importance to observe the mounting instructions of the manufacturer. In all parts of the world, a wind direction sensor needs to point due north, as this is the switchover point from 360° to 0°. Even ultrasonic sensors possess a marking on the case that needs to be pointed north. Figure 13.10 shows several examples of wind sensors as discussed above.

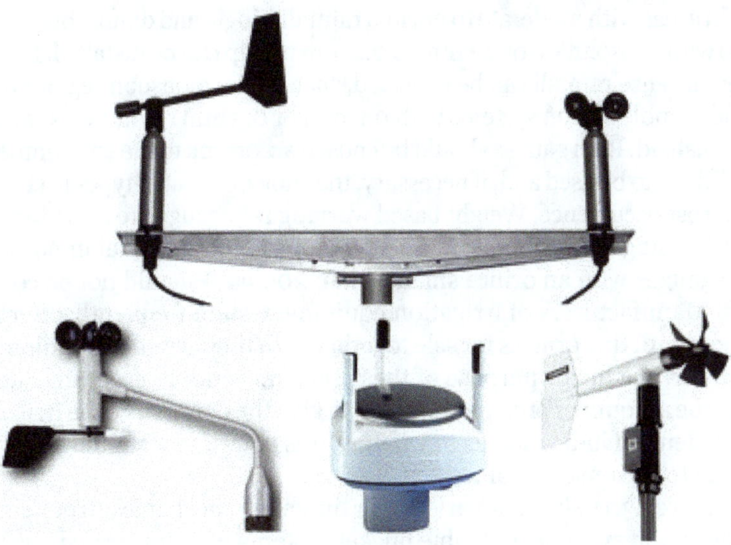

Figure 13.10: Various wind sensors, including propeller (bottom right), wind cap and ultra–sonic sensor (bottom center); (Source: Pacher).

Precipitation

The need for accurate and reliable rain gauges for agricultural purposes cannot be overstated. Whether for rain–fed or irrigated agriculture, it is of paramount importance to properly monitor precipitation.

Rain gauges come in all sizes and materials and with a large variety of measuring principles, ranging from single tipping buckets to sound pattern detection and laser technology. For obvious reasons, mechanical designs are predominant in agrometeorology. They offer sufficient accuracy at relatively low prices and require very little power.

The most common type of rain sensor is the tipping bucket sensor with a pulse output. The principle is simple: a defined amount of water fills a spoon or bucket until it tips over. In doing so, it not only empties itself out, it also moves a magnet past a reed switch, thus closing or opening an electrical contact and creating a pulse that can be measured by a data logger.

Both single as well as double tipping bucket designs can be found. They differ mostly in their ability to cope with high–intensity rain events. In such cases, the amount of water pouring through the funnel of the rain gauge onto the measuring element can be so great that water is lost during the tipping process. However, this not only applies to single spoon systems, but also to high–resolution double–bucket systems, when rain intensity exceeds the capacity of the rain gauge as quoted by the manufacturer (usually given in mm/hour or tips per minute). It is therefore extremely important to choose the proper rain gauge for the type of rainfall events predominant at the installation site.

In areas with moderate to normal rainfall, single and double bucket systems with a resolution of 0.1 mm to 0.2 mm per tip can be installed. In areas where intense rainfall can be expected, such as the monsoon regions of the world, double bucket systems with 0.2 mm or 0.5 mm resolution should be used instead. Rain gauges should be chosen according to the environmental conditions to be used and, if necessary, the choice of heated types in climates with frost occurrence. Weight based working rain gauge are also used.

Choosing the proper orifice of the rain gauge is of equal importance. Rain gauges with an orifice smaller than 200 cm^2 should not be considered. Manufacturers of irrigation equipment quite frequently offer rain sensors with tiny orifices for sale together with their irrigation equipment. However, the major purpose of these sensors is not to accurately determine the amount of precipitation, but to give the operator of the irrigation equipment an indication as to whether this equipment should be turned off due to the onset of rain.

Figure 13.11 shows a variety of rain gauge mechanics, from simple single spoon systems to double bucket systems with integrated levelling features.

Figure 13.11: Various rain gauge (tipping bucket) mechanics (Source: Pacher).

As with most sensors, proper installation is the key to representative results. It is obvious that a tipping bucket system, which relies heavily on the well–balanced action of the spoons, must not be tilted sideways, but needs to be installed absolutely horizontally. This is also of importance because for every millimeter that a rain gauge's orifice is tilted sideways, its capacity to properly catch rain is further reduced.

Particularly in agricultural environments, it is highly recommendable to install a bird deterrent from the very beginning. While for scientific purposes a bird deterrent might minimally alter readings, for practical reasons such inaccuracies (if any) are secondary to the benefit of such a device. Bird droppings might become a serious problem, clogging the funnel, and birds of prey like to feed while sitting on rain gauges. Figure 13.12 shows a typical bird protector, while Figure 13.13 shows a large bird of prey resting on a rain gauge.

Figure 13.12: Bird protector for rain gauge (Source: Pacher).

Figure 13.13: Bird resting on rain gauge (Source: Pacher).

Soil moisture

In recent years, many companies have introduced a wide range of various soil moisture sensors, which are now available for just about any budget, soil and crop, and in many different technologies. There is no ideal sensor, one that does it all, one that is equally suited for sand and clay, for blueberries and grapes, turf and pecans. But within the range of available products everyone will be able to find one that meets the criteria for the task at hand.

The idea behind direct soil moisture monitoring is simple: rather than determining the amount of irrigation by processing (meteorological elements) to monitor the actual soil water status at the plant's active root zone, the measurement of soil moisture directly assess the water needs of the plant.

Soil moisture sensors, however, offer much more insight into what is happening in the soil and the root zone than just measuring soil moisture. They indicate the infiltration rates of irrigation and natural precipitation, they allow us to asses water table fluctuations, they assist in determining agronomical thresholds such as refill point/wilting point and field capacity, and last but not least they help us to properly determine the dimension of the active root zone and its variability over time in the various phenological phases. As is known, soil water content is also a sign of agricultural drought.

When looking at the measurement of soil moisture, we need to distinguish between the various measurement methods used to determine soil moisture levels, between single and multi–level sensing, and between the frequencies at which readings are taken.

One of the first accurate methods to monitor soil moisture in the field was the Neutron moderation method, developed in the 1950s. As the equipment needed was not only very expensive but also required a very skilled operator authorized to handle nuclear material, it was necessary to serve as many clients and take measurements in as many fields as possible with one instrument. Thus usually only one reading could be obtained per week, with the gaps having to be filled by interpolation methods.

Recent developments, however, have brought the advance of less costly equipment. In conjunction with a data logger it has now become affordable to leave an instrument permanently on site and to take a reading every 15 to 30 minutes. This means that if the logger is also connected to a telemetry device, data can then be sent to the meteorological service or to the farm manager in almost real time. In the case of needing to adjust one's daily irrigation practice flexibly to changing conditions, a permanent monitoring system in combination with a telemetry device is clearly the preferred option. Permanent monitoring significantly reduces the time needed to establish the most important agronomical indicators of every field, namely field capacity and wilting point. Furthermore, daily observation of plant behaviour is very helpful to better understand its varying physiology during the different phenological phases.

Rapidly advancing sensor development has led to the multi–level probe, developed in the early 1990s, which allows the comfortable installation of several sensors on one site without the need to disturb the root zone by digging a profile. Such sensors are usually installed through an access tube which is installed into the soil, and can be spaced at various sensor intervals, up to a placement density of one sensor every 10 cm. Such sensors can deliver accurate readings right after installation, without the need to wait for several weeks or months until the soil has settled again from digging in several individual sensors. However, since multi–level sensors of the access–tube kind necessarily determine soil water content without being in direct contact with the medium, they are also subject to some limitations. Pricewise, single level sensors are less expensive than multi–level sensors and are usually installed for crops with shallow roots.

Soil moisture content with in–situ sensors can be determined using:

- volumetric methods and
- soil water potential monitoring.

Most sensors used these days employ some form of *volumetric methods* (Fig. 13.14). They give a direct reading of soil water status (units of water per unit of soil, e.g. ml of water/cm^3 of soil, or simply expressed in %), but shed no light on the ability of a plant to extract water from the soil.

The technology is based on determining the dielectric constant of soil, indicating the capacity of a non–conductor (i.e., soil as insolator) to transmit an electromagnetic wave, such as the one emitted by a soil moisture sensor. As the dielectric of dry soil and that of water are known, a curve can be developed correlating the sensors' response to the moisture content of the soil, given that this content is very low in dry soil and very high in saturated soil.

Two kinds of in–situ measurement systems are available today:

- Frequency domain reflectometry (FDR);
- Time domain reflectometry (TDR).

The capacitance or FDR technique determines the dielectric permittivity of a medium by measuring the charge time of a capacitor, which uses that medium as a dielectric. In contrast, time domain reflectometry (TDR) determines the dielectric permittivity of a medium by measuring the time it takes for an electromagnetic wave to propagate along a transmission line that is surrounded by the medium. TDR measurements are theoretically less susceptible to soil and environmental conditions compared to capacitance sensors. However, the interpretation of TDR output can be a considerable source of error when high salinity diminishes the reflectance waveform or temperature changes the endpoint (https://www.metergroup.com/environment/articles/tdr–capacitance–missing–point/).

Soil water potential monitoring or tension (see Chapter 7) is measured by the matrix potential or soil suction method often referring to the force a plant needs to exert to extract water from the soil. The unit of measure is kilopascals. Such sensors are rather inexpensive by comparison, but require more attention and maintenance than volumetric sensors. Their common denominator is the use of a porous media such as gypsum that allows water to penetrate into the instrument. Their most common representatives are tensiometers, gypsum blocks and Watermark sensors

(Fig. 13.15). These sensors differ in the measurable range on soil water potential, and need, especially for measuring absolute soil water potentials, regular calibration to the relevant soil types. Depending on the material used, the lifetime of these sensors is limited, especially for the low cost gypsum blocks.

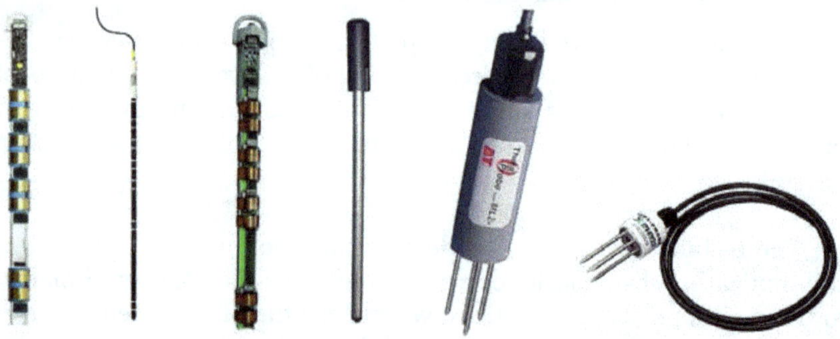

Figure 13.14: Various FDR/TDR soil moisture probes from different producers. From left to right: Sentek EnviroSmart, Delta-T Profile Probe, Agrilink C-Probe, AquaCheck, Delta-T Profile Probe and Stevens Hydra Probe, (Source: Pacher).

Figure 13.15: Soil water potential sensor for agrometeorological use (example: Watermark), (Source: Pacher).

Fig. 13.16 illustrates the output of a soil moisture sensor. In this specific example, moisture is monitored at five different layers, with one sensor every 10 cm, and each sensor visualized by the time series of a different colour. For better optical recognition of each layer, the individual lines are separated by a slight offset on the Y–axis so as to spread the information over a larger area. The top of the graph indicates the sums of the moisture of all five sensors. The two major agronomical lines, full point and refill point, have already been established, with the area between them highlighted in red. The art of the irrigator is to maintain the moisture sums within those two lines, thus avoiding both overwatering, which would result in excessive drainage and runoff, and under–watering, which would cause great stress to plants.

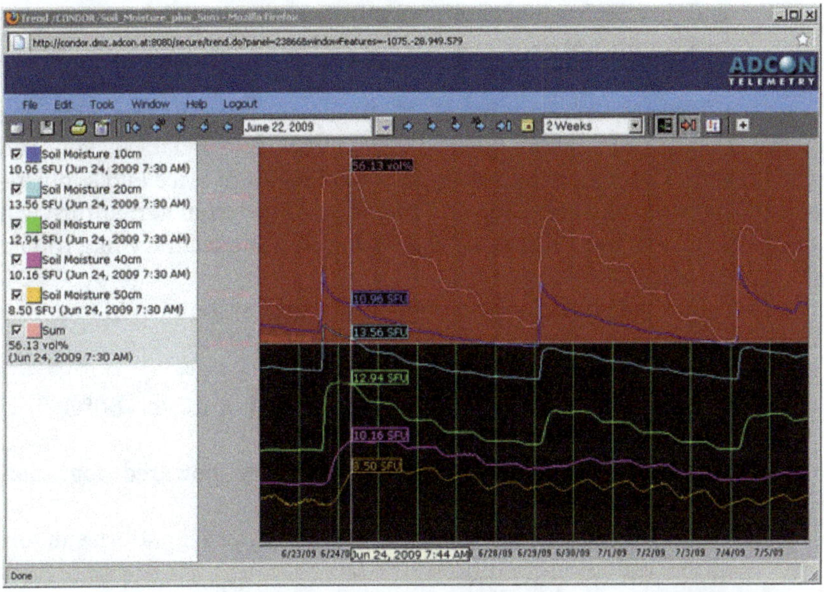

Figure 13.16: Monitored soil moisture graphs, (Source: Pacher).

A discussion about which sensor is best for which crop and which soil would go far beyond the scope of this book. Vendors certainly have good arguments and supporting information explaining why their products would fit best, and there is abundant literature on the topic. So what can a farmer get out of it? The list of potential benefits is long:

- save water: savings of up to 80% as compared to standard flood irrigation have been achieved, but even with more accurate systems like drip and pivot significant savings of another 50% or more is possible;
- save energy: water not pumped is energy not consumed;
- save fertilizer: over–irrigation washes out nutrients, while under–irrigation will not get the fertilizer to the active root zone, where it is needed;
- increase crop quality: For many crops, proper irrigation can help in improving the quality of the final product, be it by avoiding stress or by applying desirable amounts of stress (controlled deficit irrigation, partial root zone drying);
- increase crop quantity: most farmers are not aware of the disastrous effects of over–irrigation, which are about as severe as the effects of under–irrigation. Very few species of crop can cope with waterlogging, the excessive application of water beyond field capacity. The effect is akin to suffocating the plant, replacing oxygen in the soil with water, which slows down its water uptake

and thus its "metabolism", effectively hampering the growth cycle to a significant extent.

Data loggers: In order to automatically collect and store data, the above–mentioned sensors need to be connected to a data logger. The market offers a huge variety of loggers at all levels of price and sophistication. As with sensors, the prime selection criterion in agriculture will not be precision, but "fitness for the job"–sufficient accuracy in a robust package at an affordable price.

Therefore, a list of requirements for a logger system could look like the following:

- robust, at least IP–65 rated case;
- wide operational temperature range, from -40°C to +60°C;
- case resistant to agro–chemicals;
- low power consumption, allowing battery–powered operation with compact solar panels;
- at least three weeks of operation off the internal batteries in low global radiation conditions;
- maintenance–friendly connector system to allow even non–technical personnel to replace defective components or routinely swap one sensor element for a new or a recalibrated one;
- 10–bit resolution or higher;
- two weeks of memory;
- simple user interface;
- a telemetry option that is easy to install and configure.

As stated above, it thus seems more recommendable to trade ultimate accuracy and a wealth of features for robustness and ease of operation. High accuracy comes at a price, and it is wiser to invest available funds in equipment fit for the job and make accruals for ongoing maintenance than to buy the most accurate gear available and then never perform the required recalibrations because they are too expensive. On the other hand, it is equally unwise to buy the cheapest logger available and then to compromise data quality.

Figure 13.17 shows a variety of data loggers, most of which need an additional protective enclosure to prevent damage from environmental impact.

Date transfer: While data as such can be collected by a logger system as described above, they will have to be retrieved on a daily basis either by visiting the site and collecting data with a laptop, or by connecting a telemetry device to the logger. Experience shows that manual data retrieval does not work very well over time. In the days immediately after the purchase and installation of a weather station, data may be retrieved

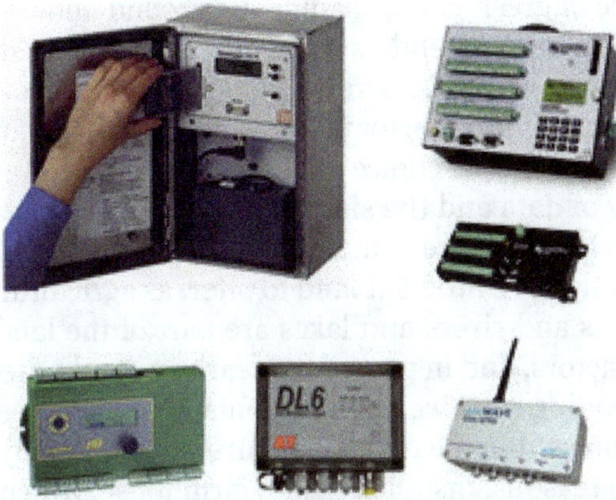

Figure 13.17: A variety of data loggers from different producers, (Source: Pacher).

daily. As the season progresses and people get busier and as the weather changes to cold and rainy, the intervals between site visits tend to get longer and longer.

It is thus highly advisable to purchase from the very beginning a telemetry device that will automatically transmit data from the weather station back to the base station. This not only provides a continuous stream of data, but also a "technical condition" of the station that allows one to check if the station and its sensors are still operational.

13.2.3 Installation of an agrometeorological weather station

A typical installation site in agriculture is very different from a meteorological or hydroglogical application. Weather stations are normally installed on the perimeters of a field (usually in the case of arable crops) or directly inside the crop (mostly in tree crops and vineyards). Therefore, a weather station should occupy as little space as possible so as not to obstruct the daily crop management routine, such as spraying, cropping, pruning, irrigating, etc. As a result, the required space for an ideal installation is frequently unavailable, and a suitable compromise needs to be found.

Some of the requirements listed above in the section on data loggers make more sense now, when regarded from an installation perspective. A small solar panel offers less exposure to wind pressure, thus reducing the forces that need to be absorbed by the pole of the station, while large solar panels might require guy wires to securely anchor the station in the soil. Small solar panels are also less visible, and thus less attractive to vandals and thieves and less likely to be shot at (yes, this *is* a problem!).

In order to comply with the requirements of evapotranspiration models, the ideal installation site is described by the FAO (Allen *et al.*, 1889) as "an extensive surface of green grass, shading the ground and not short of water". Other sources) recommend a surface area of 225 to 300m^2 of flat land, "mown at least once per week". There are numerous other recommendations, but in most cases the limiting factor is the availability of land for the purpose.

Nonetheless, the proper choice of the installation site is a crucial factor for the usability of data and the significance of the computation results. This is of especial importance in mountainous areas, which, like all alpine regions, frequently have little flat land to offer, as agricultural land alternates with forests and rivers and lakes are part of the landscape. All of these external factors, and in particular nearby water bodies, can largely influence the readings of a weather station and simulate environmental conditions not present in the crop area. Furthermore, many stations are, for convenience reasons, installed near a farmhouse, which not only obstructs wind, but quite frequently alters all other readings as well.

Figure 13.18a shows an ETo station. The ET_0 weather station provides continuous monitoring of air temperature, solar radiation, rainfall, relative humidity, and wind speed and direction. Using these measurements, a computer program calculates reference evapotranspiration ET_0 using the Penman–Monteith equation. It is set up according to FAO56 on a large patch of grass, with the rain gauge on a separate pole, and a telecommunications antenna at the tip of the tower. In Fig. 13.18b a Bowen Ratio station is shown, which estimates estimates the vertical fluxes of sensible and latent heat at the local surface. Flux estimates are calculated from observations of net radiation, soil surface heat flux, and the vertical gradients of air temperature and relative humidity.

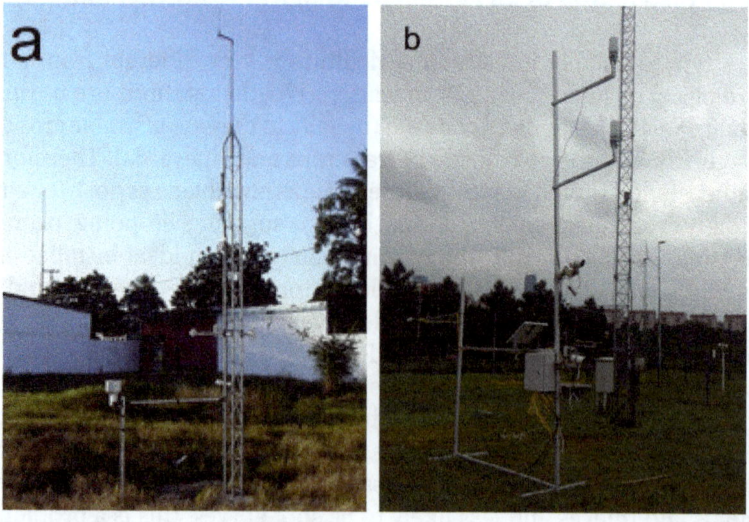

Figure 13.18: ETo station with telemetry (a, left) and Bowen–Ratio station (b, right); (Source: Pacher (a); Saylan (b)).

It is highly recommended to take photographs of the installation at a distance of about 20 m to the station from all four directions. A very desirable result of having a weather station is the ability to get time series of site–specific weather data, enabling a grower to correlate crop events to weather events. This requires the ability to identify at any given time the original condition of the site when the weather station was first installed; photographs thus enable the station's operator to restore the original site conditions.

Often agrometeorological weather stations have to be placed within crop fields or within canopy in orchards for measuring accurately micro-climatic conditions (e.g. for disease monitoring, irrigation scheduling or specific other applications); (Fig. 13.19 and Fig. 13.20).

In such cases the temporal change of the growing canopy should be considered and reported (by regular measurements of Leaf area index or canopy height etc. or by observations by means of notes or photos) for data interpretation. For example, if canopy height or density is changing during measurement period, the vertical gradients and behaviour of microclimatic parameters can be affected significantly.

While taking pictures it is also very advisable to take the GPS coordinates of the weather station–not only to be able to link the data to an exact location, but also to allow such mundane tasks as finding the equipment again!

Figure 13.19: Agrometeorological weather stations in orchards for disease monitoring (Source: PFNS).

Figure 13.20: Agrometeorological weather stations (with variety of sensors) in an alfalfa field for crop water balance estimations (Source: BOKU–Met).

14

Remote Sensing Technologies

14.1 Introduction

Remote sensing (either by ground, aircraft, drones or satellite–based sensors) is a technique to gather information about a target without "touching" it. It may be split into "active" and "passive" remote sensing. Passive sensors gather radiation that is emitted or reflected by the object or surrounding areas. While, active sensors provide their own source of energy (electromagnetic) to illuminate the objects they observe. An active sensor emits radiation in the direction of the target to be investigated. The sensor then detects and measures the radiation that is reflected or backscattered from the target. However, it is notable to mention that for signal interpretation and calibration, in many cases ground truth data (e.g. biomass, soil water content, etc., directly measured by conventional ground–based methods) are required.

Some of the remote sensing applications can include:

- cartographic mapping of regions/world;
- thematic mapping;
- identification of substances, structures and regional interrelationships;
- recognition of processes and trends;
- monitoring of regional and seasonal developments;
- agroforestry mapping;
- mapping of soil characteristic, crop condition assessment and drought monitoring weather forecasting.

It is, then, partly geometric and partly radiometric information which is of interest. While geometric and mapping information is provided by a black–and–white or color photograph, radiometric information is contained in:

- spectral reflectance;
- scatter / backscatter varieties;
- emissivity as a function of materials and substances;
- temporal variations;
- structural / textural features.

Satellite–based remote sensing can use these electromagnetic bands for earth surface observations, for which the atmosphere is transparent (see Fig. 2.4 in chapter 2 and Tab. 14.1). Sensor type is roughly divided into two: Optical sensor and Microwave sensor. Optical sensors observe visible lights and infrared rays (Visible/near infrared, thermal infrared). Microwave sensors receive microwaves, which is longer wavelength than visible light and infrared rays; and observation is not affected by day, night or weather.

The "raw" data of satellite measurements need to be processed to obtain specific information from the earth's surface, which is then made ready for further applications.

Optical bands:	**Wavelength (micron – µm)**	
UV and visible		
Near infrared (NIR)	0.30–0.70	
Middle infrared (MIR)	0.70–3.00	
Thermal infrared (TIR)	3.00–6.00	
Far infrared (FIR)	6.00–20.00	
	Longer than 20.00	
Microwavebands:	**Frequency (GHz)**	**Wavelength (cm)**
P–band	0.31	100.00 – 30.00
L–band	1 – 2	15.00 – 30.00
S–band	2 – 4	7.50 – 15.00
C–band	4 – 8	3.75 – 7.50
X–band	8 – 12	3.80 – 2.50
K–band	18 – 27	1.10 – 1.70
Q–band	33 – 50	0.60 – 0.90
V–band	40 – 75	0.40 – 0.75
W–band	75 – 110	0.27 – 0.40
KU	12 – 18	1.70 – 2.50
KA	27 – 40	0.75 – 1.10

Table 14.1: Nomenclature of spectral bands used in remote sensing[4].

[4] https://www.nasa.gov/directorates/heo/scan/communications/outreach/funfacts/txt_band_designators.html – https://www.everythingrf.com/tech–resources/frequency–bands.

The most relevant variables that can be measured over land are:

Solar radiation, daily global albedo (the fraction of solar radiation reflected back from the Earth's surface and clouds), vegetation indices, LAI, land surface temperature, soil surface moisture, rainfall, terrestrial water storage, cloud cover, fires and burnt areas, snow cover inboth hemispheres, digital elevation maps of ice sheet surfaces, glacier evolution and land cover.

An ever greater quantity of useful spectral information has been supplied to geoscientists from sophisticated space sensors with increased detection capability. Therefore, these data are a very attractive source of geospatial data, that needs more exploration, especially for areas with sparse/absence of ground measurements. Because the quality of spectral information depends on its sensor characteristics, sensor development is the key to remote sensing techniques.

In this context, Tab. 14.2 reveals that the new generation of satellite sensors (e.g. MODIS on TERRA platform, VEGETATION on SPOT, SEVIRI on MSG, AVHRR–3 on EPS/NOAA, and recently Sentinel 1+2) have brought an upgraded level of remote–sensed information to the user community. Thanks to improved spatial, temporal, spectral and angular sampling of the radiative fields emerging from the surface of the Earth potentials applications strongly may increase. The time resolution and global coverage provided by the new satellites/instruments, together with the extensive sampling in both the spectral and angular domains, have paved the way for a broad spectrum of novel applications, specifically within the scope of land surface processes and land–atmosphere interactions.

For example, several vegetation indices, such as VCI, VHI and TCI, are used in single or in combination to detect drought conditions in agriculture (Tab. 14.2).

Climatic variables (examples)	Biophysical variables (examples)
Surface temperature	**NDVI (see below)**
Precipitation	SAVI (Soil adjusted vegetation indices)
Snow cover	LAI (Leaf Area Index)
Solar radiation	VCI (Vegetation Condition Index)
Albedo	VHI (Vegetation Health Index)
Cloud cover and other cloud products	TCI (Temperature Condition Index)
Synthetic Aperture Radar (SAR) products (precipitation, soil moisture, snow)	EVI (Enhanced Vegetation Index)
	Soil moisture
	Vegetation cover
Air stability	Land cover
Storm detection	Evapotranspiration
Ozone content	Degree days
Sea ice, sea wind	Burned area

Table 14.2: Meteorological and biophysical variables surveyed by satellite sensors.

Moreover, satellite measurements are available from several observation systems, then a logical next step is to combine the best features from each approach into a single estimate. The products of this type are often called: multi–sensor, blending or multi–source fusion.

14.2 Remote sensing for phenology and vegetation dynamics

The most advantages and benefits of using satellite data is that, they cover large portions of the earth continuously. Hence, time series of remotely sensed data are an important source of information for understanding land cover dynamics. Vegetation dynamics can be defined over several time scales. In the short term, plant communities have seasonally changing phenological phases which typically follow annual cycles.

Between years, phenological markers (e.g., onset of greenness, length of growing season) may respond differently; these changes are affected by short–term climate fluctuations (e.g., temperature, rainfall) and/or anthropogenic forcing (e.g., groundwater extraction, urbanization) (Elmore *et al.*, 2000). Over a longer time period, annual shift of the beginning of phenological phases may be a result of climate changes and large–scale anthropogenic disturbance.

Differentiation of annual, inter–annual, and long–term phenological patterns are an important component of the monitoring and modelling of global ecosystems and may lead to a better understanding of how and why land cover changes over time. The most common measure and the most used satellite–derived biophysical of the photosynthetic 'greenness' of vegetated land cover used to determine the phenological phases is the normalized difference vegetation index (NDVI) (e.g. Leeuwen *et al.*, 2006; Peters *et al.*, 2002); it is described in more detail below.

The Normalized Difference Vegetation Index (NDVI). The NDVI is an index of plant "greenness" which is defined as the ratio of the difference between the near infrared and red reflectances to their sum (Fig. 14.1); it is written as:

$$\text{NDVI} = (\text{near infrared} - \text{red}) / (\text{near infrared} + \text{red}) \qquad (14.1)$$

The NDVI index is called normalized because it is divided by the sum of radiances and thus normalizes somewhat for differences in solar spectral irradiances. Calculations of NDVI for a given pixel show a number that ranges from minus one (-1) to plus one (+1). Negative values are sometimes found when the red reflectance is higher than the near infrared, as for certain types of dry soils. The value of the NDVI is mainly determined by the difference between the near infrared response–which increases with increasing vegetation on the scene–and the red response–which decreases with decreasing vegetation.

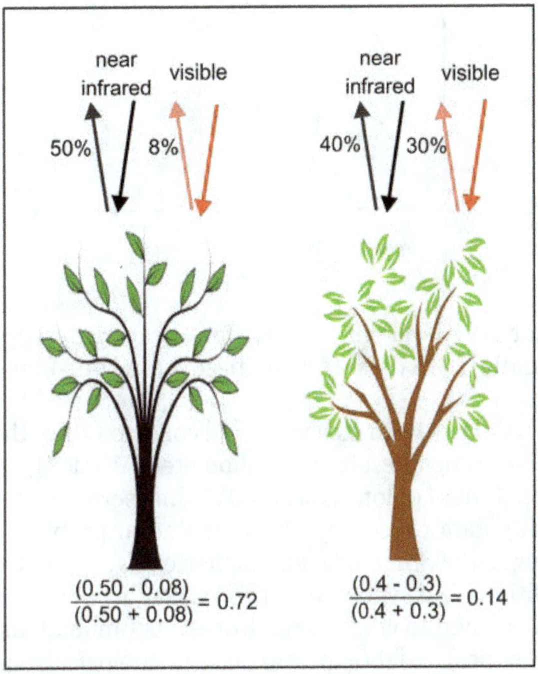

Figure 14.1: Scheme of NDVI detection (schematic, based on: https://earthobservatory.nasa.gov/Features/MeasuringVegetation/measuring_vegetation_2.php).

Field measurements have established that there is a relationship between green biomass and the NDVI up to a certain value of the biomass cover, above which the NDVI remains constant (this is the saturation level) (Fig. 14.2 and 14.3).

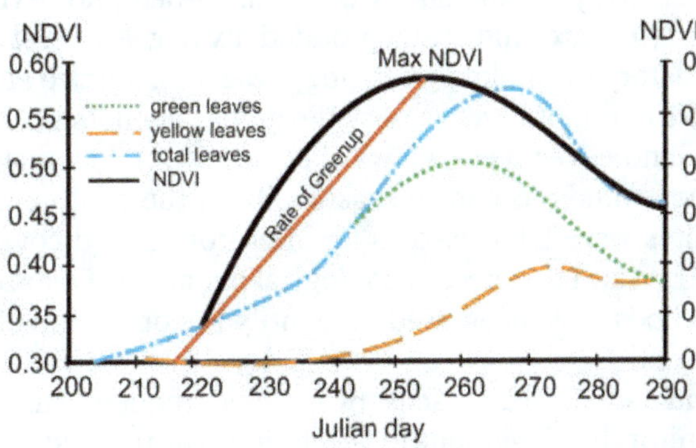

Figure 14.2: NDVI development for sorghum plots in Mali (schematized, based on Bartholomé, 1987).

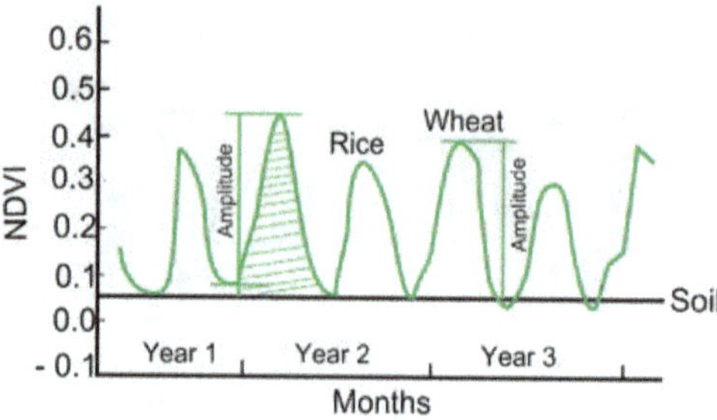

Figure 14.3: NDVI development curve for wheat–rice cropping system in the Punjab (India) (GVI data) (schematized, based on Malingreau, 1986).

Global NDVI have been measured and collected since the early 1980s by Advanced Very High Resolution Radiometer (AVHRR) satellites. However, the full potential of long–term NDVI time series is often hampered by poor quality data caused by instrumentation problems, changes in the sensor angle, the Sun angle, atmospheric (e.g., clouds and haze)and ground conditions (e.g., snow cover) and the aging of satellite detectors. These problems tend to create data drop–outs (anomalously low NDVI values in time series) or data gaps and make phenological markers difficult to identify (Reed *et al.*, 1994).

Overlarge scales, it is extremely difficult to obtain consistent field phenology observations across land covers which represent ecosystem activity rather than species–level phenology. To overcome this difficulty, numerous studies have used high–frequency coverage of the terrestrial biosphere by the National Oceanic and Atmospheric Administration (NOAA) Advanced Very High Resolution Radiometer (AVHRR) to quantify ecosystem vegetation phenology.

The NDVI has been related to several biophysical parameters including chlorophyll density, percent canopy cover, absorbed photosynthetically active radiation, leaf area index and productivity (e.g. Kogan *et al.*, 2003). Early in the history of satellite phenology research, Justice *et al.* (1985) used the NDVI to qualitatively assess the global phenological dynamics of numerous land cover types. Goward *et al.* (1985) demonstrated that the NDVI corresponds to known seasonality in the continental United States. Satellites were later used as an indicator of land cover changes in South America and to detect phenological dynamics. Quantitatively, a variety of methods have been used to identify the onset and offset dates from satellite data.

Visible and near–infrared sensors are most frequently used to determine and monitor the actual state of vegetation. The most advantages and benefits of using satellites isthat, they can cover near–globally the earth surface and atmosphere and are equipped with different type of sensors.

They also observe long–term time periods with frequent measurements. Another solution, which is currently still being developed, is the use of microwave sensors. Below are listed the most extensively used data series.

Satellite sensors used for phenological studies:

- AVHRR 1981–present; (8 km resolution) global coverage, 1989–present; (1 km resolution);
- SPOT Vegetation 1998–present; 1 km resolution;
- MODIS 2000–present; 250m, 500m, 1 km resolution;
- TM/ETM Landsat 1984–2003 16 days, 10–30 m resolution;
- SMMR data from 1978–1987, and SSM/I data from 1987 to 2005;
- SENTINEL 1 and 2 products (since 2015), high resolution (larger than10 m).

Numerous studies have been conducted to detect and estimate vegetation phenology analogues at continental or regional levels; these include:

- the start date of a growing season (SGS);
- the end date of a growing season (EGS);
- the length of a growing season (LGS).

The analysis temporal variability of vegetation indices allows for the determination of the metrics to be used to characterize the phenological cycle and its individual components (Fig. 14.2).

The most frequently used indices retrieved from temporal curves of NDVI are listed in Tab. 14.3. Their biological meaning and recognized limitations are also presented.

Index	Type of measure	Definition	Biological meaning	Comments
Integrated NDVI	Overall productivity and biomass	Sum of positive NDVI values over a given period	Annual production of vegetation	Not relevant when resource quality is *at least as* important as quantity (e.g. highly selective foragers)
Annual Maximum NDVI	Overall productivity and biomass	Maximum value of NVDI over a year	Annual production of vegetation	Sensitive to false highs and noise correction
Relative annual range of NDVI	Inter–annual variability in productivity	(Maximum NDVI value– Minimum NDVI value)/ INDVI	Enables inter–annual comparisons of vegetation biomass	Sensitivity of range definition to outliers in both directions

Rate of increase/ decrease of NDVI	Phenology	Slope between two NDVI values at two defined dates, slopes of fitted logistic curves to NDVI time–series	Greening up (spring) or senescence (fall) phases	Sensitive to false highs and noise correction
Dates of start or end of growing season	Phenology	Dates estimated from threshold models or moving average procedures	Start of green–up	Accuracy is linked to temporal scale of Time series considered (with the problem that higher temporal resolution leads to more contaminated data)
Length of 'green' season	Phenology	Number of days where NDVI = 0; number of days between estimated date of green–up and end of growing season	In seasonal environments, number of days when food is available	Sensitive to false highs and noise correction
Timing of annual maximum NDVI	Phenology	Date when maximum NDVI value occurs within a year	Timing of maximum availability of vegetation	Sensitive to false highs and noise Correction

Table 14.3: Indices retrieved from temporal variation of NDVI and their relation to the vegetation cycle (Pettorelli *et al.*, 2005).

Other vegetation indices

The NDVI is a vegetation index that has demonstrated its usefulness in many ecological and agronomic studies. Hence, in some situations other vegetation indices might be more appropriate. The relationship between the NDVI and vegetation can be biased in sparsely vegetated areas (e.g. arid to semi–arid zones in Australia) and dense canopies (e.g. Amazonian Forest). In sparsely vegetated areas with a leaf area index (LAI) of less than 3, the NDVI is influenced mainly by soil reflectance, whereas in densely vegetated areas (i.e., LAI arger than 6), the relationship between the NDVI and NIR becomes saturated (Asrar *et al.* 1984; Xie *et al.*, 2008).

Therefore, in sparsely vegetated areas, the *soil–adjusted vegetation index* (SAVI) is recommended instead of the NDVI. However, the SAVI requires *local calibration* because it is difficult to predict how soil effects are manifested within large pixel areas: with aggregate soils and vegetation of many different types, each requires, in principle, separate calibration.

Another index that has appeared with MODIS is the *Enhanced Vegetation Index* (EVI) (Huete *et al.*, 2002). This index provides complementary information about the spatial and temporal variations of vegetation, while minimizing many of the contamination problems present in the NDVI, such as those associated with canopy background and residual aerosol influences. Whereas the NDVI is chlorophyll sensitive and responds mostly to RED variations, the EVI is more NIR sensitive and responsive to canopy structural variations, including LAI, canopy type, and architecture. This index is thus meant to take full advantage of the new state–of–the–art measurement capabilities of MODIS and other sophisticated sensors.

Additionally, the EVI does not become saturated as easily as the NDVI when viewing rainforests and other areas of the Earth with large amounts of green material. However, the EVI has been developed on MODIS data and so data are only available from 2000 onwards. Available MODIS–based indices related to phenological cycle are listed in Tab. 14.4.

MODIS Product	Spatial resolution	Description
MOD09GQK	250 m	Daily surface reflectance computer from MODIS bands 1 (620–670 nm) and 2 (841–876 nm).
MOD13Q1 NDVI	250 m	Normalized difference vegetation index (NDVI) computed from MOD09GQK and composited from 16 days of data.
MOD13Q1 EVI	250 m	Enhanced vegetation index (EVI) computed from MOD09GQK and composited from 16 days of data.
MOD15A2 LAI	1000 m	Leaf area index (LAI, one–sided) computed from 1 km surface reflectance and land cover definition using radiative transfer or empirical (backup) methods, 8–day composite.
MOD15A2 FPAR	1000 m	Fraction of photosynthetically active radiation (FPAR) absorbed by vegetation computed from 1 km surface reflectance and land cover definition using radiative transfer or empirical (backup) methods, 8–day composite.
MOD43B4	1000 m	Nadir BRDF–adjusted reflectance composited from 16 days of data.

Table 14.4: MODIS data products useful for phenological studies.

NDVI data sets are generally well–documented, quality–controlled data sources that have been pre–processed to reduce many of the problems mentioned. However, some noise is still present in the downloadable data sets, meaning that NDVI time series need to be smoothed over before being used. Such noise is mainly due to remnant cloud cover, water, snow or shadow, sources of errors that tend to decrease NDVI values.

Use of satellite data in phenology analysis, its benefits and limitations can be summarised as follows:

- seasonal and inter–annual variability;
- global monitoring, 25+ years;
- spatially integrated climate change impact;
- gaps in atmospheric disturbances;
- diagnostic: no information about future.

Figures 14.4 – 14.6 present some practical examples of satellite data use based on past observations and following events. Figure 15.5 shows the development of the summer droughts and heat waves over Europe (France) using NDVI deviations indicating drought and heat stress for vegetation in 2003, 2005 and 2006, showing the highest impact in the year 2003.

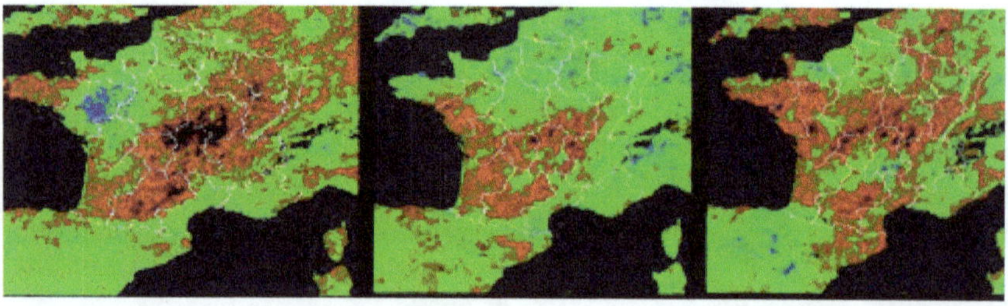

Figure 14.4: Summer droughts of 2003 (left), 2005 (center) and 2006 (right) in France. NDVI deviations (August/mean 2002–2004) based on VEGETATION/SPOT 5 data (by CNES, processed by O. Hagolle, IN: Nejedlik and Orlandini, 2008).

In comparison, Figure 14.5 shows the impact of the drought and heat wave in 2003 in terms of July surface temperature anomaly to the reference year 2002.

An example of observing crop phenology or agricultural management is shown in Fig. 14.6, indicating areas of harvested crops and/or bare soils over south–eastern Europe using NDVI.

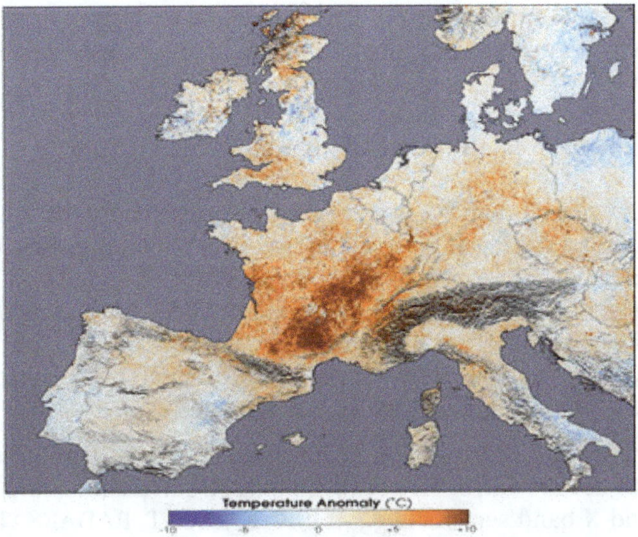

Figure 14.5: The 2003 heatwave on European scale (July 2003/July 2002). Surface temperature map based on MODIS data (by NASA Earth Observatory - VISIBLE EARTH); (Nejedlik and Orlandini, 2008).

Figure 14.6: Atmospherically corrected NDVI map of Hungary. NOAA/AVHRR 10–day composite image of 1–10 September 1997 (Nejedlik and Orlandini, 2008).

NDVI–based monitoring is used nowadays, for example, by the FAO and other organizations for predicting yield failures over the world and to provide early warning for hunger crises.

14.3 Microwave satellites for soil moisture estimation

Other important and complementary information for this application is provided by the satellite–based monitoring of soil moisture by radar satellites (using microwave sensors). In the past two decades, microwave remote sensing has proven successful missions for estimating dielectric properties of soil based on land surface emissivity, leading to soil moisture estimation. Several satellite soil moisture products are available from microwave, optical and thermal sensors (Brocca *et al.* 2017).

In the microwave range, low frequencies (X, C, and L bands) have typically been used to detect bare or vegetated soil surface moisture content. The C and X band sensors (e.g., AMSR–E, ASCAT, RADARSAT, WindSAT) onboard various satellites have shown promise for global surface water content estimates. However, the spatial resolution of the currently available satellite products is still toocoarse (25 km)when applied to local basins and regions and presents limits to practical applications in agriculture. It should also be borne in mind that one limit of remotely– sensed soil moisture data is that it provides information to a depth of only a few centimeters below the surface.

Although several uncertainties regarding data retrieval, validation, and climate–specific bias persist, microwave remote sensing of soil moisture has improved over the past few years (Mohanty *et al.*, 2017). While in the past, plant–available water was estimated exclusively by in–situ measurements or model simulations, in recent years remote sensing has played an increasingly important role in receiving spatial information on soil surface conditions (Wagner *et al.*, 2013).

Satellite and radar data are being applied to agriculture in several ways:

- agricultural water management (irrigation scheduling) by including remotely–sensed soil moisture status as a boundary condition for soil hydrology and crop growth models on various spatiotemporal scales (e.g. Thaler *et al.*, 2018);
- weather and climate forecasting by assimilating regional– and global–scale soil moisture into numerical weather and climate models;
- earth water balance estimates (evapotranspiration, groundwater recharge, surface runoff) by linking fluxes to surface and root–zone soil water content, resulting in improved drought or flood forecasting, and

- radar sensors can provide an accurate overview of the acreage being cultivated, while distinguishing between crop types and determining their health and maturity.

Further examples and information on microwave–based soil moisture monitoring and actual global maps of soil moisture can be found at: http://rs.geo.tuwien.ac.at/data–viewers/

14.4 Application examples of remote sensing in agricultural practice

The use of remote sensing data is steadily increasing for agricultural applications. Data is gathered either by satellites or by drones or tractors equipped with sensors, which are used increasingly for precision farming techniques. Just a few examples of such applications are briefly described below.

Satellite–based applications. Satellite sensors provide coverage of vast land areas and are especially useful for monitoring agricultural land use and mapping, field sizes and crop status, for calculating losses from severe weather events, and for conducting yield assessments. In Europe, for example, farmers report actual land use in order to receive subsidies. Land use is controlled by satellite, and farmers have web–based access to high–resolution aerial photographs or satellite pictures of their farm area and fields.

NDVI data are used for the crop yield forecasting system of the European MARS (Monitoring Agricultural Resources System; https://ec.europa.eu/jrc/en/mars),in combination with crop modeling and statistical data. Satellite data such as the NDVI are also used for agricultural drought monitoring systems (see Fig. 14.6 and the drought monitoring links discussed in Chapter 9), for observing phenology of vegetation or crops (Fig. 14.6) and for measuring soil surface temperature (Fig. 14.5).

One of the most significant developments in agriculture in recent years has been the rise of smart farming. It encompasses observing, measuring and responding to field and crop variability with the goal of optimizing returns on inputs while preserving resources. Combining space–based systems and terrestrial resources for smart farming has great potential for such aspects as yield optimization, irrigation management, and regulatory compliance (https://business.esa.int/projects/showcases/smart–farming–satellites).For example, the availability of high spatial resolution data allows field–based applications such as crop irrigation scheduling(based on MODIS and Sentinel 2 data) to be carried out operationally (Vuolo *et al.*, 2015).

However, many potential applications to practical agriculture are mostly limited by the spatial or temporal resolution of satellite remote–sensing data and by the inability to directly predict developments for actual decision making in crop management at the farm level. New high–resolution sensors (such as those in SENTINEL satellites) or methods (e.g. Haboudane *et al.*, 2004) may partly overcome such limitations, but problems still exist with regard to effective data management and storage, delayed data retrieval, and costs for tailored service products. Today, exist many farm applications for real–time decision making, therefore increasingly rely on precision farming technologies, which provide real–time data and are very flexible in changing the target area.

Precision farming applications. Ground sensors are handheld, mounted on tractors and combines, or free–standing in a field. Common uses for these include evaluating nutrient levels for more specific chemical and nutrient application and measuring weather or soil moisture content. Farmers find the most useful information closer to the visible spectrum, as color can be used to measure a plant's chlorophyll levels and provide insight into a plant's health and growth status. Simple red–green–blue sensors can provide color information, but more sophisticated sensors are available that peer into the near–infrared and short–wave infrared spectral bands (https://agfundernews.com/remote–sensing–powers–precision–agriculture.html).

Conventional applications are based on the visible and near–infrared portions of the spectrum. Increasingly, electromagnetic wavelengths are used in the range from the ultraviolet to microwave of the spectrum, enabling advanced applications such as light detection and ranging (LIDAR), fluorescence spectroscopy, and thermal spectroscopy. Also, the spectral bandwidth has decreased, which hyperspectral remote sensing, allows to improved analysis of specific compounds, molecular interactions, crop stress, and crop biophysical or biochemical characteristics. It enables the acquisition of data simultaneously in hundreds of spectral bands. A variety of spectral indices now exist for various precision agriculture applications (Mulla, 2013).

However, the more sophisticated the sensor, the higher the cost. Farmers must always check the potential for increased yield against the capital investment for each sensing platform. Sensing technologies are evolving rapidly. It is often up to farmers to use trial and error to determine what off–the–shelf products will be economically useful. Agricultural drones are high–tech systems that can do things a farmer can't: conduct soil health scans, monitor crop health, apply fertilizers, water fields, and even track weather and estimate yields. After collecting and analyzing data, these systems provide information for prompt action. Drones, in short, can mechanize every step of farming, eliminating the costs of human error and enabling farmers to react quickly to threats (such as drought conditions and pests). In the end, drones help farmers maximize income and returns on their investment (http://www.businessinsider.de/commercial–drone–uses–agriculture–business–military–2017–8?r=US&IR=T).

Drones are also starting to be used to apply nutrients and pesticides to plants, instead of using workers on foot or tractors. Aerial spraying is much faster and cheaper than using traditional methods, such as tractors. Crop dusting drones have been in operation in Japan for over two decades and are finally starting to make their appearance in precision farming globally(http://www.droneomega.com/drones–in–agriculture/).

Currently, the most used imaging sensors on drones for practical applications are:

- RGB (Red, Green Blue): for plant counting, elevation modeling and visual inspection; they can further detect the area and extent of crop damage from pests, hail, frost, etc.;
- NIR (Near infra–red): for water management, erosion analysis, plant counting, soil moisture analysis and crop health;
- RE (red edge): for plant counting, water management (water stress status detection) and crop health (e.g. nitrogen status);
- thermal infrared: for irrigation scheduling, plant physiology and yield forecasting; they produce high–resolution crop stress indices, etc.

Numerical examples

In the next 20 pages you can find numerical examples related to first seven chapters of the book called the basics. This chapter is also related to the numerical examples in Excel that are free available for download online at the address http://serbiaforexcell.com/education/book/.

Multiplication Factor	Short	Prefix	Symbol
1 000 000 000 000 000 000	10^{18}	Exa	E
1 000 000 000 000 000	10^{15}	Peta	P
1 000 000 000 000	10^{12}	Tera	T
1 000 000 000	10^{9}	Giga	G
1 000 000	10^{6}	Mega	M
1 000	10^{3}	Kilo	k
100	10^{2}	Hecto	h
10	10^{1}	Deka	da
0.1	10^{-1}	Deci	d
0.01	10^{-2}	Centi	c
0.001	10^{-3}	Mili	m
0.000 001	10^{-6}	Micro	μ
0.000 000 001	10^{-9}	Nano	n
0.000 000 000 001	10^{-12}	Pico	p
0.000 000 000 000 001	10^{-15}	Femto	f
0.000 000 000 000 000 001	10^{-18}	atto	a

Table E1: SI prefixed and symbols for multiplication and submultiplication of units.

E1 Origin and composition of the atmosphere

E1.1 Units

Pressure. In the SI system, pressure is measured in pascals; one pascal equals one newton per square meter. Atmospheric pressure, *p*, is the pressure exerted by the force of air on the Earth's surface. It is also classified as barometric pressure. With the help of a mercury barometer, atmospheric pressure can be determined. Atmospheric pressure is expressed in several different systems of units: millimeters of mercury (mmHg), standard atmospheres (atm), pascals (Pa), millibars (mb) and several others not mentioned here. Standard sea–level atmospheric pressure is referred to here as standard air pressure. Relations between units describing standard atmospheric pressure at sea–level are

$$760 \text{ mmHg} = 1 \text{ atm} = 101325 \text{ Pa} = 1013.25 \text{ mb}$$

Example:
If the partial pressure of the CO_2 inside the leaf tissue is 20 Pa convert it to milibars.

$$20 \text{ Pa} = 20\times10^{-2} \text{ hPa or } 0.2 \text{ mb}$$

Concentration. Concentration is expressed in a number of ways. The most common are parts per million (ppm) and parts per billion (ppb), which mean volumes of gas in 10^6 and 10^9 volumes of air, respectively. Converted to SI units, the volume ratio is equal to the mole fractions μmol mol^{-1} and nmol mol^{-1}, respectively.

$$1 \text{ ppm} = \frac{\text{gas volume}}{10^6 \text{ air volume}} = \frac{\text{mol}}{\text{mol}}\times10^{-6} \quad \text{(E1.1)}$$

$$1 \text{ ppb} = \frac{\text{gas volume}}{10^9 \text{ air volume}} = \frac{\text{mol}}{\text{mol}}\times10^{-9} \quad \text{(E1.2)}$$

Concentration can be expressed in pressure units. It is described by Dalton's law, which states that the total pressure of a mixture of gases is equal to the sum of the partial pressures of the individual component gases. The partial pressure is the pressure that each gas would exert if it alone occupied the volume of the mixture at the same temperature.

$$p = \sum_{i=1}^{n} p_i, \quad \text{(E1.3)}$$

where $p_1, p_2, \ldots, p_n$ represent the partial pressures of each component in air mixture. From there partial pressure of the specific gas is $p_i = n_i \cdot p$, where n_i is the mole fraction of the ith component in the total mixture of n components.

> ***Example:***
>
> If the concentration of CO_2 in the air canopy space is 380 ppm, express it's partial pressure, p_{CO2} in all pressure units if the standard air pressure is considered.
>
> $C_{CO2} = 380\,ppm = 380\times10^{-6} mol\,mol^{-1}, if\,p = 101325\,Pa, then\,for\,CO_2,$
> $p_{CO2} = 380\times10^{-6}\cdot101325\,Pa = 38.5\,Pa = 0.385\,mb,\ 0.289\,mmHg$

Photosynthesis can be modelled depending on how complex a model is required. In the past 30 years, photosynthesis models have developed from simple empirical (Monteith, 1977) to complex biochemical (Farquhar *et al.* 1980) models. Basically, the process of photosynthesis can be described through a gas exchange model where photosynthesis depends on the transpiration rate at the leaf level (Bierhuizen and Slatyer, 1965)

$$\frac{A_{LAI}}{E_{LAI}} = \frac{k_1}{VPD_{LAI}} \tag{E1.4}$$

where A_{LAI} is the net assimilation rate of CO_2 per unit of leaf area, E_{LAI} is the rate of evaporation per unit leaf area, k_1 is a constant for leaves of a given crop, and VDP_{LAI} is the leaf to canopy air space vapour pressure deficit. During CO_2 intake, the stomata are open, so plant is releasing water in the same time trough the process of transpiration. These two processes, assimilation and transpiration, are usually combined in photosynthesis models.

The authors of this simple model explain that from the leaf– to the canopy–level calculation, it is simply necessary to change the coefficient k_1. Then, equation C1.4 becomes

$$\frac{B_M}{T_M} = \frac{k_c}{VPD} \tag{E1.5}$$

where B_M is crop biomass production (kg), T_M is the crop transpiration rate, expressed in the amount of water (kg) and k_L is the canopy–level constant for a given crop (Pa) and *VDP* is the air vapour pressure deficit (the replacement of VDP_{LAI} with *VDP* can be made only if the temperature of the leaf is close to the air temperature). The relation between crop biomass production and transpiration rate is usually called the transpiration–use efficiency and describes the generated biomass per unit of transpired water. k_L is mostly attained from the relationship between CO_2 concentration inside and outside of the leaf.

Example:
If the transpiration–use efficiency for wheat is 4.93 g kg^{-1}, and the vapour pressure is 10 mb at 20 °C, calculate coefficient k_L for this cultivar.

First we need to express the vapour pressure deficit, which represents the difference between maximal vapour pressure (e_s) and vapour pressure (e) at given temperature. To do this, see example in C_4. Here, we will use the calculated value VDP = e_s -e = 23.29 -10 = 13.29 hPa = 1329 Pa.

$$k_L = VPD\frac{B_M}{T_M} = 1329 \cdot 4.93 \times 10^{-3} = 6.551\,Pa$$

E2 Energy balance of the Atmosphere

E2.1 Units

	MJ m^{-2} day^{-1}	J cm^{-2} day^{-1}	cal cm^{-2} day^{-1}	W m^{-2}
1 MJ m^{-2} day^{-1}	1	100	23.9	11.6
1 cal cm^{-2} day^{-1}	4.1868 10^{-2}	4.1868	1	0.485
1 W m^{-2}	0.0864	8.64	2.06	1

Table E2.1: Conversion factors for energy (J) received on a unit surface per unit time.

Commonly, the irradiance (radiation intensity) of a light source is given in W m^{-2}. However, in the case of photosynthesis rate calculation, it is expressed in quanta mol m^{-2} s^{-1} as the photosynthetic photon flux density (PPFD). The conversion between these two units is based on Eqs. E2.1 and E2.2. Therefore, the energy of the photon of the specific wavelength can be calculated as

$$E = 6.6262 \cdot 10^{-34}\,\frac{2.998 \cdot 10^{8}}{\lambda}\,. \qquad \text{(E2.1)}$$

This means that photons on the smaller wavelength have more energy. If the wavelength is in the PAR spectrum, for example 550 nm, then the energy of each photon, or quantum, is 3.61395×10^{-19} J. If the photon irradiance is 1500 µmol m^{-2} s^{-1}, then 1500 µmol of 550 nm light is falling on 1 m² every second. Multiplying that number by Avogadro's number (6.022×10²³ quanta mol⁻¹), we acquire the number of quanta m⁻² s⁻¹, and by multiplying it further by the already calculated energy of the quantum, we acquire the irradiance in J s⁻¹ m⁻² = W m⁻². Therefore, the conversion factor is expressed for the average wavelength of the PAR, not for all wavelengths between 400 and 700 nm (Campbell and Norman, 1998). In the literature, the conversion factor is usually given in the form of

$$1\, W m^{-2} \approx 4.6\, \mu mol\, m^{-2}\, s^{-1} \tag{E2.2}$$

If irradiance of a surface is 450 W m⁻² the value of PPFD is

$$PPFD = 450\, J\, m^{-2} s^{-1} \cdot 4.6 \cdot 10^{-6}\, mol\, J^{-1} = 2070 \cdot 10^{-6}\, mol\, m^{-2} s^{-1}$$

Other conversions:

$$MJ\, m^{-2}\, day^{-1} = W m^{-2} \cdot 3600\, s\, h^{-1} \cdot 24\, h\, day^{-1} \cdot 10^{-6}\, MJ\, W^{-1} = 0.0864 \cdot W m^{-2}$$

$$mol\, m^{-2}\, day^{-1} = \mu mol\, m^{-2}\, s^{-1}\, 3600\, s\, h^{-1} \cdot 24\, h\, day^{-1} \cdot 10^{-6}$$

$$mol\, \mu mol^{-1} = 0.0864 \times \mu mol\, m^{-2}\, s^{-1}$$

E2.1 Transfer of measured solar radiation (W m⁻²) in daily energy budget

Method 1. Daily sum of the energy received from the global radiation can be calculated as an integral of the function representing hourly measured solar radiation intensity. For this purpose, *numerical integration is used* is used as follows

$$G_d = h\left(\frac{G_0 + G_N}{2} + \sum_{i=1}^{N-1} G_i\right), \tag{E2.3}$$

where h is the ime step between two measurements in seconds (if the measures are made on every hour, h=3600s), N is the number of measurements made, and G_o and G_N are values of the first and last measurements. The intensity of the global radiation is measured by a pyranometer.

Example:

Measurements of incoming global radiation are made by pyranometers for the location Prospect Hill Tract, Harvard Forest (42°32'N, 72°10'W), USA on 26th June 2006 (Fitzjarrald and Sakai, 2009). Calculate the daily sum of the global radiation with the values given in the table below.

Time	G_d (W m^{-2})
5:30	13.4
6:30	34.6
7:30	104.1
8:30	144.8
9:30	353.5
10:30	372.9
11:30	365.8
12:30	207.8
13:30	252.8
14:30	409.1
15:30	216.1
16:30	240.2
17:30	175.3
18:30	97.5
19:30	25.2

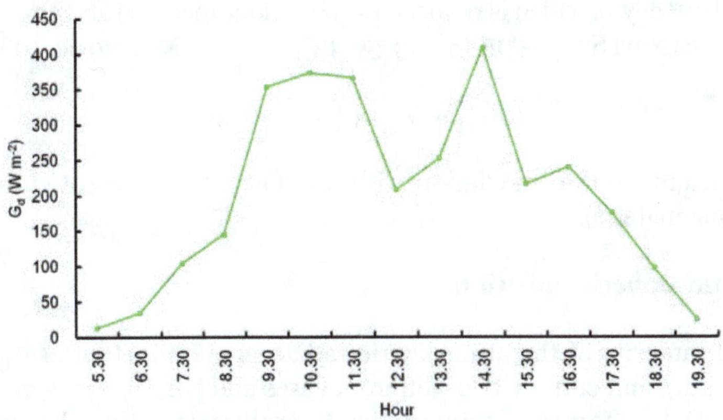

Figure E2.1: Daily course of global radiation.

$$G_d = 3600\left(\frac{13.4+25.2}{2} + (34.6+104.1+144.8+353.5+372.9+365.8+207.8+252.8+409.1+216.1+240.2+175.3+97.5)\right) = 1077.78\times10^4\,Jm^{-2}day^{-1}$$

Method 2. The daily global solar radiation (G_d, MJ m^{-2} day^{-1}) can be calculated with the Ångstrom–Prescott method, which considers the solar radiation that reaches the top of the atmosphere (G_a, MJ m^{-2} day^{-1}) and the relative sunshine duration or relative duration of direct solar radiation ($S_a \cdot S_p^{-1}$)

$$G_d = \left(a + b\frac{S_a}{S_p}\right)G_a \qquad \text{(E2.4)}$$

where S_a is the actual direct solar radiation duration (hour), S_p is the potential (maximal) direct radiation duration (hour). Coefficients a and b are empirical and represent the ratio of solar radiation that reaches the Earth's surface on a completely clear day and the solar radiation that reaches the top of the atmosphere, respectively. The values of the coefficients are changing depending on geographical location. This method is used if the solar radiation measurements are missing. In addition, if there is no actual radiation data available coefficients a = 0.25 and b = 0.5 can be used. The coefficients can be found in the literature since this method is widely used.

Method 3. The *Coulson method* is an additional method for estimating the daily global radiation ($G_{d,}$ MJ m^{-2} day^{-1}). This method is based on the daily global radiation that reaches the Earth's surface on a cloudless day ($G_{do,}$ MJ m^{-2} day^{-1}), which is reduced by daily cloudiness (n) and empirical constant ξ (Novi Sad, 45°15'18", 19°50'42", ξ = 1.89, Mihailović, 1988)

$$G_d = G_{d0}\left[1 + (1 - \xi)n\right] \tag{E2.5}$$

An example of a calculation with this formula is presented in the electronic material.

E2.3 Atmospheric radiation

The intensity of the atmospheric radiation is a function of the air temperature and emissivity of different gases and particles present (see Chapter 2). For the quantification of the atmospheric radiation, the Brunt's empirical relation (Brunt, 1932) is frequently used

$$B_a = \sigma T_a^4\left(0.625 + 0.056 \cdot \sqrt{e}\right) \tag{E2.6}$$

In this formula, the water vapour is the gas responsible for the atmospheric radiation, and its concentration is given in the form of vapour pressure (e) in hPa at absolute air temperature T_a in K.

Example:

Using Brunt's empirical formula, calculate the atmospheric radiation when the air temperature is 20°C and the vapour pressure is 5 mb.

$$B_a = 5.67 \cdot 10^{-8}\,(20 + 273.15)^4\left(0.625 + 0.056 \cdot \sqrt{5}\right) = 314.146 \text{ W m}^{-2}$$

To calculate the energy that the Earth's surface receives every day per cm^2 from atmospheric radiation in J cm^{-2} day^{-1} we need to multiply the result by 8.64 as given in Tab. 2.5.

$$B_{ad} = 314.146 \cdot 8.64 = 2714.22 \text{ J cm}^{-2} \text{ day}^{-1}$$

E2.4 Outgoing terrestrial radiation

The intensity of the outgoing terrestrial radiation is a function of the surface temperature and emissivity. The equation used for the calculation is in Chapter 2 Eq. 2.3, and an additional example is given in electronic material.

> ***Example:***
> If the soil surface temperature is 15°C and the soil emissivity is 0.95 calculate the intensity of the longwave radiation.
>
> $$B_E = 0.95 \cdot 5.67 \cdot 10^{-8}(15 + 273.15)^4 = 371.35\ W\ m^{-2}$$

E3 Soil and air temperature

E3.1 Units

	Kelvin	Celsius	Fahrenheit
Kelvin (K)	–	1K–273.15	(1K–273.15) 9/5 + 32
Celsius (°C)	1°C + 273,15	–	(1°C·9/5) + 32
Fahrenheit (F)	(1F–32) 5/9 + 273.15	(1F–32) · 5/9	–

Table E3.1: Conversion between Kelvin–Celsius–Fahrenheit.

$$1\ °C = 1 + 273.15\ K = (1 \cdot 9/5) + 32\ F$$

Heat capacity at constant pressure, C_p (J kg^{-1} K^{-1}), and heat capacity at constant volume, C_v (J kg^{-1} K^{-1}), are the commonly used constants. Their relationship depends on the gas, and for the ideal gas, the relationship is

$$\gamma = \frac{C_p}{C_v} = \frac{5}{3} \tag{E3.1}$$

E3.2 Daily, annual temperature, their variation and extremes

The calculation of daily and annual temperature variation are given in the electronic material. Temperature extremes like freezing days are defined with temperature above or below certain threshold (Tab. E3.2).

	Maximum temperature (°C)	Minimum temperature (°C)
Frost day	–	≤ 0 °C
Freez day	≤ 0 °C	–
Strong frost day	–	≤ -10 °C
Summer day	≥ 25 °C	x
Tropical day	≥ 30 °C	–
Tropical night	–	≥ 20 °C

Table E3.2: Daily temperature extremes.

E3.3 Calculating Accumulated Degree Days

Air temperature method called degree–days (DD) are widely used in agriculture as a method for predicting plant phonological phases and pest and disease incubation period, since it is more suitable than calendar time. Sum of degree days over particular period usually called accumulated or growing degree days (GDD) represent the accumulation of temperature above a given threshold. Threshold temperatures are predefined in laboratory experiments and here are given in Chapter 3 Table 3.4.

To calculate DD for specific day different methods are developed.

Method 1. Averaging method (McMaster and Wilhelm, 1997):

$$DD = T_{average} - T_t = \frac{T_{\max} + T_{\min}}{2} - T_t \quad \text{(E3.2)}$$

where DD are the degree–days, T_{max} and T_{min} are the daily maximum and minimum temperatures, respectively, and T_t is the temperature threshold. If the result is negative, it is assumed that there is no accumu-lation, and DD=0.

This method is commonly used in practice, but it can introduce error into the calculation since it can underestimate DD when temperatures fluctuate above and below the minimum threshold. In that time, plants will experience temperatures above the threshold, but this will not be accounted in the calculated DD.

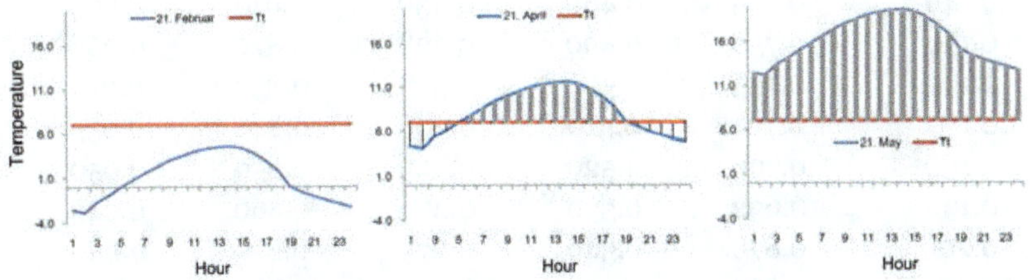

Figure E3.1: Temperature above threshold (T_t) included into the GDD calculation.

To compensate for this error McMaster and Wilhelm (1997) are using following rules:

if $T_{min} > T_t$ then

$$DD = \frac{T_{max} + T_{min}}{2} - T_t, \tag{E3.3}$$

else if $T_{max} < T_t$

$$DD = 0, \tag{E3.4}$$

else if $T_{max} > T_t > T_{min}$

$$DD = \varepsilon\, f(R) \tag{E3.5}$$

where $\varepsilon = \frac{T_{max} - T_{min}}{2}$ and $R = \frac{T_t - T_{min}}{T_{max} - T_{min}}$, while function $f(R)$ is taken from table based on calculated R.

R	$f(R)$	R	$f(R)$	R	$f(R)$
0.000	1.000	0.340	0.495	0.670	0.167
0.010	0.981	0.350	0.483	0.680	0.159
0.020	0.962	0.360	0.471	0.690	0.152
0.030	0.944	0.370	0.459	0.700	0.144
0.040	0.927	0.380	0.448	0.710	0.137
0.050	0.910	0.390	0.436	0.720	0.130
0.060	0.893	0.400	0.425	0.730	0.123
0.070	0.876	0.410	0.413	0.740	0.116
0.080	0.859	0.420	0.402	0.750	0.109
0.090	0.843	0.430	0.391	0.760	0.102
0.100	0.827	0.440	0.381	0.770	0.096
0.110	0.811	0.450	0.370	0.780	0.090
0.120	0.796	0.460	0.359	0.790	0.084
0.130	0.780	0.470	0.349	0.800	0.078

0.140	0.765	0.480	0.339	0.810	0.072
0.150	0.750	0.490	0.328	0.820	0.066
0.160	0.735	0.500	0.318	0.830	0.061
0.170	0.721	0.510	0.308	0.840	0.055
0.180	0.706	0.520	0.299	0.850	0.050
0.190	0.692	0.530	0.289	0.860	0.045
0.200	0.678	0.540	0.279	0.870	0.040
0.210	0.664	0.550	0.270	0.880	0.036
0.220	0.650	0.560	0.261	0.890	0.031
0.230	0.636	0.570	0.251	0.900	0.027
0.240	0.622	0.580	0.242	0.910	0.023
0.250	0.609	0.590	0.233	0.920	0.019
0.260	0.596	0.600	0.225	0.930	0.016
0.270	0.583	0.610	0.216	0.940	0.013
0.280	0.570	0.620	0.208	0.950	0.010
0.290	0.557	0.630	0.199	0.960	0.007
0.300	0.554	0.640	0.191	0.970	0.004
0.310	0.532	0.650	0.183	0.980	0.002
0.320	0.519	0.660	0.175	0.990	0.001
0.330	0.507			1.000	0.000

Table E3.3: Values of f(R).

Method 2. The Baskerville–Emin (BE) method requires a curve (sine or cosine function) for the maximum and minimum temperature. The curve will then give values of the air temperature for every hour and the area of the curve above the base temperature is the DD.

Method 3. Use of measured data is the most accurate method, since they represent the real field conditions and are not mathematically reproduced. After assimilation of measured air temperature data, DD is calculated by subtracting base temperature from the measured data.

Values of DD for every day are summed up in the particular period to give GDD.

Example:

Calculate the degree day for maize with threshold temperature of 10 °C calculate DD for days when T_{max} was 15 °C and 11 °C, and T_{min} 8 °C and 9 °C respectively.

$$DD1 = \frac{15+9}{2} - 10 = 12.5 - 10 = 2.5$$

$$DD2 = \frac{11+8}{2} - 10 = 9.5 - 10 = -0.5, \quad DD2 = 0$$

E3.4 Approximation of hourly temperatures

If hourly temperature data are missing, there are several methods in use that can reproduce hourly temperatures from some time specific values such as temperature extremes. The Cessarici temperature model is based on the maximum (T_{max}) and the minimum (T_{min}) temperature of the air during the day, the temperature of the air (T_0) at the time of sunset (H_0) and the time of the sunrise on the next day.

Symbols	Sunrise (time of the T_{min})	Time of the T_{max}	Sunset	Sunrise (time of the sunrise the next day)
Time	H_{min1}	H_{max}	H_0	H_{min2}
Temperature	T_{min1}	T_{max}	T_0	T_{min2}

$$T(t) = \begin{cases} T_{min1} + \alpha \sin\left[\frac{\pi}{2}\left(\frac{t - H_{min1}}{H_{max} - H_{min1}}\right)\right] & H_n < t \le H_x \\ T_0 + R\sin\left[\frac{\pi}{2} + \frac{\pi}{2}\left(\frac{t - H_{max}}{4}\right)\right] & H_x < t \le H_0 \\ T_0 + b\sqrt{t - H_0} & H_0 < t \le H_p \end{cases} \quad \text{(E3.6)}$$

$$T_0 = T_{max} - 0.39(T_{max} - T_{min2}) \quad \text{(E3.7)}$$

Coefficients α, R and b are calculated from the given data:

$$\alpha = T_{max} - T_{min}, \; R = T_{max} - T_0, \text{ and } b = \frac{T_{min2} - T_0}{\sqrt{H_0 - H_{min2}}} \quad \text{(E3.8).}$$

Example:

On the 17th and 18th of May 2007 on the Rimski Sancevi measurement site extreme temperature values were measured. Hourly values are missing. For the purpose of data filling and frost forecast, calculate the missing data with the Cessarici method.

Symbols	Sunrise (time and temperature T_{min})	Time and value of the T_{max}	Sunset (time and temperature)	Sunrise – the next day (time and temperature)
Time	H_{min1}	H_{max}	H_0	H_{min2}
Temperature	T_{min1}	T_{max}	T_0	T_{min2}
Time	5:15	16:10	20:10	5:13
Temperature	13.4	19.6	16.7	10.5

$$\alpha = T_x - T_n = 19.6 - 13.4 = 6.2\ ^\circ C\ ,$$

$$R = T_x - T_0 = 19.6 - 16.7 = 2.9\ ^\circ C,$$

$$b = \frac{T_p - T_0}{\sqrt{H_0 - H_p}} = \frac{10.5 - 16.7}{\sqrt{20.1 - 5.13}} = \frac{-6.2}{3.869} = -1.6\,.$$

Date	Hour	$H_{min1} < t \le H_{max}$	$H_{max} < t \le H_0$	$H_0 < t \le H_{min2}$
17.5.2007	6	14.2		
17.5.2007	7	15.0		
17.5.2007	8	15.9		
17.5.2007	9	16.7		
17.5.2007	10	17.4		
17.5.2007	11	18.0		
17.5.2007	12	18.6		
17.5.2007	13	19.0		
17.5.2007	14	19.3		
17.5.2007	15	19.5		
17.5.2007	16	19.6		
17.5.2007	17		19.4	
17.5.2007	18		18.8	
17.5.2007	19		17.9	
17.5.2007	20		16.8	
17.5.2007	21			15.2
17.5.2007	22			14.5
17.5.2007	23			14.0
17.5.2007	24			13.5
18.5.2007	1			13.2
18.5.2007	2			12.8
18.5.2007	3			12.5
18.5.2007	4			12.2
18.5.2007	5			11.9

E3.5 Frost prediction

There are several methods for predicting the frost appearance. The most used method in practice is the FAO model (Snyder and de Melo–Abreu, 2005) for predicting the minimal air temperature during the night and frost duration. This method was developed in the framework of active frost protection with sprinkle spraying. Therefore, the method also gives the mm of water that need to be sprinkled over the 1h period in order to save plant from damage.

The function for the minimal air temperature prediction is derived from the observations. The first step is to calculate linear regression coefficients (a, b) from the temperature measured 2 hours after sunset (T_{2h}) and the observed minimal air temperatures (T_{min}) during the frost nights, without wind and precipitation. These coefficients than can be used to calculate initial minimal temperature for all values of air temperature observed

$$T_p^1 = b_1 T_{2h} + a_1 \tag{E3.9}$$

Then, factor R_1 is calculated as the difference between T_{2h} and T_{min} and used in a second linear regression coefficient calculation together with the dew point temperature (T_d)

$$T_p^2 = b_2 T_d + a_2 \tag{E3.10}$$

At the end, the minimal temperature can be predicted according to the following equation

$$T_p = T_p^1 + T_p^2 \tag{E3.11}$$

If dew point temperature is not observed, only Eq. C3.9 can be used for the minimal temperature prediction but with larger error.

The air temperature trend calculation uses a square root function from two hours after sunset until sunrise (minimal temperature) the next morning. The factor b' is calculated from predicted minimum temperature (T_p), T_{2h} and the hours between these two temperatures $h = (24 - \tau_{2h}) + \tau_p$

$$b' = \frac{T_p - T_{2h}}{\sqrt{h}} \tag{E3.12}$$

$$T_i = T_{2h} + b'\sqrt{\tau_i} \tag{E3.13}$$

where τ_i is any hour after τ_{2h} given as ordinal number. With this formula, we can determine when temperature will go below 0 °C, when the temperature will reach a critical point and how long will stay under.

Example:

Table E3.4 presents meteorological measurements during the winter and early spring in 2015 at Rimski Sancevi in Serbia. Following the FAO frost model described above, predict the minimal air temperature when the measured temperature 2 hour after the sunset was 1 °C and the dew temperature was 0.1 °C, and calculate the temperature trend in the following night. Indicate in which hour frost appeared and when temperature reached a critical value of -1.7 °C. Assume that time of critical temperatures will end on the sunrise next day.

T_{min} (°C)	T_{2h} (°C)	e (hPa)	T_d (°C)	wind (m s^{-1})	precipitation (mm)
-6.2	-3.2	4.5	-3.7	0	0
-1.8	1.2	6.2	0.2	0.8	0
-2.3	4.6	7.4	2.3	0.8	0
-1.2	0.2	5.9	-0.4	0.0	1
-3.5	2.1	6.7	1.1	0.8	0
-5.5	-0.4	5.2	-1.9	0.8	0
-1.9	1.6	6.2	0.2	0.8	0
-3.5	1.2	6	-0.2	0.8	0
-0.2	8.2	7.3	2.2	0.8	0
-2.7	4.0	6.9	1.5	0.0	0
-3.6	3.2	7.1	1.8	0.8	0
-2.4	5.0	8.1	3.5	0.8	0
-2.0	3.2	7.7	2.8	0.8	0
-2.8	0.7	6.4	0.6	0.8	0
-2.0	-1.6	5.3	-1.7	0.0	0
-1.0	-0.4	5.9	-0.4	0.8	0
-0.3	2.8	7.5	2.5	0.0	0
-1.5	1.2	6.7	1.1	0.8	0
-2.0	0.6	6.4	0.6	0.8	0
-4.9	-4.9	3.7	-5.9	0.8	0

Table E3.4: Observed values of meteorological parameters during the frost episodes in 2015 on the location Rimski Sancevi.

$$T_p^1 = 0.3228T_{2h} - 3.419$$
$$T_p^2 = 0.1821T_d + 0.2371$$

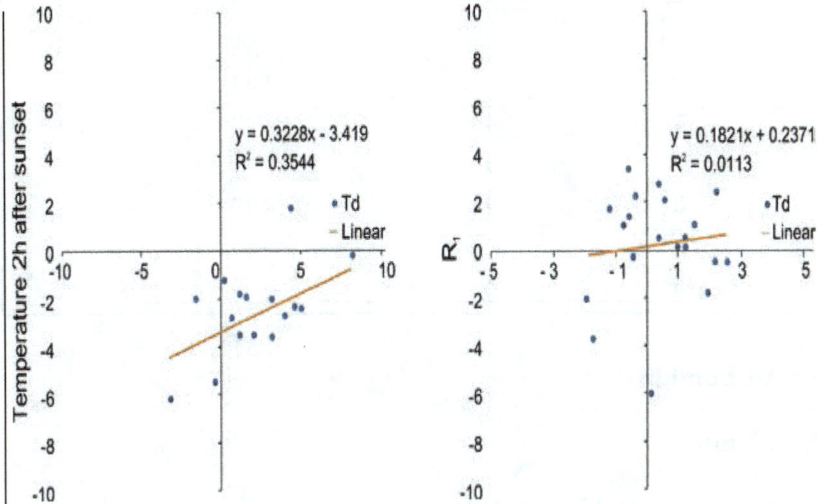

Figure E3.2: Linear regression coefficients for observed minimal temperature (left) and dew point temperature (right).

$$T_p = 0.3228T_{2h} - 3.419 + 0.1821T_d + 0.2371$$

Then, if t_{2h} =1 °C and t_d = 0.1 °C, then the t_p calculated from the previous equation is -2.8 °C.

$$b' = \frac{2.8 - 1}{\sqrt{(24-21)} + 5} = -1.3435\ ^\circ C\, h^{-0.5}$$

$$T_i = 1 - 1.3435\sqrt{1} = -0.34^\circ C$$

By changing the ordinal number of hours you can fill in the table below. From that table, we can see that in the first hour after t_{2h} the temperature drops below 0 °C, and the critical temperature started at 2 hour after midnight.

Hour	Ordinal number	Temperature
21	0	1.00
22	1	-0.34
23	2	-0.90
24	3	-1.33
1	4	-1.69
2	5	-2.00
3	6	-2.29
4	7	-2.55
5	8	-2.80

E4 Air humidity

E4.1 Units

e – pressure units: hPa (or mb), Pa
e_s – pressure units: hPa (or mb), Pa
VPD – pressure units: hPa (or mb), Pa
T_d – temperature units: °C, K, F
r – unitless or in percentage %
q –given in kg kg^{-1} (or usually as g kg^{-1})
a – given in kg m^{-3} (or usually as g m^{-3})

E4.2 Air humidity quantification

The Clausius–Clapeyron equation gives water vapour at or near standard atmospheric conditions (temperature and pressure)

$$\frac{de_s}{dT} = \frac{L_v(T)e_s}{R_v T^2}, \tag{E4.1}$$

after derivation

$$e_s = \frac{L_v(T)}{R_v}\left(\frac{1}{T} - \frac{1}{T_0}\right) \tag{E4.2}$$

where e_s is saturation vapour pressure, T temperature, L_v specific latent heat of evapour ation of water, R_v gas constant of water vapour with value 461 J K^{-1} kg^{-1} and T_0 is 273.15 K. Derivation of given formula gives e_s as a temperature function. Since L_v is also temperature dependant

$$L_v(T) = 2501 - 2.361 \cdot (T + 273.15) \quad \text{(E4.3)}$$

when T is in °C for the simplification several approximations are developed.

The special form of Clausius–Clapeyron equation called Tetens' empirical equation is widely used for computing e_s from temperature (Campbell and Norman, 1998)

$$e_s = a\exp\left(\frac{bT}{T+c}\right) \quad \text{(E4.4)}$$

where a, b, c are constants calculated for various ranges of data and t is temperature in °C.

Tetens' formula for t > 0 °C, over water is

$$e_s = 6.1078\exp\left(\frac{17.27 \cdot T}{T + 237.3}\right) \quad \text{(E4.5)}$$

Tetens' equation for t ≤ 0 °C, over ice is

$$e_s = 6.1078\exp\left(\frac{21.875 \cdot T}{T + 265.5}\right) \quad \text{(E4.6)}$$

where temperature t is in °C and e_s is in hPa (mb).

The FAO (Allen *et al.*, 1998) uses a slightly different formula because of practical application. In most cases, the only measured temperatures during the day are maximal and minimal temperatures. From these values, the daily average of saturation vapour pressure can be calculated based on the same equation as above but using T_{max} and T_{min}

$$e_{s,\max} = 6.1078\exp\left(\frac{17.27 \cdot T_{\max}}{T_{\max} + 237.3}\right) \quad and \quad e_{s,\min} = 6.1078\exp\left(\frac{17.27 \cdot T_{\min}}{T_{\min} + 237.3}\right) \text{(E4.7)}$$

$$e_s = \frac{e_{s,\max} + e_{s,\min}}{2} \quad \text{(E4.8)}$$

The actual vapour pressure can be determined from the difference between the dry (T_{dry}) and wet bulb (T_{wet}) temperatures measured with a psychrometer. Psychrometers consist of two thermometers. One thermometer measure air temperature and the other is wet, and measures the wet bulb temperature.

The relationship is expressed by the following equation:

$$e = e_{s,wet} - p\gamma\left(T_{dry} - T_{wet}\right) \tag{E4.9}$$

where p is the atmospheric pressure, γ is the thermodynamic psychrometer constant whit a value of $6.66 \cdot 10^{-4}$ °C^{-1}. However, since the psychrometer is not an adiabatic system, γ can differ from theoretical values. FAO defines values of depending on the type of phihrometer used: 0.000662 for ventilated (Asmann type) psychrometers, with an air movement of some 5 m/s, 0.000800 for natural ventilated psychrometers (about 1 m/s), 0.001200 for non–ventilated psychrometers installed indoors (Allen *et al.*, 1998). For calculating $e_{s,wet}$ we use Eq. 4.5 and 4.6.

Other humidity elements mostly used in practise are vapour pressure deficit:

$$VPD = e_s - e \tag{E4.10}$$

and relative humidity.

$$r = \frac{e}{E} \cdot 100\% \tag{E4.11}$$

Both are a function of vapour pressure and maximal vapour pressure.

Specific humidity (q) is the mass of water vapour per unit mass of humid air (g kg^{-1}).

$$q = \frac{m_v}{m_d + m_v} \cdot \frac{1/V}{1/V} = \frac{\rho_v}{\rho} = \frac{R_d}{R_v} \frac{e}{p - \left(1 - \frac{R_d}{R_v}\right)e} \approx \frac{R_d}{R_v} \frac{e}{p} \tag{E4.12}$$

Absolute humidity (a) is the mass of water vapour per unit volume of humid air (kg m^{-3}).

$$a = \frac{m_v}{V} = \rho_v = \frac{e}{R_v T} \tag{E4.13}$$

In Eq. E4.12 and E4.13 m_v is the mass of water vapour, m_d is the mass of dry air, V is volume, p is atmospheric pressure, and R_d and R_v are the gas constant of water vapour and dry air in J K^{-1}kg^{-1}, respectively. They are mostly given as unitless fractions

$$\varepsilon = \frac{R_d}{R_v} = \frac{287}{461} = 0.622 \quad \text{(E4.14)}$$

Example:

Using the Tetens' empirical equation and psychrometric equation, calculate the humidity parameters if the air temperature is 10 °C and the temperature of the wet bulb is 5 °C. Have in mind that the atmospheric pressure is 1013.25 hPa.

a. Maximal vapour pressure

$$\mathrm{T_{dry}} = 10\ °\mathrm{C},\ \mathrm{T_{wet}} = 5\ °\mathrm{C}$$

$$e_s = 6.1078\exp\left(\frac{17.27 \cdot T_{dry}}{T_{dry} + 237.3}\right) = 6.1078\exp\left(\frac{17.27 \cdot 10}{10 + 237.3}\right) = 12.28\ hPa,$$

$$e_{s,wet} = 6.1078\exp\left(\frac{17.27 \cdot T_{wet}}{T_{wet} + 237.3}\right) = 6.1078\exp\left(\frac{17.27 \cdot 5}{5 + 237.3}\right) = 8.72\ hPa$$

b. Vapour pressure

$$e = e_{s,wet} - p\gamma(T_{dry} - T_{wet}) = 8.72 - 1013.25 \cdot 6.66 \cdot 10^{-4}(10 - 5) = 5.35\ hPa$$

c. Relative humidity

$$r = \frac{e}{e_s} \cdot 100\% = \frac{5.35}{12.28} \cdot 100\% = 43.57\%$$

d. Vapour pressure deficit

$$VPD = e_s - e = 12.28 - 5.35 = 6.93\ hPa$$

e. Specific humidity

$$q = 0.622\frac{5.35}{1013.25} = 0.003284\ kg\ kg^{-1} = 3.284\ g\ kg^{-1}$$

f. Absolute humidity

$$a = \rho_v = \frac{e}{R_v T} = \frac{5.35 \cdot 10^2}{461(10 + 273.15)} = 0.0041\ kg\ m^{-3} = 4.1\ g\ m^{-3}$$

E5 Clouds and precipitation

E5.1 Units

Precipitation is defined as the volume of water per meter squared. In practise, it is mostly expressed in mm. One mm of rain means that 1 l (10^{-3} m^3) of water has fall on the 1 m^2 of the ground.

$$1\ mm = 1\ l\ m^{-2} = 10^{-3}\ m^3\ m^{-2}$$

E5.2 Impact of cloudiness on energy balance

One of the examples of the impact of clouds on energy that the surface receives from the Sun is presented in empirical relations, such as the Coulson method described by Equation E2.4. Another example of how clouds can affect the energy balance of both the surface and atmosphere is found in the measurements at Harvard forest given in Figure E5.1.

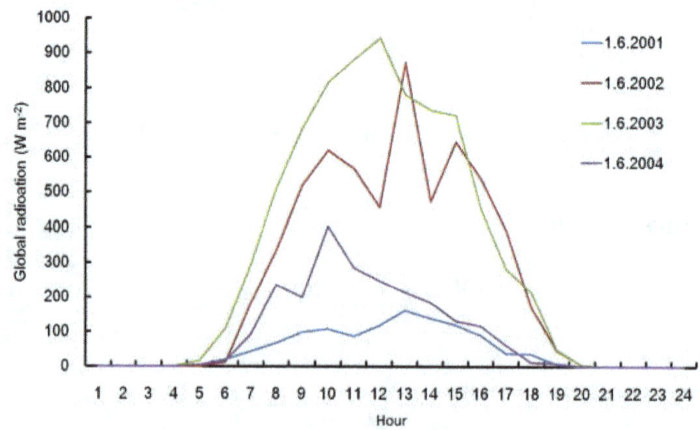

Figure E5.1: Global radiation measured at Harvard forest Fisher Meteorological Station, all at 1st June trough different years.

E5.3 Measurement of precipitation

The precipitation is a sum of water amount from precipitation during one day, week, month or year. The average amount of precipitation can be expressed only from long time series data by averaging the sum of precipitation for a given time step (5 days, week, month, and year) for all years. Therefore, the average amount of precipitation for a time step of j days and a time series of k years is

$$H_j^k = \frac{1}{k}\sum_{i=1}^{k} H_j^i \qquad \text{(E5.1)}$$

E6 Atmospheric circulations and winds

E6.1 Units

Wind is defined by its direction, speed or force. In the SI system wind speed is given as m s^{-1}, but in practise we have wide range of units in use.

	m s^{-1}	knots	km h^{-1}	Mph (mile per h)
1 m s^{-1}	–	1.9	3.6	2.2
1 knots	0.5	–	1.9	1.2
1 km h^{-1}	0.3	0.5	–	0.6
1 Mph (mile per h)	0.4	0.9	1.6	–

Table C6.1: Conversion factors for wind speed units.

Unlike other meteorological elements wind has a descriptive unit called Beaufort which is a unit of wind intensity. The Beaufort wind force scale is an empirical measure that is a function of wind speed.

Wind speed on the Beaufort scale is based on the empirical relationship

$$v = 0.836 \cdot B^{\frac{3}{2}} \qquad \text{(E6.1)}$$

Beaufort number	Name of wind	Wind speed			Description on land surface
		m s^{-1}	knots	kph	
0	calm	<0.3	<1	<1	Smoke rises vertically.
1	light air	0.3–1.5	1–3	1–5	Direction shown by smoke drift but not by wind vanes.
2	light breeze	1.6–3.3	4–6	6–11	Wind felt on face; leaves rustle; wind vane moved by wind.
3	gentle breeze	3.4–5.4	7–10	12–19	Leaves and small twigs in constant motion; light flags extended.

4	moderate breeze	5.5–7.9	11–16	20–28	Raises dust and loose paper; small branches moved.
5	fresh breeze	8.0–10.7	17–21	29–38	Small trees in leaf begin to sway; crested wavelets form on inland waters.
6	strong breeze	10.8–13.8	22–27	39–49	Large branches in motion; whistling heard in telegraph wires; umbrellas used with difficulty.
7	moderate gale (or near gale)	13.9–17.1	28–33	50–61	Whole trees in motion; inconvenience felt when walking against the wind.
8	fresh gale (or gale)	17.2–20.7	34–40	62–74	Twigs break off trees; generally impedes progress.
9	strong gale	20.8–24.4	41–47	75–88	Slight structural damage (chimney pots and slates removed).
10	whole gale (or storm)	24.5–28.4	48–55	89–102	Seldom experienced inland; trees uprooted; considerable structural damage.
11	storm (or violent storm)	28.5–32.6	56–63	103–114	Very rarely experienced; accompanied by widespread damage.
12–17	hurricane	32.7–36.9	64 and above	117 and above	Devastation.

Table E6.2: The Beaufort scale of wind (WMO 2008).

E6.2 Wind rose

The wind rose is graphical representation of percentage of the time the wind blows from each direction during the observation period, together with strengths or speed of this wind and the percentage of the time that calm air or light winds is observed. A wind rose usually comprises 4, 8, 16 or 32 directions, where main are: N (North), E (East), S (South) and W (West).

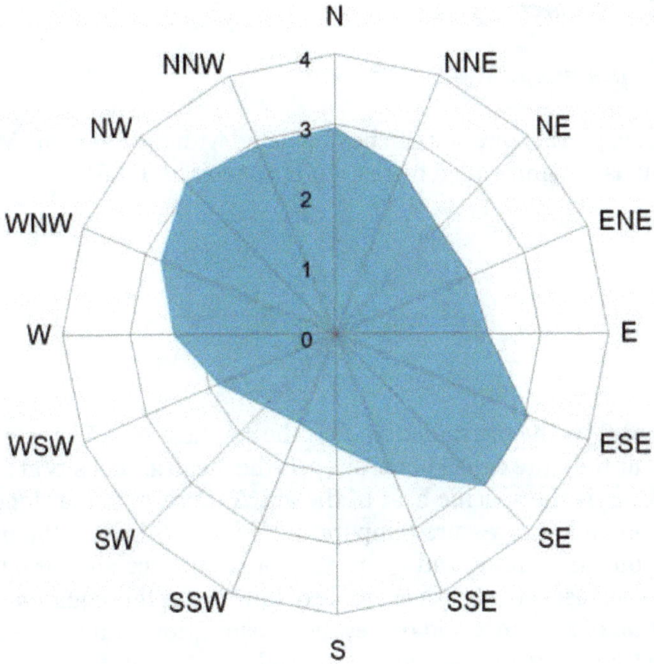

Figure E6.1: Wind rose.

An example of how to make a wind rose is given in the electronic material.

E6.3 Wind energy and power

Wind is a source of energy that can be exploited by wind turbines to create electrical current. To estimate whether a is region adequate for the establishment of a "wind farm" (large number of wind turbines in one location) it is important to calculate wind energy and power. With a density of the air ρ going normal through the surface *A* with the speed *v* during the time *t* energy is calculated as

$$E_w = \frac{1}{2}mv^2 = \frac{1}{2}(Avt\rho)v^2 = \frac{1}{2}(At\rho)v^3, \qquad \text{(E6.2)}$$

from that power can be calculated as

$$P_w = \frac{E}{t} = \frac{1}{2}(A\rho)v^3. \qquad \text{(E6.3)}$$

E7 Soil and water

E7.1 Evapotranspiration

The evapotranspiration can be calculated with the Penman–Monteith equation as recommended by the FAO (Allen *et al.* 1998)

$$\lambda ET = \frac{\Delta(R_n - G) + \rho_a C_p \frac{VPD}{r_a}}{\Delta + \gamma\left(1 + \frac{r_s}{r_a}\right)} \tag{E7.1}$$

where R_n is the net radiation, G is the soil heat flux, VPD represents the vapour pressure deficit of the air, ρ_a is the mean air density at constant pressure, c_p is the specific heat of the air, Δ represents the slope of the saturation vapour pressure temperature relationship, γ is the psychrometric constant, and r_s and r_a are the (bulk) surface and aerodynamic resistances, respectively. An example of how to use this equation and the calculation of reference and real evapotranspiration with FAO method is given in Chapter 7 and electronic material.

When the constants in Eq. E7.1 are replaced with the actual values (the procedure is present at http://www.fao.org/docrep/X0490E/x0490e00.htm#Contents) we get formula for the reference evapotranspiration (Eq. E7.2, see Chapter 7), which in combination with crop coefficient gives

$$ET_0 = \frac{0.408\Delta(R_n - G) + \gamma \frac{900}{t + 273} u_2 VPD}{\Delta + \gamma(1 + 0.34u_2)} \tag{E7.2}$$

where R_n is transferred with 0.408 to the unit mm day^{-1} and the value of wind on 2 m height is needed (u_2). In most cases the value of G is neglected, or taken as portion of the R_n. The slope of saturation vapour pressure temperature relationship is the function of the air temperature (T)

$$\Delta = \frac{4098\left[0.6108 \exp\left(\frac{17.27 \cdot T}{T + 237.3}\right)\right]}{(T + 237.3)^2} \tag{E7.3}$$

and psihrometric constant γ is a function of the atmospheric pressure (p)

$$\gamma = 0.665 \cdot 10^{-3} \cdot p. \tag{E7.4}$$

All details about transformation of different meteorological variables that can be used in the Eq. E7.2 are available online (Allen *et al.* 1998) and in electronic material.

Example. Calculate potential evapotranspiration (E_c) for the wheat with the k_c coefficient of 1.12 if air temperature is 20 °C, global net radiation 15 MJ m^{-2} day^{-1} vapour pressure deficit is 0.6 kPa, wind speed 2 m s^{-1}. Consider the standard atmospheric pressure.

$$\Delta = \frac{4098\left[0.6108\exp\left(\frac{17.27\cdot 20}{t+237.3}\right)\right]}{(20+237.3)^2} = 0.1447\ kPa\ °C^{-1}$$

$$\gamma = 0.655\cdot 10^{-3}\cdot 101.325 = 0.0674\ kPa\ °C^{-1}$$

$$ET_0 = \frac{0.408\cdot 0.1447\cdot 15 + 0.0674\left(\frac{900}{20+273}\right)2\cdot 0.6}{0.1447+0.0674(1+0.34\cdot 2)} = 4.4\ mm\ day^{-1}$$

$$ET_c = ET_0\cdot k_c = 4.4\cdot 1.12 = 4.93\ mm\ day^{-1}$$

Appendix 1

Phenological databases

After a certain decline in 1980s and 1990s phenology have undergone a renaissance in recent decades. For example, phenological observations are going on in almost all European countries (Nekovar *et al.*, 2008). Phenological activities, previously organized exclusively at national level, have entered into wide international cooperation in Europe through a program COST725 in 2004 which succeeded to establish, in practical terms, a European phenological database. International Society of Biometeorology established the Commission for Phenology in order to harmonize the methods of observation. Further databases are organized either on wider regional (Germany) or national levels (the USA, Australia). These databases are freely accessible and support and promote the research in climatology and environmental sciences. The example of such databases come from Europe and the USA.

Pan European Phenological database (PEP725)

The Pan European Phenology (PEP) project is a European infrastructure to promote and facilitate phenological research, education, and environmental monitoring. The main objective is to maintain and develop a Pan European Phenological database (PEP725) with an open, unrestricted data access for science and education. PEP725 is the successor of the database developed through the COST action 725 "Establishing a European phenological data platform for climatological applications". So far, 32 European meteorological services and project partners from across Europe have joined and supplied data collected by volunteers from 1868 to the present for the PEP725 database. The database presently holds

almost 12 million records, about 46 growing stages and 265 plant species (including cultivars).
http://www.pep725.eu/

The plant phenological online database (PPODB)

PPODB is an online database that provides unrestricted and free access to over 16 million plant phenological observations from over 8,000 stations in Central Europe between the years 1880 and 2009. Unique features are (1) a flexible and unrestricted access to a full–fledged database, allowing for a wide range of individual queries and data retrieval, (2) historical data for Germany before 1951 ranging back to 1880, and (3) more than 480 curated long–term time series covering more than 100 years for individual phenological phases and plants combined over Natural Regions in Germany. Time series for single stations or Natural Regions can be accessed through a user–friendly graphical geo–referenced interface.
www.ppodb.de

USA National Phenology Network (USANPN)

Phenology Observation Portal related to USNPN contains customized datasets of observational data from the National Phenology Database, which includes phenology data collected via the *Nature's Notebook* phenology program (2009–present for the United States), and additional integrated datasets, such as historical lilac and honeysuckle data (1955–present). Filters are available to specify dates, regions, species and phenophases of interest.
www.usanpn.org

References:
Nekovar, J (ed) 2008: COST Action 725, The history and current status of plant phenology in Europe. ISBN 978–951–40–2091–9, COST Office, 2008.

Appendix 2

Phenological models

Table A.1

Type of model	Basic principe	Functional dependence on climatic and terrestrial parameters	Principal description of the model(s) (example of use)
GDD models (growing degree day)	Start of the particular phenological phase is related to the sum of temperatures calulated over a certain treshold (base temperature) starting on a defined date. (In some modifications the base temperature depends on other parameters (photoperiod, soil moisture etc.).	R = f(T)	Miller et.al. (2001)
NGD models (number of growing days)	Start of the particular phenological phase is related to the number of days during a certain period when the mean air temperature is over a certain limit (depending on the bionom, frequently used are -5°C, 0°C, 5°C...).	R = f (T)	Botta et all. (2000)
GPT models (genotype x temperature x photoperiod)	Start of the particular phenological phase is constituted mostly by genes other than photoperiod genes and regulated by temperature. Any activity of photoperiod genes delays the process of plant development.	R = f (T,P)	Weikai and Wallace (1998)

Type of model	Basic principe	Functional dependence on climatic and terrestrial parameters	Principal description of the model(s) (example of use)
NCD–GDD two–phase phenological model (Chilling Days–Growing Degree Day model	Chilling Days represent the dormancy required for the start of a new growing season. After the dormancy demand is met the model works in GDD mode. Many variations of such models assum reduced plant`s GDD demand with the increase of chilling days.	R = f(T)	Fu *et. al.* (2014)
Parallel model	Start of the particular phenological phase is related to both chilling and thermal forcing acting in the parallel. The time spent in chilling period increases the potential of the plant organs to respond to temperature forcing.	R = f(T)	Landsberg (1974)
Sequential models	The dormancy and the start of active phase of vegetation cycle do not act in parallel but they are in a strict sequence. The periods of rest and quiescence before the start of the vegetation are adopted. The start of quiescence requires a critical state of chilling. The active phase of vegetative process starts just after a critical state of forcing is reached.	R = f(T)	Sarvas (1974)
R–rate of plant development T, P, Pc–external environmental parameters influencing the plant development T–temperature, P–photoperiod, Pc–precipitation			
Deepening rest model	This type of model concentrates on the period of dormancy. During the phase of rest a subphases of deepening rest and decreasing rest is distinguished while other models recognize only decreasing rest. During the quiescence phase which comes after the rest the rate of the development increases with the temperature.	R = f(T)	Kobayashi (1982), Kobayashi and Fuchigami 1983

Type of model	Basic principe	Functional dependence on climatic and terrestrial parameters	Principal description of the model(s) (example of use)
Four phase model	The rest phase of the dormancy is broken into three subphases (pre rest, true rest and post–rest) defined by the rate of receptivness of the plant to start to grow. During the quiescence which follows after the post–rest the plants (buds) already respond to forcing temperature.	R = f(T)	Vegis (1964)
Alternating model	The relation of the state of forcing and chilling is used to predict the start of the particular phenological phase (budburst). It is set after so called c ritical thermal time is reached. This value is not constant but declines exponentially with the state of chilling. From the state of quiescence the state of chilling increases when the daily mean temperature is below the base temperature or increases if being above.	R = f(T)	Murray *et. al.* (1989).
BBGC (Biome–BioGeochemical Cycles model)	The model accounts the temperature and precipitation for setting the critical summ of soil temperature and precipitation to start the particular phenological stage. Model distinguishes the processes for woody plants and grasses. Model is process oriented based on modelling the processes that control flux of energy and mass.	R = f(T, Pc)	Fu *et. al.* (2014)
R–rate of plant development T, P, Pc–external environmental parameters influencing the plant development T–temperature, P–photoperiod, Pc–precipitation			

References

Ch–1

Barry, R.G.; Chorley, R.J.; 2003. *Atmosphere, Weather and Climate*, 8th Ed.; Routledge Taylor & Francis Group, p. 421, ISBN 0–203–42823–4.

Bierhuizen, J.F.; Slatyer, R.O.; 1965. Effect of atmospheric concentration of water vapour and CO2 in determining transpiration–photosynthesis relationships of cotton leaves. *Agric. Meteorol.* 2, 259–270, doi: 10.1016/0002–1571(65)90012–9.

Farquhar, G.D.; von Caemmerer, S.; Berry, J.A.; 1980. A biochemical model of photosynthetic CO2 assimilation in leaves of C3 species. *Planta* 149, 78–90, doi: 10.1007/BF00386231.

Grantz, D.A.; Gunn, S.; Vu, H.–B.; 2006. O3 impacts on plant development: a meta–analysis of root/shoot allocation and growth. *Plant Cell Environ.* 29, 1193–1209, doi:10.1111/j.1365–3040.2006.01521.x.

Le Treut, H.; Somerville, R.; Cubasch, U.; Ding, Y.; Mauritzen, C.; Mokssit, A.; Peterson, T.; Prather, M. Historical overview of climate change. In: *Climate Change 2007: The Physical Science Basis. Contribution of Working Group I to the Fourth Assessment Report of the Intergovernmental Panel on Climate Change*. Solomon, S., Qin, D., Manning, M., Chen, Z., Marquis, M., Averyt, K.B., Tignor, M., Miller, H.L., Eds.; Cambridge University Press, Cambridge, United Kingdom and New York, NY, USA, 2007, pp. 95–122, ISBN 978–0–521–88009–1.

NOAA, 2018. Earth system research laboratory (ESRL), *Global Monitoring Division. Trends in atmospheric carbon dioxide*. Available online: https://www.esrl.noaa.gov/gmd/ccgg/trends/ (accessed on 28 May 2018).

Salehi, A.; Fallah, S.; Abasi Sourki, A. *Organic and inorganic fertilizer on soil* CO_2 *flux, microbial biomass and growth of* Nigella sativa L. *Int Agrophy* 2017, 31, 103–116, doi: 10.1515/intag–2016–0032.

Thompson, R.D.; 1998. *Atmospheric Processes and Systems*. Routlige, 11 New Fetter Lane, London, ISBN 978–0415171465.

van de Geijn, S.C.; Goudriaan, J.; 1996. The effects of elevated CO_2 and temperature change on transpiration and crop water use. In: *Global Climate Change and Agricultural Production: Direct and Indirect Effects of Changing Hydrological Soil and Plant Physiological*. Bazzaz, F., Sombroek, W., Eds.; Food and Agriculture Organisation of the United Nations, Rome, John Wiley & Sons, ISBN 0–471–96927–3.

Ward, J.K.; Tissue, D.T.; Thomas, R.B.; Strain, B.R.; 1999. Comparative responses of model C3 and C4 plants to drought in low and elevated CO2. *Glob. Change Biol.*, 5, 857–867, 10.1046/j.1365–2486.1999.00270.x.

WMO – *International Meteorological Vocabulary*, 2nd Ed.; Secretariat of the World Meteorological Organization, Geneva, 1992, p. 636, ISBN 92–63–02182–1.

Ch–2

Campillo, C.; Fortes, R.; del Henar Prieto, M.; 2012. *Solar Radiation Effect on Crop Production, Solar Radiation*, Babatunde, E.B. Ed., ISBN: 978–953–51–0384–4.

Eitzinger, J.; Koessler, Ch.; 2002. *Microclimatological characteristics of a miscanthus* (Miscanthus cv. giganteus) *stand during stable conditionsat night in the nonvegetative winter period*. Theor. Appl. Climatol. 72, 3-4, 245-25

Liou, K.N.; 2002. *An Introduction to Atmospheric Radiation*, 2nd Ed.; Academic Press, London, UK, p. 599, ISBN 0–12–451451–0.

Moore, K.E.; Fitzjarrald, D.R.; Sakai, R.K.; Goulden, M.L.; Munger, J.W.; Wofsy, S.C.; 1996. Sesonal variation in radiative and turbulent exchange at a deciduous forest in Central Massachusetts, *J. Appl. Meteorol.*, 35, 122–134.

Munger and Wofsy, 2017. *Canopy–atmosphere exchange of Carbon, Water and Energy at Harvard Forest EMS tower since 1991*. Harvard Forest data archive: HF004.

Thompson, R.D.; 1998. *Atmospheric Processes and Systems*. Routlige, 11 New Fetter Lane, London, ISBN 978–0415171465.

Varlet–Grancher, C.; Bonhomme, R.; Sinoquet, H.; 1993. *Crop Structure and Light Microclimate*, INRA, Paris, p. 518, ISBN 2 7380 0448 2.

Ch–3

Munger and Wofsy, 2017. *Canopy–atmosphere exchange of carbon, water and energy at Harvard Forest EMS tower since 1991*. Harvard Forest data archive: HF004.

Ferris, R.; Ellis, R.H.; Wheeler, T.R.; Hadley, P.; 1998. *Effect of high temperature stress at anthesis on grain yield and biomass of field–grown crops of wheat*. Annal. Bot. 82, 631–639.

Porter, J.R.; Gawith, M.; 1999. Temperatures and the growth and development of wheat: a review. *Eur. J. Agron.* 10, 23–36.

Wahid, A.; 2007. Physiological implications of metabolites biosynthesis in net assimilation and heat stress tolerance of sugarcane (*Saccharum officinarum*) sprouts. *J. Plant Res.* 120, 219–228.

Commuri, P.D.; Jones,R.D.; 2001. High temperatures during endosperm cell division in maize: a genotypic comparison under in vitro and field conditions. *Crop Sci.* 41,1122–1130.

Hatfield, J.H.; Prueger, J.H.; 2015. Temperature extremes: Effect on plant growth and development. *Weather Clim. Extremes* 10, 4–10, doi: 10.1016/j.wace.2015.08.001.

Ch–4

Allen, R.G.; Pereira, L.S.; Raes, D.; Smith, M.; 1998. Crop evapotranspiration – Guidelines for computing crop water requirements – FAO Irrigation and drainage paper 56, FAO – Food and Agriculture Organization of the United Nations, Rome, ISBN 92–5–104219–5.

Blain G.C.; 2012. Revisiting the probabilistic definition of drought: strengths, limitations and an agrometeorological adaptation. Bragantia, 71, 1, 132–121.

Gillette, H.P.; 1950. A creeping drought under way. *Water and Sewage Works*, 104–105.

Jacimovic, G.; 2012. Optimisation of mineral nutrition of wheat depending on weather conditions during the year. University of Novi Sad. PhD dissertation, p. 164 (In Serbian).

Monteith, J.L., Unsworth, M.H.; 2008. *Principles of Environmental Physics*. 3rd Ed. AP, Amsterdam.

Murray, F.W.; 1967. On the computation of saturation vapour pressure. *J. Appl. Meteorol.* 6, 203–204.

Munger, W.; Hadley, J.; 2018. Microclimate at Harvard Forest HEM, LPH and EMS Towers since 2005. *Harvard Forest Data Archive*: HF206.

Thornthwaite, C.W.; 1948. An approach toward a rational classification of climate. *Geogr. Rev.*, 38(1), 55–94.

Van Lanen, H.A.J.; 2006. Drought propagation through the hydrological cycle. In: *Climate Variability and Change Hydrological Impacts*. Demuth, S., Gustard,A., Planos, E., Scatena, F., Servat, E., Eds.; vol. 308, IAHS–AISH Publication, IAHS Press, Wallingford, UK, pp. 122–127.

Vidal, J.–P.; Martin, E.; Franchistéguy, L.; Habets, F.; Soubeyroux, J.–M.; Blanchard, M.; Baillon; M.; 2010. Multilevel and multiscale drought reanalysis over France with the Safran–Isba–Modcou hydrometeorological suite. *Hydrol. Earth Syst. Sci.* 14, 3, 459–478, doi: 10.5194/hess–14–459–2010.

Wilhite, D.A.; Glantz, M.H.; 1985. Understanding the Drought Phenomenon: The Role of Definitions. *Water Int.* 10, 3, 111–120.

Wilhite, D.A.; 2000. Drought as a Natural Hazard: Concepts and Definitions. In: *Drought: A Global Assessment*, Vol. I, D.A. Wilhite Ed., Drought Mitigation Center Faculty Publications. 69.

Wilhite, D.A.; 2005. The Role of Disaster Preparedness in National Planning with Specific Reference to Droughts. In: Natural Disasters and Extreme Events in Agriculture, Sivakumar, M.V., Motha, R.P., Das, H.P. Eds.; Springer, Berlin, Heidelberg, ISBN 978–3–540–22490–7.

Ch–5

Faghri, A.; Zhang, Y.; 2006. Transport Phenomena in Multiphase Systems, Elsevier Inc., p. 1030, ISBN: 978–0–12–370610–2.

Garrett, T.J.; Fallgatter, C.; Shkurko, K.; Howlett, D.; 2012. Fallspeed measurement and high–resolution multi–angle photography of hydrometeors in freefall. *Atmos. Meas. Tech.*, 5, 2625–2633, doi:10.5194/amt–5–2625–2012.

International Cloud Atlas, WMO. Available online: https://cloudatlas.wmo.int (accessed on 28 May 2018).

McDonald, J.E.; 1962. Homogeneous nucleation of vapor condensation. I. Thermodynamic Aspects. *Am. J. Phy.* 30, 870; doi: 10.1119/1.1941841.

Wood, R.; 2012. Stratocumulus Clouds. Mon. Weather Rev. 140, 2373–2423, doi: 10.1175/MWR–D–11–00121.1.

Ch–6

Caborn, J.M.; 1957. Shelterbelts and Microclimate, Forestry Commission Bulletin No. 29.

Cleugh, H.A.; 1998. Effects of the wind brakes on airflow, microclimates and crop yields. *Agroforest. Sys.* 41, 55–84.

NOAA 2018. How hurricanes form? National Ocean Service website. Available online: https://oceanservice.noaa.gov/facts/how–hurricanes–form.html (accessed on 28 May 2018).

The Climate System, EESC 2100 Spring 2007. Available online: http://eesc.columbia.edu/courses/ees/climate/lectures/atm_dyn.html (accessed on 03 June 2018).

Todey, D.; Taylor, E.; Shelterbelts, Agronomy 541. Available online: http://agron–www.agron.iastate.edu/courses/Agron541/classes/541/lesson05a/541L5.pdf (accessed on 03 June 2018).

Numerical examples

Allen, R.G.; Pereira, L.S.; Raes, D.; Smith, M.; 1998. *Crop evapotranspiration – Guidelines for computing crop water requirements* – FAO Irrigation and drainage paper 56, FAO – Food and Agriculture Organization of the United Nations, Rome, ISBN 92–5–104219–5.

Bierhuizen, J.F.; Slatyer, R.O.; 1965. Effect of atmospheric concentration of water vapour and CO2 in determining transpiration–photosynthesis relationships of cotton leaves. *Agric. Meteorol.* 2, 259–270, doi: 10.1016/0002–1571(65)90012–9.

Brunt, D.; 1932. Notes on radiation in the atmosphere. *Q. J. R. Met. Soc.* 58, 389–42, doi: 10.1002/qj.49705824704.
Campbell, G.S.; Norman, J.M.; 1998. *An Introduction to Environmental Biophysics*, 2nd Ed.; Springer, p. 306.
Farquhar, G.D.; von Caemmerer, S.; Berry, J.A.; 1980. A biochemical model of photosynthetic CO2 assimilation in leaves of C3 species. *Planta* 149, 78–90.
Fitzjarrald, D.; Sakai, R.; 2009. Radiation Measurements at Harvard Forest EMS Tower 1991–2007. *Harvard Forest Data Archive:* HF102.
McMaster, G.S.; Wilhelm, W.W.; 1997. Growing Degree–Days One Equation, Two Interpretations. *Agric. Forest Meterol.*, 87, 291–300.
Mihailovic, D.; 1988. Basics of meteorological observations and data processing, Faculty of Agriculture. Institute of Field and Vegetable Crops, Novi Sad, p. 217 (in Serbian).
Monteith, J.L.; 1977. Climate and Efficiency of Crop Production in Britain. Philosophical Transactions of the Royal Society London B, 281, 277–294.
Snyder, R.L.; de Melo–Abreu, J.P.; 2005. *Frost Protection: fundamentals, practice and economics.* Vol 1, Food and Agriculture Organization of the United Nations, Rome, p. 126.
WMO; 2008. *Guide to Meteorological Instruments and Methods of Observation.* World Meteorological Organization, 2008, WMO–No. 8, p.716, ISBN 978–92–63–10008–5.

Ch–7

References, further READING and LINKS

Allen, R.G.; Pereira, L.S.; Raes, D. und Smith, M.; 1998. *Crop evapotranspiration Guidelines for computing crop water requirements.* FAO Irrigation and Drainage Papers. Paper No 56, FAO, Rome, 326 S. Online available at: http://www.fao.org/docrep/X0490E/X0490E00.htm.
AQUACROP model: http://www.fao.org/aquacrop
Gobin, A.; Kersebaum, K.C.; Eitzinger, J.; Trnka, M.; Hlavinka, P.; Takac, J.; Kroes, J.; Ventrella, D.; Dalla Marta, A.; Deelstra, J.; Lalic, B.; Nejedlik, P.; Orlandini, S.; Peltonen–Sainio, P.; Rajala, A.; Saue, T.; Saylan, L.; Stricevic, R.; Vucetic, V.; Zoumides, C., 2017. *Variability in the Water Footprint of Arable Crop Production across European Regions.* WATER–SUI.9(2): 93; doi:10.3390/w9020093.
Hlavinka, P.; Trnka, M.; Balek, J.; Semeradova, D.; Hayes, M.; Svoboda, M.; Eitzinger, J.; Mozny, M.; Fischer, M.; Hunt, E.; Zalud, Z.; 2011. *Development and evaluation of the SoilClim model for water balance and soil climate estimates.* Agr Water Manage. 2011; 98(8): 1249-1261.
Hoyningen–Huene, J. v.; 1980. *Mikrometeorologische Untersuchungen zur Evapotranspiration von bewässerten Pflanzenbeständen.* Ber Inst Meteorol Klimatol Univ Hannover Nr. 19:1–168.
Kroes, J.G.; Wesseling, J.G. and J.C. Van Dam; 2000. *Integrated modelling of the soil–water–atmosphere–plant system using the model*

SWAP 2.0 an overview of theory and an application. Hydrological Processes, 14, 1993–2002. Available at: http://onlinelibrary.wiley.com/doi/10.1002/1099–1085(20000815/30)14:11/12%3C1993::AID–HYP50%3E3.0.CO;2–%23/abstract.

Lemon, E. R. 1963 after Shinn, J.H. and Lemon, E.R. 1963; IN: Evans, L.T. (ed); 1963. *Environmental Control of Plant Growth, Symposium Proceedings*, Canberra, Australia, Academic Press.

Marshall, T.J.; Holmes, J.W.; Rose, C.W.; 1996. *Soil Physics*. Cambridge Univ. Press, ISBN: 9780521457668; Information on this title: www.cambridge.org/9780521457668.

Moene, A.F.; Van Dam, J.C.; 2014. *Transport in the Atmosphere–Vegetation–Soil Continuum*. Cambridge Univ. Press, ISBN: 9780521195683; Information on this title: www.cambridge.org/9780521195683.

Rose, C.W.; 2004. *An Introduction to the Environmental Physics of Soil, Water and Watersheds*. Cambridge Univ. Press, ISBN: 9780521829946; Information on this title: www.cambridge.org/9780521829946.

Scheffer, F.; Schachtschabel, P.; 1982. *Lehrbuch der Bodenkunde*; Ferdinand Enke Verlag, Stuttgart. ISBN 10: 3432847718, ISBN 13: 9783432847719.

Steduto, P.; Hsiao, C.; Fereres, E.; Raes, D.; 2012. *Crop Yield Response to Water*. FAO Irrigation and Drainage Paper No. 66. ISBN: 978–92–5–107274–5. Online available at: http://www.fao.org/documents/card/en/c/c355da16–217c–555b–acbc–505d87bade00/.

SWAP model: http://www.swap.alterra.nl/.

Thaler, S.; Eitzinger, J.; Trnka, M; Dubrovsky, M.; 2012. Impacts of climate change and alternative adaptation options on winter wheat yield and water productivity in a dry climate in Central Europe. *Journal of Agricultural Science, Cambridge*. DOI: http://dx.doi.org/10.1017/S0021859612000093.

Xue, J.; Ren, L.; 2016. Evaluation of crop water productivity under sprinkler irrigation regime using a distributed agro–hydrological model in an irrigation district of China. Agricultural Water Management, 178, 350–365. http://doi.org/10.1016/j.agwat.2016.10.003.

Ch–8

References, further READING and LINKS

Anderson, J. L.; Richardson, E. A.; Kesner, C. D.; 1986. *Validation of chill unit and flower bud phenology models for 'Montmorency' sour cherry*. Acta Hortic 184: 71–78.

Bünning, E.; 1953. *Entwicklungs– und Bewegungsphysiologie der Pflanzen*. 3.Aufl. Springer, Berlin, 1953.

Chang, Jen–Hu; 1968. *Climate and Agriculture – An Ecological Survey*. Aldine Transaction. ISBN 978–0–202–36249–6

Eitzinger, J.; Kersebaum, K. C.; Formayer, H.; 2009. *Landwirtschaft im Klimawandel – Auswirkungen und Anpassungsstrategien für die Land- und Forstwirtschaft in Mitteleuropa*. http://de.agrimedia.com, 320, Agrimedia, D–29459 Clenze, Deutschland; ISBN: 978–3–86037–378–1 (in german).

Grigg, D.; 1984. *An Introduction to Agricultural Geography*. Routledge, ISBN: 0–203–41927–8.

Hodges, T.; 1990. *Predicting crop phenology*. CRC press, ISBN 9780849367458

Schwartz, M.D.; 2003. Phenology: An Integrative Environmental Science. Springer; ISBN: 978–1–4020–1580–9; DOI https://doi.org/10.1007/978–94–007–0632–3

Huglin, P; 1986. *Biologie et écologie de la vigne*. Lavoisier (Edition Tec & Doc), Paris 1986, ISBN 2–60103–019–4. S. 292 (371 S.).

Larcher, W.; 2003. *Physiological Plant Ecology – Ecophysiology and Stress Physiology of Functional Groups*. Springer. ISBN 978–3–540–43516–7.

Luedeling, E.; Zhang, M.; Luedeling, V.; Girvetz, H.; 2009. *Sensitivity of winter chill models for fruit and nut trees to climatic changes expected in California's Central Valley*. Agric Ecosyst Environ 133: 23–31 CrossRf.

Manshard, W.; 1968. *Agrargeographie der Tropen*. English. Tropical agriculture. London ; New York : Longman, 1974

Meier, U. (Ed.); 1997. *BBCH–Monograph. Growth stages of plants. Entwicklungsstadien von Pflanzen. Estadios de las plantas. Stades dedéveloppement des plantes*. Blackwell Wissenschafts–Verlag Berlin, Wien, 622 p.

Meier, U.; 2003. *Phenological growth stages*, 269–283. In (Ed.) SCHWARTZ, M. D.: *Phenology: An integrative science*. Kluver Academic Publishers, Dordrecht / Boston / London, p 564.

Meier, U.; Bleiholder, H.; Buhr, L.; Feller, C.; Hack, H.; Heß, M.; Lancashire, P.D.; Schnock, U.; Stauß, R.; van den Boom, T.; Weber, E.; Zwerger, P.; 2009. *The BBCH system to coding the phenological growth stages of plants – history and publications*. Journal für Kulturpflanzen, 61 (2). s. 41–52, 2009, Issn 0027–7479, Verlag Eugen Ulmer, Stuttgart. Online: https://www.journal–kulturpflanzen.de/artikel.dll/meier–et–al_OTAyMjUy.PDF.

Nejedlik, P, Orlandini, S. (ed.); 2008. *Impact of Climate Change and Variability on European Agriculture – Survey of agrometeorological practices and application in Europe regarding climate change impacts. Final report* COST734. Online: http://www.cost734.eu/reports–and–presentations/SURVEY%20OF%20AGROMETEOROLOGICAL%20PRACTICES%20AND%20APPLICATIONS%20IN%20EUROPE%20%20REGARDING-CLIMATE%20CHANGE%20IMPACTS%20–%20COST%20734.pdf.

Pan European Phenological database (PEP725): http://www.pep725.eu/

Richardson, E.A.; Seeley, S.D.; Walker, D.R.; 1974. *A model for estimating the completion of rest for Redhaven and Elberta peach trees*. HortScience, 9(4), 331–332.

Schnelle, F. (ed); 1963. *Frostschutz im Pflanzenbau*, vol 1: Die meteorologischen und biologischen Grundlagen der Frostschadensverhütung. Bayerischer Landwirtschaftsverlag, München.
The plant phenological online database (PPODB): www.ppodb.de
USA National Phenology Network (USANPN): www.usanpn.org
Zadoks, J.C.; Chang, T.T.; Konzak, C.F.; 1974. *A decimal code for the growth stages of cereals*. Weed Research, 14, 415–421 and Eucarpia Bulletin, 7, 49–52.

Ch–9

Anandhi, S., Perumal, P.H., Gowda, M., Knapp, S., Hutchinson, J., Harrington Jr., L., Murray, M.B., Kirkham, C.W., 2013: R*ice. Long–term spatial and temporal trends in frost indices in Kansas*, USA. *Clim. Change*, **120**, 169–181.
Baisi, F., Galligani, P.L., Pergola, V., 1999: *Corso di Agronomia ed elementi di meccanizzazione agraria*. Edagricole – Edizioni Agricole della Calderini s.r.l. – Bologna.
Basher, R., 2006: UN–ISDR Platform for the Promotion of Early Warning (PPEW), Goerresstrasse 30, 53113 Bonn, Germany, In "Global early warning systems for natural hazards: systematic and people–centred" – Phil. Trans. R. Soc. A (2006) 364, 2167–2182.
Blumenthal, C.S., Batey, I.L., Bekes, F., Wrigley, C.W., Barlow, E.W.R., 1991: Seasonal changes in wheat–grain quality associated with high temperatures during grain filling. *Australian Journal of Agricultural Research*, 42, 21 – 30.
Ceccon, P., Fagnano, M., Grignani, C., Monti, M., Orlandini, S., 2017: *Agronomia*. EdiSES S.r.l. – Napoli.
Christner, B.C., Cai, C., Morris, C.E., McCarter, K.S., Foreman, C.M., Skidmore, M.L., Montross, S.N., Sands, D.C., 2008: *Geographic, seasonal, and precipitation chemistry influence on the abundance and activity of biological ice nucleators in rain and snow*. In: Proceedings of the National Academy of Sciences, 105:18854–18859.
Edwards, D.C., McKee, T.B., 1997: *Characteristics of 20th century drought in the United States at multiple time scales*. Climatology Report Number 97–2, Department of Atmospheric Science, Colorado State University, Fort Collins.
FAO – *Frost protection: fundamentals, practice, and economics* (2005) – Volume 1 – Produced by: Natural Resources Management and Environment Department.
Garen, D.C., 1993: Revised surface-water supply index for Western United States. *Journal of Water Resources Planning and Management*, 119, 437–454.
Guttman, N.B., 1998: Comparing the palmer drought index and the standardized precipitation index". *Journal of the American Water Resources Association*, 34, 113–121.
Hatfield, J.L., Prueger, J.H., 2015: Temperature extremes: Effect on plant growth and development. *Weather and Climate Extremes*, 10, 4–10.
IPCC, 2007. Climate Change 2007: Impacts, Adaptation and Vulnerability. Contribution of Working Group II to the Fourth Assessment Report of the IPCC.

Kogan, F.N., 1995: Application of vegetation index and brightness temperature for drought detection. *Adv. Space Res.*, 5. No. 11, pp. (11)91–(1l)

Kovats, R.S., Valentini, R., Bouwer, L.M., Georgopoulou, E., Jacob, D., Martin, E., Rounsevell, M., Soussana, J.F., 2014: Europe. In: *Climate Change 2014: Impacts, Adaptation, and Vulnerability*. Part B: Regional Aspects. Contribution of Working Group II to the Fifth Assessment Report of the Intergovernmental Panel on Climate Change [Barros, V.R., C.B. Field, D.J. Dokken, M.D.

Levitt, J. 1980: "Responses of plants to environmental stresses". Vol. 1.

Li, M., Zhao, C., Li, D., Yang, X., Sun, C., Wang, Y., 2008: *Towards developing an early warning system for cucumber diseases for greenhouse in China.* IFIP International Federation for Information Processing, Volume 259; Computer and Computing Technologies in Agriculture, Vol. 2; Daoliang Li; (Boston: Springer), pp. 1375–1378.

Luo, Q., 2011: Temperature thresholds and crop production: a review. *Climatic Change*, 109: 583–598

Orlandini S., Nejedlik P., Eitzinger, J., Alexandrov, V., Toulios, L., Calanca, P., Trnka, M., Olesen, J.E., 2008: *Impacts of Climate Change and Variability on European Agriculture. Results of Inventory Analysis in COST 734 Countries.* Trends and Directions in Climate Research: Ann. N.Y. Acad. Sci. 1146: 338–353 (2008). doi: 10.1196/annals.1446.013C. New York Academy of Sciences.

Shafer, B. A. and L. E. Dezman (1982). *Development of a surface water supply index (SWSI) to assess the severity of drought conditions in snowpack runoff areas. Proceedings of the Western Snow Conference*, 164–175.

Van Loona, A.F. and Laahab, G., 2015: *Hydrological drought severity explained by climate and catchment characteristics. Journal of Hydrology*, 526, 3–14.

Watts, N., Amann, M., Ayeb–Karlsson, S., Belesova, K., Bouley, T., Boykoff, M., 2017: The Lancet Countdown on health and climate change: From 25 years of inaction to a global transformation for public health. *Lancet* 2017.

Wilhite, D.A., 2005: *Drought and Water Crises*: Science, Technology, and Management Issues.

WMO, 2010: *Guidelines on early warning systems and application of nowcasting and warning operations.*

WMO, 2012a: *Guide to Agricultural Meteorological Practices* (GAMP).

WMO, 2012b: *Standardized Precipitation Index User Guide* (M. Svoboda, M. Hayes and D. Wood). (WMO–No. 1090), Geneva.

Woodruff N. P., Siddoway F. H., 1965. *A Wind Erosion Equation.* Soil Science Society of America Journal Vol. 29, 5: 602–608.

Ch–10

Allen, B., Maréchal, A., 2017: *Agriculture GHG emissions: determining the potential contribution to the Effort Sharing Regulation.* In: *Report prepared for Transport and Environment.* Institute for European Environmental Policy, London.

Asfaw, S., Lipper, L., 2016: *Managing climate risk using climate–smart agriculture* – Food and Agriculture Organization of the United Nations.
Brooks, N., 2003: *Vulnerability, risk and adaptation: A conceptual framework*. Tyndall Centre for Climate Change Research and Centre for Social and Economic Research on the Global Environment (CSERGE) School of Environmental Sciences University of East Anglia.
Cardona, O.D., M.K. van Aalst, J. Birkmann, M. Fordham, G. McGregor, R. Perez, R.S. Pulwarty, E.L.F. Schipper, and B.T. Sinh, 2012: *Determinants of risk: exposure and vulnerability*. In: *Managing the Risks of Extreme Events and Disasters to Advance Climate Change Adaptation* [Field, C.B., V. Barros, T.F. Stocker, D. Qin, D.J. Dokken, K.L. Ebi, M.D. Mastrandrea, K.J. Mach, G.–K. Plattner, S.K. Allen, M. Tignor, and P.M. Midgley (eds.)]. A Special Report of Working Groups I and II of the Intergovernmental Panel on Climate Change (IPCC). Cambridge University Press, Cambridge, UK, and New York, NY, USA, pp. 65–108.
FAO, 2013. *Climate–Smart Agriculture: Sourcebook*. Rome, Italy: Food and Agriculture Organization of the United Nations.
Gilard, O., 2016: *Hazards, Vulnerability and Risk*. 19–29. 10.1007/978–94–017–7462–8_2. In: Pedrono, Miguel & Locatelli, Bruno & Ezzine–de–Blas, Driss Pesche, Denis Morand, Serge Binot, Aurélie. (2016). Impact of Climate Change on Ecosystem Services. 348 pp. 10.1007/978–94–017–7462–8.
IPCC Climate Change 2007 – *Impacts, Adaptation and Vulnerability Contribution of Working Group II to the Fourth Assessment Report of the IPCC* (978 0521 88010–7 Hardback; 978 0521 70597–4 Paperback).
Stephenson, D.B., 2008. *Definition, diagnosis, and origin of extreme weather and climate events*. In H. Diaz R. Murnane (Eds.), Climate Extremes and Society (pp. 11–23). Cambridge: Cambridge University Press doi:10.1017/CBO9780511535840.004
WMO, 2007: *Managing Weather and Climate Risks in Agriculture. Sivakumar*, M.V.K., Motha, R.P. (eds), Springer, Berlin, Heidelberg DOI https://doi.org/10.1007/978–3–540– 72746–0
WMO, 2016: *Guidelines on the definition and monitoring of extreme weather and climate events*. Task Team on Definitions of Extreme Weather and Climate Events (TT–DEWCE).

Ch–11

Literature and recommended READING and LINKS

Abrahamsen, P.; Hansen, S.; 2000. *Daisy: an open soil–crop–atmosphere system model*. Environmental Modelling & Software 15/3, 313–330.
Arnold, J. G.; Srinivasan, R.; Muttiah, R. S.; Williams, J.R.; 1998. *Large-area hydrologic modeling and assessment*: Part I.Model development. J. American Water Resour. Assoc. 34(1):73-89.

Asseng, S.; van Keulen, H.; Stol, W.; 2000. *Performance and application of the APSIM Nwheat model in the Netherlands*, European Journal of Agronomy, Volume 12, P: 37– 54.

Austrian drought monitoring system for agriculture: https://warndienst.lko.at/winterweizen+2500++1073299 (Austria, in german).

Bellocchi, G.; Confalonieri, R.; Donatelli, M.; 2006. *Crop modelling and valiation: Intergration of Irene–DLL in the warm environment*. Italian Journal of Agrometeorology 3.

Brisson, N.; Gary, C.; Justes, E.; Roche, R.; Mary, B.; Ripoche, D.; Zimmer, D.; Sierra, J.; Bertuzzi, P.; Burger, P.; Bussiere, F.; Cabidoche, Y.M.; Cellier, P.; Debake, P.; Gaudillere, J.P.; Henault, C.; Maraux, F.; Seguin, B.; Sinoquet, H.; 2003. *An overview of the crop model STICS*. Eur. J. Agron.,18, 309–332.

Cabelguenne, M.; Debaeke, P.; Bouniols, A.; 1999. *EPIC phase, a version of the EPIC model simulating the effects of water and nitrogen stress on biomass and yield, taking account of developmental stages: validation on maize, sunflower, sorghum, soybean and winter wheat*. Agricultural Systems 60: 175–196.

Connor, D.J.; Loomis, R.S.; Cassman, K.G; 2011. *Crop Ecology – Productivity and Management in Agricultural Systems*. Cambridge Univ. Press, ISBN: 9780521744034; Information on this title: www.cambridge.org/9780521744034

COST734 crop modelling survey results – presentation: http://www.cost734.eu/reports–and–presentations/6th–management–committee–meeting–in–oscarsborg/oscarsborg_p6.pdf

Donatelli, M.; Magarey, R.D.; Bregaglio, S.; Willocquet, L.; Whish, J.P.M.; Savary, S.; 2017. *Modelling the impacts of pests and diseases on agricultural systems*. Agricultural Systems 155: 213–224.

Examples of operational drought monitoring systems worldwide: https://combirisk.boku.ac.at/links.html

Gijsman, A.J.; Jagtap, S.S.; Jones, J.W.; 2002. *Wadding through a swamp of complete confusion: how to choose a method for estimating soil water retention parameters for crop models*. Euro. J. Agron. 18, 77–106.

Grassini, P.; Lenny, G.J.; van Bussel; van Wart, J.; Wolf, J.; Claessens, L.; Yang, H.; Boogaard, H.; de Groot, H.; van Ittersum, M.K.; Cassman, K.G.; 2015. *How good is good enough? Data requirements for reliable crop yieldsimulations and yield–gap analysis*. Field Crops Research 177, 49–63.

Jamieson, P.D.; Semenov, M.A.; Brooking, I.R.; Francis, G.S.; 1998. *Sirius: a mechanistic model of wheat response to environmental variation*. European Journal of Agronomy, 8: 161–179.

Jones, J.W.; Hoogenboom, G.; Porter, C.H.; Boote, K.J.; Batchelor, W.D.; Hunt, L.A.; Wilkens, P.W.; Singh, U.; Gijsman, A.J.; Ritchie, J.T.; 2003. *The DSSAT cropping system model*, European Journal Of Agronomy, Volume 18 (3–4), P: 235–265.

Keating, B.A.; Carberry, P.S.; Hammer, G.L.; Probert, M.E.; Robertson, M.J.; Holzworth, D.; Huth, N.I.; Hargreaves, J.N.G.; Meinke, H.; Hochman, Z.; McLean, G.; Verburg, K.; Snow, V.; Dimes, J.P.; Silburn, M.; Wang, E.; Brown, S.; Bristow, K.L.; Asseng, S.; Chapman, S.; McCown, R.L.; Freebairn, D.M.; Smith, C.J.; 2003. *An overview of APSIM, a model designed for farming systems simulation.* Eur. J. Agron. 18: 267–288.

Kersebaum, K.C.; Boote, K.J.; Jorgenson, J.S.; Nendel, C.; Bindi, M.; Frühauf, C.; Gaiser, T.; Hoogenboom, G.; Kollas, C.; Olesen, J.E.; Rötter, R.P.; Ruget, F.; Thorburn, P.J.; Trnka, M.; Wegehenkel, M.; 2015. *Analysis and classification of data sets for calibration and validation of agro–ecosystem models.* Env. Model. Software 72, 402–417.

Kersebaum, K.Ch.; Hecker, J.–M.; Mirschel, W.; Wegehenkel, M. (ed.); 2007. *Modelling Water and Nutrient Dynamics in Soil–Crop Systems. Springer*, ISBN:978–1–4020–4478–6.

Kersebaum, K.Ch.; 2007. *Modelling nitrogen dynamics in soil–crop systems with HERMES.* Nutr Cycl Agroecosyst (2007) 77: 39. https://doi.org/10.1007/s10705–006–9044–8

Lalic, B.; Francia, M.; Eitzinger, J.; Podraščanin, Z.; Arsenić, I. 2016. *Effectiveness of short–term numerical weather prediction in predicting growing degree days and meteorological conditions for apple scab appearance.* Meteorol. Appl. 23: 50–56; DOI: 10.1002/met.1521

Martre, P.; Jamieson, P.D.; Semenov, M.A.; Zyskowski, R.F.; Porter, J.R.; Triboi, E.; 2006. *Modelling protein content and composition in relation to crop nitrogen dynamics for wheat.* Eur. J. Agron. 25: 138–154.

Matthews, K.B.; Sibbald, A.R.; Craw, S.; 1999. *Implementation of a spatial decision support system for rural land use planning: integrating geographic information system and environmental models with search and optimisation algorithms.* Comput. Electron. Agric. 23(1):9–26.

Matthews, R.B.; 2006. *The People and Landscape Model (PALM): towards full integration of human decision–making and biophysical simulation models.* Ecol. Modelling 194(4):329–343.

Mihailovic, D.T.; Koci, I.; Lalic, B.; Arsenic, I.; Radlovic, D.; Balaz, J. 2001. *The main features of BAHUS – biometeorological system for messages on the occurrence of diseases in fruits and vines.* Environ. Modell. Softw. 16: 691–696; DOI: 10.1016/S1364–8152(01)00032–9

Nendel, C. M.; Berg, K.; Kersebaum, K.Ch.; Mirschel, W.; Specka, X.; Wegehenkel, M.; Wenkel, K.O.; Wieland, R.; 2011.*The MONICA model: Testing predictability for crop growth, soil moisture and nitrogen dynamics.* Ecological Modelling 222/9, 1614–1625.

Nejedlik, P.; Orlandini, S. (ed.); 2008. *Impact of Climate Change and Variability on European Agriculture – Survey of agrometeorological practices and application in Europe regarding climate change impacts. Final report COST734*; online: http://www.cost734.eu/reports–and–presentations/SURVEY%20OF%20AGROMETEOROLOGICAL%20 PRACTICES%20AND%20APPLICATIONS%20IN%20EUROPE%20

%20REGARDINGCLIMATE%20CHANGE%20IMPACTS%20-%20 COST%20734.pdf.

Oerke, E-C.; Gerhards, R.; Menz, G.; Sikora, R.A.; 2010. *Precision Crop Protection – the Challenge and Use of Heterogeneity*. Springer, ISBN: 978-90-481-9276-2 https://doi.org/10.1007/978-90-481-9277-9.

Parton, W.J.; Hartman, M.; Ojima, D.; Schimel, D.; 1998. *DAYCENT and its land surface submodel: description and testing. Global and Planetary Change* 19:35–48.

Penning de Vries, F.W.T.; Jansen, D.M.; ten Berge, H.F.M.; Bakema, A.; 1989. *Simulation of ecophysiological processes of growth in several annual crops*. PUDOC, Wageningen. ISBN 90-220-1000-7. Available online: http://edepot.wur.nl/108856.

Rötter, R.P.; Appiah, M.; Fichtler, E.; Kersebaum, K.Ch.; Trnka, M.; Hoffmann, M.P.; 2018. *Linking modelling and experimentation to better capture crop impacts of agroclimatic extremes – a review*. Field Crops Research 221, 152–156.

Sadras, V.; Calderini, D.; 2014. Crop Physiology. *Applications for Genetic Improvement and Agronomy*. ELSEVIER, ISBN: 9780124171046.

Stockle, C.O.; Donatelli, M.; Nelson, R.; 2003. *CropSyst, a cropping system simulation model*. Europ. J. Agron. 18: 289–307.

Supit, I.; Hooijer, A.A.; van Diepen, C.A. (Eds.); 1994. *System description of the WOFOST 6.0 crop simulation model implemented in CGMS*. Volume 1: Theory and Algorithms. Joint Research Centre of the European Commission, EUR 15956 EN, Luxembourg.

van Ittersum, M.K.; Leffelaar, P.A.; van Keulen, H.; Kropff, M.J.; Bastiaans, L.; Goudriaan, J.; 2003. *On approaches and applications of the Wageningen crop models*. Europ. J. Agron. 18: 201–234.

Willmott, C.J.; Ackleson, S.G.; Davis, R.E.; Feddema, J.J.; Klink, K.M.; Legates, D.R.; O'Donnell, J.; Rowe, C.M.; 1985. *Statistics for the evaluation of model performance*. Journal of Geophysical Research 90(C5): 8995–9005.

Willmott, C.J.; Matsuura, K.; 2005. *Advantages of the mean absolute error (MAE) over the root mean square error (RMSE) in assessing average model performance*. Climate Research 30: 79–82.

Wu, L.; Kersebaum, K.-C.; 2008. *Modeling water and nitrogen interaction responses and their consequences in crop models*. In: Ahuja, L.R.,, Reddy, V.R., Saseendran, S.A., Yu, Q. (eds.) *Response of crops to limited water: understanding and modeling water stress effects on plant growth processes. Advances in Agricultural Systems Modeling 1*: 215–249; Madison, WI, USA (ASA, CSSA, SSSA).

Yang, J.; Greenwood, D.J.; Rowell, D.L.; Wadsworth, G.A.; Burns, I.G.; 2000. *Statistical methods for evaluating a crop nitrogen simulation model*, N-ABLE. Agricultural Systems 64, 37–53.

Ch–12

References, recommended literature and links

Alexandrov, V.; Eitzinger, J.; Cajic, V.; Oberforster, M.; 2002. *Potential impact of climate change on selected agricultural crops in north–eastern Austria.* Global Change Biology, 8, 4, 372–389.

Alexandrov, V.; Eitzinger, J.; Hoogenboom, G.; 2011. *Climate variability and change and related impacts on agroecosystems in southeast and central Europe as well as southeast USA.* COST, European Science Foundation, 232, BOLID–INS, Sofia, Bulgaria; ISBN: 978–954–394–055–4.

APCC; 2014. *Österreichischer Sachstandsbericht Klimawandel 2014 (AAR14).* Austrian Panel on Climate Change (APCC), Verlag der Österreichischen Akademie der Wissenschaften, Wien, Österreich, 1096 Seiten. ISBN 978–3–7001–7699–2.

Behringer, W.; 2010. *A Cultural History of Climate.* Verlag Polity, ISBN 0745645291.

Bryant, C.R.; Sarr, M.A.; Delusca, K. (eds).; 2016. *Agricultural Adaptation to Climate Change.* Springer, ISBN 978–3–319–31392–4.

Dinar, A.; Mendelsohn, R. (ed.); 2011. *Handbook on Climate Change and Agriculture.* Edward Elgar Publishing, ISBN: 978 1 84980 116 4.

Durand, J.–L.; Delusca, K.; Boote, K.; Lizaso, J.; Manderscheid, R.; Weigel, H.–J.; Ruane, A.C.; Rosenzweig, C.; Jones, J.W.; Ahuja, L.; Anapalli, S.; Basso, B.; Baron, C.; Bertuzzi, P.; Biernath,C.; Deryng, D.; Ewert, F.; Gaiser, T.; Gayler, S.; Heinlein, F.; Kersebaum, K.C.; Kim, S.–H.; Müller, C.; Nendel, C.; Olioso, A.; Priesack, E.; Ramirez Villegas, J.; Ripoche, D.; Rötter, R.P.; Seidel, S.I.; Srivastava, A.; Tao, F.; Timlin, D.; Twine, T.; Wang, E.; Webber, H.; Zhao, Z.; 2017. *How accurately do maize crop models simulate the interactions of atmospheric CO2 concentration levels with limited water supply on water use and yield?* Europ. J. Agron. (in press). Doi: 10.1016/j.eja.2017.01.002.

EEA; 2018. *Greenhouse gas emissions and trends, data, maps, trends*: online at: https://www.eea.europa.eu/data–and–maps/indicators/greenhouse–gas–emission–trends–6

EEA; 2016. *Agriculture and climate change*: online at: https://www.eea.europa.eu/signals/signals–2015/articles/agriculture–and–climate–change

Eitzinger, J.; Kersebaum, K. C.; Formayer, H.; 2009. *Landwirtschaft im Klimawandel – Auswirkungen und Anpassungsstrategien für die Land– und Forstwirtschaft in Mitteleuropa.* http://de.agrimedia.com, 320, Agrimedia, D–29459 Clenze, Deutschland; ISBN: 978–3–86037–378–1 (in german).

Eitzinger, J.; Trnka, M.; Semerádová, D.; Thaler, S.; Svobodová, E.; Hlavinka, P.; Siska, B.; Takáč, J,; Malatinská, L.; Nováková, M.; Dubrovský, M.; Zalud, Z.; 2013. *Regional climate change impacts on agricultural crop production in Central and Eastern Europe – hotspots, regional differences and common trends.* The Journal of Agricultural Science 151(6): 787–812.

European Environmental Agency (EEA); 2005. *Vulnerability and adaptation to climate change in Europe*. EEA Technical Report No. 7. Copenhagen: EEA.

European Environmental Agency (EEA); 2007. *Climate change: the cost of inaction and the cost of adaptation*. EEA Technical Report No. 13. Copenhagen: EEA.

Eurostat information on greenhouse gas emissions from agriculture: available online at: http://ec.europa.eu/eurostat/statistics-explained/index.php?title=Archive:Agriculture_-_greenhouse_gas_emission_statistics&oldid=273928

EPA; 2018. Global greenhouse gas emissions shares and trends: https://www.epa.gov/ghgemissions/global-greenhouse-gas-emissions-data

Fuhrer, J.; 2003. *Agroecosystem responses to combinations of elevated CO2, ozone and global climate change*. Agriculture, Ecosystems and Environment 97, 1–20.

Global Yield Gap Atlas, 2018: www.yieldgap.org.

Hoekstra A.Y.; Chapagain A.K.; Aldaya M.M.; Mekonnen M.M.; 2009. *Water Footprint Manual: State of the Art 2009*. Water Footprint Network, Enschede, The Netherlands.

IEEP; 2017. *Agriculture GHG emissions: Determining the potential contribution to the effort sharing regulation*. Available online at:https://ieep.eu/uploads/articles/attachments/6882b42e-9769-43be-bb10-f83d06510b7d/Agriculture_GHG_emissions_determining_the_potential_contribution_to_the_Effort_Sharing_Regulation._IEEP_2017_.pdf?v=63664510042

IPCC; 2012. Summary for Policymakers. In: *Managing the Risks of Extreme Events and Disasters to Advance Climate Change Adaptation. A Special Report of Working Groups I and II of the Intergovernmental Panel on Climate Change* (Cambridge University Press).

IPCC; 2013. Climate Change 2013: *The Physical Science Basis*. Contribution of Working Group I to the Fifth Assessment Report of the Intergovernmental Panel on Climate Change [Stocker, T.F., D. Qin, G.-K. Plattner, M. Tignor, S.K. Allen, J. Boschung, A. Nauels, Y. Xia, V. Bex and P.M. Midgley (eds.)]. Cambridge University Press, Cambridge, United Kingdom and New York, NY, USA, 1535 pp.

Kersebaum, K.C.; Nendel, C.; 2014. *Site-specific impacts of climate change on wheat production across regions of Germany using different CO2 response functions*. Eur. J. Agronomy 52: 22–32.

Klik, A.; Eitzinger, J.; 2010. *Impact of climate change on soil erosion and the efficiency of soil conservation practices in Austria. Journal of Agricultural Science*, Cambridge 148, 529–541.

Kroes J. *et al.*; 2016. *Water Footprint of agricultural products: progress, challenges and solutions*. Final EURO-Agriwat conference. Abstracts online access: http://edepot.wur.nl/380126.

Lalic, B.; Francia, M.; Eitzinger, J.; Podrascanin, Z.; Arsenic, I.; 2015. *Effectiveness of short-term numerical weather prediction in predicting growing degree days and meteorological conditions for Apple Scab appearance*. Meteorological Applications.

Lalic, B.; Firanj Sremac, A.; Eitzinger, J.; Stričević, R.; Thaler, S.; Maksimović, I.; Danicic, M.; Perišić, D.; Dekić, L.; 2018. *Seasonal forecasting of green water components and crop yield of summer crops in Serbia and Austria*. J AGR SCI–CAMBRIDGE.

Lobell, D.B.; Field, C.B.; 2007. *Global scale climate–crop yield relationships and the impacts of recent warming*. Environmental Research Letters.

Lobell, D.B.; Cassman, K.G.; Field, C.B.; 2009. *Crop yield gaps: their importance, magnitudes, and causes*. Annual Review of Environment and Resources34, 179–204.

Louise E.B.; Lassoie, J.P.; Fernandes, E.C.M.; 1998. *Agroforestry in Sustainable Agricultural Systems*. CRC Press, ISBN 9781566702942.

Maharjan, K.L.; Joshi, N.P.; 2013. *Climate Change, Agriculture and Rural Livelihoods in Developing Countries Springer*. ISBN 978–4–431–54343–5.

Manderscheid, R.; Erbs, M.; Weigel, H–J.; 2014. *Interactive effects of free–air CO2 enrichment and drought stress on maize growth*. European Journal Of Agronomy 52, 11–21.

Metzger, M.J.; Bunce, R.G.H.; Jongman, R.H.G.; Mücher, C.A.; Watkins, J.W.; 2005. *A climatic stratification of Europe*. Global Ecology and Biogeography, 14, 549–563.

Mueller, N.D.; Gerber, J.S.; Johnston, M.; Ray, D.K.; Ramankutty, N.; Foley, J.A.; 2012. *Closing yield gaps through nutrient and watermanagement*. Nature 490, 254–257. doi:10.1038/nature11420.

Nejedlik, P.; Orlandini, S. (Eds.); 2008. *Survey of agrometeorological practices and applications in Europe regarding climate change impacts*, 320 pages; COST 734 report, "Impact of climate change and variability on European agriculture – CLIVAGRI"; (available at : www.cost734.eu), Florence, Italy. Accessible as pdf online: http://www.cost734.eu/reports–and–presentations/.

Olesen, J.E.; Trnka, M.; Kersebaum, K.C.; Skjelvag, A.O.; Seguin, B.; Peltonen–Sainio, P.; Rossi, F.; Kozyra, J.; Micale, F.; 2011. *Impacts and adaptation of European crop production systems to climate change*. – European Journal of Agronomy.43 (2): 96–112.

Orlandini, S.; Nejedlik, P.; Eitzinger, J.; Alexandrov, V.; Toulios, L.; Calanca, P.; Trnka, M.; Olesen, J. E.; 2008. *Impacts of climate change and variability on European agriculture: results of inventory analysis in COST 734 countries*. Annals of the New York Academy of Sciences1146, 338–353.

Pfister, C.; Brazdil, R.; Glaser, R.; 1999. Climatic Variability in Sixteenth–Century Europe and Its Social Dimension. Springer. DOI 10.1007/978–94–015–9259–8.

Peltonen–Sainio, P.; Jauhiainen, L.; Trnka, M.; Olesen, J. E.; Alanca, P.; Eckersten, H.; Eitzinger, J.; Gobin, A.; Kersebaum, K. C.; Kozyra, J.; Kumar, S.; Dalla Marta, A.; Micale, F.; Schaap, B.; Seguin, B.; Skjelvag, A.O.; Orlandini, S.; 2010. *Coincidence of variation in yield and climate in Europe. Agriculture, Ecosystems and Environment* 139, 483–489.

Peltonen–Sainio, P.; Rötter, R.; Iglesias, A.; Orlandini, S.; Dubrovsky, M.; Hlavinka, P.; Balek, J.; Eckersten, H.; Cloppet, E.; Calanca, P.; Gobin, A.; Vucetic, V.; Nejedlik, P.; Kumar, S.; Lalic, B.; Mestre, A.; Rossi, F.; Kozyra, J.; Alexandrov, V.; Semeradova, D.; Žalud, Z.; 2011. *Agroclimatic conditions in Europe under climate change. Global Change Biology*, 17: 2298–2318. doi: 10.1111/j.1365–2486.2011.02396.x.

Rötter, R.P.; Höhn, J.; Trnka, M.; Fronzek, S.; Carter, T.R.; Kahiluoto, H.; 2013. *Modelling shifts in agroclimate and crop cultivar response under climate change*. Ecol Evol 3: 4197–4214.

Rötter, R.P.; Appiah, M.; Fichtler, E.; Kersebaum, K.C.; Trnka, M.; Hoffmann, M.P.; 2018. *Linking modelling and experimentation to better capture crop impacts of agroclimatic extremes – a review*. Field Crops Research 221, 152–156. Doi: 10.1016/j.fcr.2018.02.023.

Thaler, S.; Eitzinger, J.; Trnka, M.; Dubrovsky, M.; 2012. *Impacts of climate change and alternative adaptation options on winter wheat yield and water productivity in a dry climate in Central Europe* J AGR SCI 150(5): 537–555.

Torquebiau, E. (Ed.); 2016. Climate Change and Agriculture Worldwide. Springer, ISBN 978–94–017–7462–8

Trnka, M.; Eitzinger, J.; Semeradova, D.; Hlavinka, P.; Balek, J.; Dubrovsky, M.; Kubu, G.; Stepanek, P.; Thaler, S.; Mozny, M.; Zalud, Z.; 2011. *Expected changes in agroclimatic conditions in Central Europe*. CLIMATIC CHANGE 108(1–2): 261–289.

Troll, C.; 1955. *Der jahreszeitliche Ablauf des Naturgeschehens in den verschiedenen Klimagürteln der Erde*. Studium Generale 8, 713–733.

Walter, H.; Lieth, H.; 1967. Klimadiagram–Weltatlas, VEB Gustav Fischer Verlag, Jena.

White, J.W.; Hoogenboom, G.; Kimball, B.A.; Wall, G.W.; 2011. *Methodologies for simulating impacts of climate change on crop production*. Field Crops Research 124, 357–368.

Ch–13

Further READING and LINKS

Darnhofer, T.; 1985. *Meteorological elements and their observation*. Working paper Nr.14. 1st Revision. Pdf download: http://www.worldagroforestry.org/downloads/Publications/PDFS/58_Working_paper_no_14.pdf

Drlatief, A.; Raihana, H.K.; Sabah, P.; Syed, S.M.; 2017. *Experimental Agrometeorology: A Practical Manual*. 10.1007/978–3–319–69185–5.

Emeis, S.; 2010. *Measurement Methods in Atmospheric Sciences – In situ and remote*. Borntraeger Science Publishers. ISBN 978–3–443–01066–9

Foken, T.; 2017. Micrometeorology. Springer, ISBN 3642254403, 9783642254406

Gelo, B.; Penzar, I., Poje, D. *et al.*; 2005: *Meteorological Glossary and Multilingual Dictionary (Croatian, English, German and French)*, Croatian Meteorological and Hydrological Service, HINUS, Zagreb, p. 655. (In Croatian)

Harrison, G.; 2015. *Meteorological Measurements and Instrumentation. Wiley*, ISBN: 978–1–118–74580–9. Info: https://www.wiley.com/en–us/Meteorological+Measurements+and+Instrumentation–p–9781118745809

Otorepac, S.; 1980: *Agrometeorology*, Nolit, Beograd, p. 231. (In Serbian).

Penzar, I.; Penzar B.; 2000. *Agrometeorology*. Školska knjiga, Zagreb, p. 228. (In Croatian).

Strangeways, I.; 2003. *Measuring the Natural Environment*. Cambridge Univ. Press. Link: https://books.google.at/books?id=oTSJ6NGFSuQC&dq=book+measurement+techniques+in+meteorology&hl=de&source=gbs_navlinks_s.

TDR and FDR. *Capacitive soil water measurement method*: http://www.environmentalbiophysics.org/tdr–versus–capacitance–or–fdr/.

Vučetić, M.; Vučetić, V.; 2013: Adriatic Weather – *Meteorology for Sailers*, Fabra Press, Zagreb.

WMO; 2008. *Guide to Meteorological Instruments and Methods of Observation*. ISBN 978–92–63–10008–5 Pdf download: http://www.posmet.ufv.br/wp–content/uploads/2016/09/MET–474–WMO–Guide.pdf.

WMO; 2011. WMO standards and best practices: presentation: https://www.wmo.int/pages/prog/www/OSY/Meetings/GCW–IM1/Doc7.1_BestPractices.pdf.

Ch–14

References, recommended READING and LINKS

Asrar, G.; Fuch, M; Kanemasu, E.T. *et al.*; 1984. *Estimating absorbed photosynthetic radiation and leaf area index from spectral reflectance in wheat*. Agron J, vol. 76, 300–306.

Baghdadi, N.; Zribi, M.; 2016. *Land Surface Remote Sensing in Agriculture and Forest*. ELSEVIER, ISBN: 9781785481031

Bartholome, E.; 1987. *Radiometric measurements and crop yield forecasting: some observations over millet and sorghum experimental plots in Mali*, EarselSymp, Noordwijkerhout; IN: Toselli, F. (ed) (1988). Applications of Remote Sensing to Agrometeorlogy. Proceedings; ISPRA corses on Remote Sensing, Luwer, ISBN 0–7923–0020–3.

Binayak, P.; Mohanty, M.; Lakshmi, C.V.; Montzka, C.; 2017. *Soil Moisture Remote Sensing:State–of–the–Science*. Vadose Zone J., doi:10.2136/vzj2016.10.0105

Brocca, L.; Crow, W.T.; Ciabatta, L.; Massari, C; de Rosnay, P.; Enenkel, M; Hahn, S.; Amarnath, G; Camici, S; Tarpanelli, A. and Wagner, W.; 2017. *A review of the applications of ASCAT soil moisture products*. IEEE Journal of Selected Topics in Applied Earth Observations and Remote Sensing 10, 2285–2306.

Elmore, A. J., J. F.; Mustard, S.; Manning J. and Lobell, D.B.; 2000. *Quantifying vegetation change in semiarid environments: Precision and accuracy of spectral mixture analysis and the normalized difference vegetation index*. Remote Sens. Environ.,73,87–102.

Haboudane, D.; Miller, J.R.; Pattey, E.; Zarco–Tejada, P.J.; Strachan, I.; 2004. *Hyperspectral vegetation indices and novel algorithms for predicting green LAI of crop canopies: modeling and validation in the context of precision agriculture*. Remote Sens. Environ., 90, pp. 337–352.

Huete, A.; Didan, K.; Miura, T.; Rodriguez, E.P.; Gao, X.; Ferreira, L.G.; 2002. *Overview of the radiometric and biophysical performance of the MODIS vegetation indices*. Remote Sensing of Environment 83 (2002) 195–213.

Kogan, F. N.; Gitelson, A.; Edige, Z.; Spivak, l.; Lebed L.; 2003. *AVHRR–Based Spectral Vegetation Index for Quantitative Assesment of Vegetation State and Productivity: Calibration and Validation*. Photogramm. Engg. & Remote Sensing, 69(8), pp. 899–906.

Krishna, K.R.; 2018. *Agricultural Drones: A Peaceful Pursuit*. Apple Academic Press, ISBN–10: 1771885955

Leeuwen, W.; Orr, B. J.; Marsh, S. E.; Herrmann, S. M.; 2006. *Multisensor NDVI data continuity: uncertainties and implications for vegetation monitoring applications, Remote Sens. Environ.*, vol. 100, pp. 67–81.

Links and examples of satellite remote sensing products for agrometeorology:
www.esa.int/eo
www.ceos.org
www.ceos.org/pages/CEOSResponse_1010A.pdf
www.eumetsat.int
www.nesdis.noaa.gov,
www.orbit.nesdis.noaa.gov,
www.ncdc.noaa.gov
www.cr.usgs.gov/pub/imswelcome/
http://vegetation.cnes.fr
http://rs.geo.tuwien.ac.at/data–viewers/
https://glovis.usgs.gov/
https://landsat.gsfc.nasa.gov/
https://landsat.usgs.gov/

Link on NDVI and EVI: https://earthobservatory.nasa.gov/Features/MeasuringVegetation/

Links on remote sensing applications in agriculture: https://agfundernews.com/4–things–know–remote–sensing–agriculture.html

Links on drones in agriculture: http://www.droneomega.com/drones–in–agriculture/

Malingreau, J.–P.; 1986. *Global vegetation dynamics – Satellite observations over Asia*. International Journal of Remote Sensing. 7. 10.1080/01431168608948914.

Mulla, D.J.; 2013. *Twenty five years of remote sensing in precision agriculture: Key advances and remaining knowledge gaps*. Bio–systems Engineering 114/4, 358–371.

Pettorelli, N.; Vik, J.O.; Mysterud, A.; Gaillard, J.M.; Tucker, C.J.; Stenseth, N.C.; 2005. *Using the satellite–derived NDVI to assess ecological responses to environmental change.* Trends EcolEvol. 2005 Sep;20(9):503–10. Epub 2005 Jun 9. DOI: 10.1016/j.tree.2005.05.011

Peters, A. J.; Walter–Shea, E. A.; Ji, L.; Vina, A.; Hayes, M; Svoboda, M.D.; 2002. *Drought Monitoring with NDVI–Based Standardized Vegetation Index*, Photogrammetric engineering and remote sensing, vol. 68(1), pp. 71–76.

Reed, B.C.; Brown, J.F.; Zee, D.V.; Loveland, T.R.; Merchant, J.W.; Ohlen, D.O.; 1994. *Measuring phenological variability from satellite imagery.* Wiley Online Library; https://doi.org/10.2307/3235884

Thaler, S.; Eitzinger, J.; Trnka, M.; Možný, M.; Hahn, S.; Wagner, W.; Hlavinka, P; 2018. *The performance of Metop Advanced SCATterometer soil moisture data as a complementary source for the estimation of crop–soil water balance in Central Europe.* The Journal of Agricultural Science. https://doi.org/10.1017/S0021859618000011

Thenkabail, P.; Lyon, J.G.; Turral, G.; Biradar, C.; 2018. *Remote Sensing of Global Croplands for Food Security.* CRC Press, ISBN 9781138116559

Vuolo, F.; Urso, D.; De Michele, G.; Bianchi, C.; Cutting, B.; 2015. *Satellite–based irrigation advisory services: A common tool for different experiences from Europe to Australia.* AGR WATER MANAGE 147: 82–95.

Wagner, W.; Hahn, S.; Kidd, R; Melzer,T.; Bartalis, Z.; Hasenauer, S.; Figa–Saldana, J.; De Rosnay, P.; Jann, A.; Schneider, S.; Komma, J.; Kubu, G.; Brugger, K.; Aubrecht, C.; Züge, J; Gangkofer, U.; Kienberger, S.; Brocca L.; Wang, Y.; Blöschl, G.; Eitzinger, J.; Steinnocher, K.; Zeil, P.; Rubel, F.; 2013. *The ASCAT soil moisture product: a review of its specifications, validation results, and emerging applications.* Meteorologische Zeitschrift 22, 5–33.

Xie, Y.; Sha, Z.; Yu, M.; 2008. *Remote sensing imagery in vegetation mapping: a review.* Journal of Plant Ecology, Volume 1, Issue 1, Pages 9–23, https://doi.org/10.1093/jpe/rtm005

Numerical examples

References, recommended READING

Allen, R.G.; Pereira, L.S.; Raes, D.; Smith, M.; 1998. *Crop evapotranspiration - Guidelines for computing crop water requirements* - FAO Irrigation and drainage paper 56, FAO - Food and Agriculture Organization of the United Nations, Rome, ISBN 92-5-104219-5.

Bierhuizen, J.F.; Slatyer, R.O.; 1965. *Effect of atmospheric concentration of water vapour and CO2 in determining transpiration–photosynthesis relationships of cotton leaves.* Agric. Meteorol. 2, 259–270, doi: 10.1016/0002-1571(65)90012-9.

Brunt, D.; 1932. *Notes on radiation in the atmosphere*. Q. J. R. Met. Soc. 58, 389-42, doi: 10.1002/qj.49705824704.

Campbell, G.S.; Norman, J.M.; 1998. *An Introduction to Environmental Biophysics*, 2nd Ed.; Springer, p. 306.

Farquhar, G.D.; von Caemmerer, S.; Berry, J.A.; 1980. *A biochemical model of photosynthetic CO2 assimilation in leaves of C3 species*. Planta 149, 78–90.

Fitzjarrald, D.; Sakai, R.; 2009. *Radiation Measurements at Harvard Forest EMS Tower 1991-2007*. Harvard Forest Data Archive: HF102.

McMaster, G.S.; Wilhelm, W.W.; 1997. *Growing Degree-Days One Equation, Two Interpretations*. Agric. Forest Meterol., 87, 291-300.

Mihailovic, D.; 1988. *Basics of meteorological observations and data processing*. Faculty of Agriculture. Institute of Field and Vegetable Crops, Novi Sad, p. 217 (in Serbian).

Monteith, J.L.; 1977. *Climate and Efficiency of Crop Production in Britain*. Philosophical Transactions of the Royal Society London B, 281, 277-294.

Snyder, R.L.; de Melo-Abreu, J.P.; 2005. *Frost Protection: fundamentals, practice and economics*. Vol 1, Food and Agriculture Organization of the United Nations, Rome, p. 126.

WMO; 2008. *Guide to Meteorological Instruments and Methods of Observation*. World Meteorological Organization, 2008, WMO-No. 8, p.716, ISBN 978-92-63-10008-5.

List of the authors and contributors

Bernhard Pacher. Born in Austria in 1961, he acquired a doctoral degree of law at the University of Salzburg. Since 1999 he was active in agrometeorology and hydrology as sales manager for an internationally renowned manufactured of related technology (hard– and software), of which he became managing director in 2004. For many years he focussed on developing meteorological projects in sub–saharan Africa and Latin America, in collaboration with numerous UN agencies and local governments. In 2015, after a successful sale of the company, he established his own consultancy business.

Branislava Lalic. Born in Serbia in 1967. Study of Physics at Faculty of Science, University of Novi Sad (UNS), magisterium in dynamical meteorology, Faculty of Physics, University of Belgrade, PhD in 2006 in Meteorology and environmental modelling at UNS. 1993–2007 Univ. assistant, since 2012 associated professor in Meteorology at Faculty of Agriculture (UNS). In 2014 Fulbright scholar at ASRC, SUNY (NY, USA). Teacher for Undergraduates and graduates in various lectures, including Meteorology, Biophysics, Crop modeling, supervision of master and PhD students. Coordinator or partner of past and ongoing national and international research projects. H–Index 15, 40 SCI Publications, ORCID number 0000–0001–5790–7533.

Josef Eitzinger. Born in Austria in 1962. Study of Agronomy at University of Natural Resources and Life Sciences Vienna (BOKU), including PhD in 1991. 1991–2003 Univ. assistant in Agrometeorlogy at the Institute of Meteorology. 1998–1999 working as Visiting Scientist at Natural Ressource Ecology Laboratory,Colorado State University on a project titled „Improvement and validation of an agroecological model for specified climatological regions“ supported by the Austrian Science Foundation (FWF), (Schroedinger scholarship). 2003 Habilitation and Professor for Agrometeorology at the Institut of Meteorology (BOKUMet). Since 2013 Head of Institute of Meteorology at BOKU. Teacher for Undergraduates and graduates in various lectures, including agronomy, agrometeorology and climatology, supervision of master and PhD students. Study program supervisor of international Master program Danube AgriFood and participation in two other international master study programs. Coordinator or partner of several past and ongoing national and international research projects. H–Index 23, 66 SCI Publications, >300 publications (status 04/2018), ORCID number 0000–0001–6155–2886.

Anna Dalla Marta. She holds a PhD in Soil Science and Climatology. She works on agronomy and agricultural meteorology, environmental sustainability and precision farming, included water management. At present, she is researcher at the University of Florence (Department of Agrifood Productions and Environmental Sciences) and lecturer at the courses of Agronomy and Soil Management, supervisor of Magister degree thesis in Agricultural Sciences and supervisor of PhD students. She is member of the editorial board of the Italian Journal of Agrometeorology and of the Italian Journal of Agronomy, she is reviewer for many peer–reviewed journals. She participates to numerous national and international projects (COST, JPI, H2020) concerning modelling of crop production, assessment of impacts of climate and climate variability on crop yield and quality, precision farming. She is author of more than 40 scientific papers on peer–reviewed international journals. ORCID number 0000–0002–4606–7521.

Simone Orlandini. Full Professor of Agronomy and Head of the Department of Agrifood Production and Environmental Sciences of the University of Florence. He is member of the Senato Accademico of University of Florence. He is graduated in Agricultural Sciences and PhD in Agrometeorology. He carries out teaching activity in Agronomy at the University of Florence for under–graduated students of the Faculty of Agriculture. He also teaches for the students of the PhD in Agricultural and Environmental Sciences and master courses (Precision Agriculture – University of Teramo; Tropimundo – University of Florence). He is member of the Scientific Council of Precision Agriculture Master, Coordinated by the University of Teramo. He is presently Director of the Interdepartmental Centre for. He is Accademico Ordinario of the Accademia dei Georgofili di Firenze and member of the Accademic Council. He is President of the Climate and Sustainability Foundation. ORCID number 0000–0001–6290–9752.

Ana Firanj Sremac. She was born 1983 in Sombor where she finished primary and secondary scool. Faculty and PhD studies she has finished at Unifersity of Novi Sad. In 2015 she got her PhD title in Meteorology and Environmental modeling. From 2008 she has been employed by University of Novi Sad, first at Faulty of Science and after at Faculty of Agriculture where she currently work in the Division of Meteorology, Agrometeorology and Biophysics. She has several papers in peer rewievd journals published and she has participated in several international and national projects. ORCID number 0000–0003–0948–0189.

List of contributors BOKU

Name	Affiliation	Country
Bernhard Pacher	Consultant	Austria
Piotr Strutzig	Institute of Meteorology and water Management	Poland
Christian Kersebaum	Leibniz–Centre for Agricultural Landscape Research, Muencheberg, ZALF	Germany
Pavol Nejedlik	Slovak Hydrometeorological Institute	Slovakia
Mirek Trnka	Mendel University Brno	Czech Republic
Aliyeh Salehi	Universität für Bodenkultur, Vienna, BOKU	Austria
Gerhard Kubu	Universität für Bodenkultur, Vienna, BOKU	Austria
Levent Saylan	Istanbul Technical University	Turkey
Erich Mursch–Radlgruber	Universität für Bodenkultur, Vienna BOKU	Austria
Sabina Thaler	Universität für Bodenkultur, Vienna BOKU	Austria

List of contributors UNIFI

Name	Affiliation	Country
Leonardo Verdi	Department of Agrifood Production and Environmental Sciences University of Florence	Italy
Carolina Fabbri	Department of Agrifood Production and Environmental Sciences University of Florence	Italy

Permissions

All chapters in this book were first published in AMC, by Firenze University Press; hereby published with permission under the Creative Commons Attribution License or equivalent. Every chapter published in this book has been scrutinized by our experts. Their significance has been extensively debated. The topics covered herein carry significant information for a comprehensive understanding. They may even be implemented as practical applications or may be referred to as a beginning point for further studies.

The contributors of this book come from diverse backgrounds, making this book a truly international effort. We would like to thank all the contributing authors for lending their expertise to make the book truly unique. They have played a crucial role in the development of this book. Without their invaluable contributions this book wouldn't have been possible. They have made vital efforts to compile up to date information on the varied aspects of this subject to make this book a valuable addition to the collection of many professionals and students.

This book was conceptualized with the vision of imparting up-to-date and integrated information in this field. To ensure the same, a matchless editorial board was set up. Every individual on the board went through rigorous rounds of assessment to prove their worth. After which they invested a large part of their time researching and compiling the most relevant data for our readers.

The editorial board has been involved in producing this book since its inception. They have spent rigorous hours researching and exploring the diverse topics which have resulted in the successful publishing of this book. They have passed on their knowledge of decades through this book. To expedite this challenging task, the publisher supported the team at every step. A small team of assistant editors was also appointed to further simplify the editing procedure and attain best results for the readers.

Apart from the editorial board, the designing team has also invested a significant amount of their time in understanding the subject and creating the most relevant covers. They scrutinized every image to scout for the most suitable representation of the subject and create an appropriate cover for the book.

The publishing team has been an ardent support to the editorial, designing and production team. Their endless efforts to recruit the best for this project, has resulted in the accomplishment of this book. They are a veteran in the field of academics and their pool of knowledge is as vast as their experience in printing. Their expertise and guidance has proved useful at every step. Their uncompromising quality standards have made this book an exceptional effort. Their encouragement from time to time has been an inspiration for everyone.

The publisher and the editorial board hope that this book will prove to be a valuable piece of knowledge for students, practitioners and scholars across the globe.

Index

www.ingramcontent.com/pod-product-compliance
Lightning Source LLC
LaVergne TN
LVHW012320171026
850499LV00003B/56